Springer Series in
Surface Sciences
3

Springer Series in **Surface Sciences**

Editors: Gerhard Ertl and Robert Gomer

Volume 1: **Physisorption Kinetics**
By H. J. Kreuzer, Z. W. Gortel

Volume 2: **The Structure of Surfaces**
Editors: M. A. Van Hove, S. Y. Tong

Volume 3: **Dynamical Phenomena at Surfaces, Interfaces and Superlattices**
Editors: F. Nizzoli, K.-H. Rieder, R. F. Willis

Volume 4: **Desorption Induced by Electronic Transition, DIET II**
Editors: W. Brenig, D. Menzel

Dynamical Phenomena at Surfaces, Interfaces and Superlattices

Proceedings of an International Summer School at the Ettore Majorana Centre, Erice, Italy, July 1–13, 1984

Editors
F. Nizzoli, K. H. Rieder and R. F. Willis

With 194 Figures

Springer-Verlag
Berlin Heidelberg New York Tokyo

Professor Dr. Fabrizio Nizzoli
Dipartimento di Fisica, Università di Modena, Via Campi 213/A, I-41100 Modena, Italy

Dr. Karl-Heinz Rieder
IBM Forschungslaboratorium, Saeumerstraße 4, CH-8803 Rueschlikon, Switzerland

Dr. Roy F. Willis
Department of Physics, Cavendish Laboratory, University of Cambridge,
Madingley Road, Cambridge, CB3 0HE, Great Britain

Guest Editor

Professor Dr. Manuel Cardona
Max-Planck-Institut für Festkörperforschung, Heisenbergstraße 1
D-7000 Stuttgart 80, Fed. Rep. of Germany

Series Editor

Professor Dr. Gerhard Ertl
Institut für Physikalische Chemie, Universität München
D-8000 München, Fed. Rep. of Germany

Professor Robert Gomer
The James Franck Institute, The University of Chicago, 5640 Ellis Avenue,
Chicago, IL 60637, USA

ISBN-13: 978-3-642-82537-8 e-ISBN-13: 978-3-642-82535-4
DOI: 10.1007/978-3-642-82535-4

Softcover reprint of the hardcover 1st edition 1985

2153/3130-543210

Preface

This volume contains summaries of lectures and contributed papers delivered at an International Summer School on "Dynamical Phenomena at Surfaces, Interfaces and Superlattices" held at the Ettore Majorana Centre for Scientific Culture, Erice (Sicily), Italy, July 1-13, 1984. The School was organized under the auspices of the Surfaces and Interfaces Section of the Condensed Matter Division of the European Physical Society as the sixth course in the series on Materials Science and Technology. Approximately 60 participants from all regions of Europe, the United States, and further afield - Hong Kong, China, India - were able to take part in a program of 45 lectures and 11 contributed talks, which covered most of the solid-state aspects of the subject.

In recent years, there has been an explosion of interest in the properties of carefully prepared surfaces, interfaces, and multilayer thin films. This advance in research has received its impetus from the technological relevance of surfaces and interfacial phenomena associated with heterogeneous catalysis, corrosion, and, particularly, new developments in microelectronics. One of the most important developments to emerge over the past decade has been our ability to prepare ultra-thin structures at the submicron level, i.e., to engineer low-dimensional solids at the atomic-scale level. The excitement surrounding these advances in materials preparation lies in the unique physical characteristics of these "synthetic structures" and their application to novel devices. Examples include semiconducting, metallic (magnetic), and organo-metallic superlattices in which ultra-thin films of one kind of material are carefully grown sandwiched between layers of another material. However, with this new skill in our ability to manipulate the growth and properties of these "superlattice" solids, problems have arisen which can be traced to the control of the clean surface and single interface boundaries. It was an opportune moment, therefore, to bring together specialist research workers from the respective sub-fields of surfaces, interfaces, and superlattice-film physics. Accordingly, these Proceedings reflect the wide-ranging nature of current research and discussion in these fields.

The book is conveniently divided into six sections: Part I, dealing with atomic structure and characterization; Part II, treating lattice dynamics, followed by lattice soft-mode phase-transition phenomena in Part III; Part IV places emphasis on electronic states and transport behavior, Part V on optical properties, followed finally by Part VI on magnetic properties. Each section introduces a lecture topic, and is terminated with short contributions mainly from student participants to reflect to some degree the nature of particularly recent activities.

The School was sponsored by various organizations: besides the basic support to the Ettore Majorana Centre from the Italian Ministry of Education, the Italian Ministry of Scientific and Technological Research, and the Sicilian Regional Government, specific financial contributions were received from the European Research Office of the U.S. Army (London), the "Gruppo Nazionale Struttura della Materia" and the "Comitato Fisica" (Roma) of Italy's National Research Council (CNR), IBM Italia (Roma), Italtel (Milano), Olivetti (Ivrea), SGS (Milano), British Petroleum Company (U.K.), and General Electric Company (U.K.). In addition, we are greatful to the European Physical Society for providing student attendance bursaries, particularly to participants from Eastern Europe. This financial aid was a great contribution to the international flavor of the School, and was much appreciated by all participants.

We wish to extend our thanks to the Director of the Ettore Majorana Centre, Prof. A. Zichini, to the secretaries Dr. A. Gabriele and Dr. Pinola Savalli, and to the technical staff for their truly excellent organization, hospitality, and assistance. We should also like to express our appreciation to Prof. Minko Balkanski, the overall Director of the Schools on Materials Science and Technology, for his kindness and help with the organization of the program.

Finally, the success of a School is dependent on its lecturers and participants. It was our real pleasure to serve our "good companions" in this venture, and we accord them all our grateful thanks for their good-humored and stimulating participation.

Erice (Sicily), Italy
December 1984

F. Nizzoli · K.-H. Rieder
R.F. Willis

Contents

Part I Structure

Part IV Electronic States

Part VI Magnetic Properties

Part I

Structure

Structural Determination of Surfaces and Overlayers with Diffraction Methods

K.H. Rieder

IBM Zürich Research Laboratory, CH-8803 Rüschlikon, Switzerland

Abstract. A brief introduction to the newer developments of surface structural research using diffraction methods is given. Selected examples of results for clean unreconstructed and reconstructed surfaces, as well as on adsorbate overlayers and thin films, are presented to illustrate the current possibilities of the different methods discussed and the various research areas in which they are applied.

1. Introduction

Surface crystallography aims for the identificaton of the atomic species at and near the surface, the positions of the atoms relative to each other, the bond lengths between them, and, finally, the character of bonding [1,2]. In no real systems do atoms at and near a surface exhibit the same arrangement as in the bulk. The simplest rearrangement is surface relaxation, whereby the topmost layers retain the bulk symmetry, but the atomic distances perpendicular to the surface are different from the respective bulk value. Surface reconstruction is a stronger disturbance giving rise to rearrangements of the topmost layers into symmetries different from the respective bulk truncation. Adsorption of foreign species at the surface usually changes the geometrical and electronic arrangement of the substrate surface. Furthermore, the formation of (coverage-dependent) regular adsorbate overlayers [3] in the region up to one monolayer, owing to direct or substrate-mediated adatom-adatom interactions, appears more to be the rule rather than the exception. In many areas of physical and chemical technology, like crystal growth, catalysis, corrosion, hardening and passivation treatments, in the formation of heterojunctions, thin-film barriers and superlattices, the properties of the crystal surface are of dominant influence. Thus, it is obvious that the necessity to gain experimental knowledge about surface structures, and to understand the microscopic mechanisms leading to them, have given rise to a wealth of surface crystallographic methods. These Lecture Notes attempt to outline the basic physical principles of the different diffraction methods, and to illustrate their application by (a very personal choice of) characteristic examples of recent results. Limitations in space do not allow a description of the other important classes of methods like ion scattering [4], surface-sensitive extended X-ray absorption fine structure (SEXAFS) [5] and the different high-resolution microscopy methods [6]. Nevertheless, it is hoped that the reader can acquire a feeling as to the respective possibilities, merits and limitations of the various methods discussed, as well as to get some insight into the different classes of structures occurring on clean surfaces, as well as on surfaces with overlayers of foreign species.

2. Surface Diffraction Kinematics

In all diffraction techniques, a beam of well-defined wavelength λ ($\lesssim 1$ Å) impinges on the solid surface, and the diffracted intensities are measured by appropriate movable detectors. The necessary condition for observation of sharp diffraction beams is a sufficiently well-ordered surface. The Laue condition for diffraction from a two-dimensional periodic array relates the incoming wavevector $k_i = (\underline{K}, \underline{k}_{iz})$ ($|\underline{k}_i| = 2\pi/\lambda$) with the outgoing $\underline{k}_G = (\underline{K}_G, \underline{k}_{Gz})$ via

$$\underline{K} - \underline{K}_G = \underline{G}. \tag{1}$$

The wavelength remains unchanged during diffraction, so that $k_i^2 = k_G^2$ (elastic scattering). The reciprocal-lattice vectors $\underline{G} = j\underline{b}_1 + l\underline{b}_2$ are specified by the following relationship between the direct ($\underline{a}_i$, i = 1,2) and the reciprocal ($\underline{b}_i$, i = 1,2) unit-cell vectors: $\underline{a}_i \cdot \underline{b}_k = 2\pi\delta_{ik}$. This implies that $\underline{b}_1$ is normal to $\underline{a}_2$, and $\underline{b}_2$ normal to $\underline{a}_1$. The lengths of b_i are determined by $b_1 = 2\pi/(a_1 \sin\gamma)$, $b_2 = 2\pi/(a_2 \sin\gamma)$ with γ denoting the angle between $\underline{a}_1$ and $\underline{a}_2$; $\underline{a}_1$ is conventionally chosen to be the shorter unit-cell vector. The Laue condition can be graphically represented by the Ewald construction shown schematically for a one-dimensional case in Fig. 1. Surface diffraction can occur in all those directions, where the Ewald sphere cuts a reciprocal-lattice rod (notice the difference from the three-dimensional case,where diffraction does not necessarily occur as the Laue condition is more stringent in this case: it requires the Ewald sphere to go through a reciprocal-lattice point); thus, the measurement of the angular locations of the diffracted beams allows the determination of the dimensions of the surface unit cell as well as its orientation relative to the incoming beam.

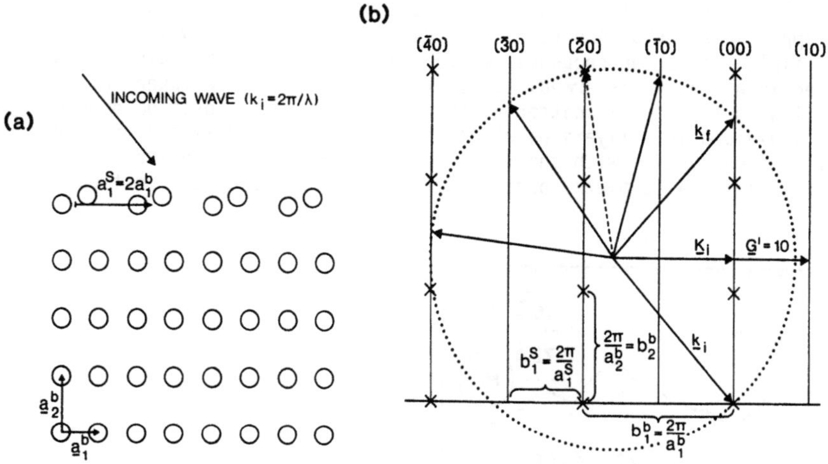

Fig. 1a and b. (a) Hard-sphere model of a (two-dimensional) crystal exhibiting a surface reconstruction with twice the corresponding bulk periodicity. The bulk and surface unit-cell vectors are indicated. (b) The Ewald construction for bulk and surface diffraction. For the case plotted, bulk diffraction occurs only in one direction (dashed arrow) where the Ewald sphere goes through a reciprocal-lattice point (the reciprocal-lattice points are indicated as crosses). In contrast to this, surface diffraction can occur in all directions in which the Ewald sphere cuts the two-dimensional lattice rods. The Miller indices of the surface reciprocal-lattice rods correspond to the surface unit cell

Determination of the distribution of the scattering centers within the unit cell requires measurement and analysis of the intensities I_G of the Bragg peaks. Analysis is complicated by the fact that the scattering equations contain the (complex) scattering amplitudes A_G, whereas only intensities $I_G \propto |A_G|^2$ can be measured, hence the information on the beam phases is lost. This is the reason why in almost all cases of intensity analysis, up to now, trial and error procedures have had to be used.

3. Low-Energy Electron Diffraction (LEED)

LEED is without any doubt the most important and fruitful surface-structural method up to now. Commercial equipment [1] is available from several manufacturers and hardly any ultra-high vacuum system dedicated to single-crystal surface research lacks a LEED retarding field analyzer [1,2]. Besides the routinely performed quick characterizations of surface order or the visual tracing of the development of regular overlayers upon adsorption, careful LEED spot-shape analysis can be used for characterization of surface disorder [7], and in some cases also for determination of the site symmetry of adsorbed particles [8].

Furthermore, most of the crystallographically complete surface-structure determinations by intensity analysis are based on LEED data [3,9]. As in the case of X-ray diffraction, the scattering centers of electrons are the regions of high electronic density, i.e. the atom cores, so that a complete structure analysis can yield the locations and mutual distances of all atom cores. Compared to X-rays, the scattering cross-section for electrons is about three orders of magnitude larger. This fact has two important consequences. On the one hand, it causes the high surface sensitivity of low-energy electrons (20 - 500 eV) owing to their small penetration depths of 5 - 8 Å, but on the other hand it gives rise to a severe complication for exact intensity calculations: whereas in the case of X-rays, a kinematical theory [10] in which multiscattering processes can be neglected, is applicable, LEED requires a full dynamical theory, in which the many different multiple-scattering sequences must be added up. Layer-by-layer treatments are used in which the intermediate scattering processes are taken into account by allowing the atomic scattering cross-sections to be position-dependent; the crystal is divided into layers parallel to the surface, and within one layer, all atoms of a given species with the same local surrounding are equivalent. As there is no translational symmetry perpendicular to the surface, the unit cell has

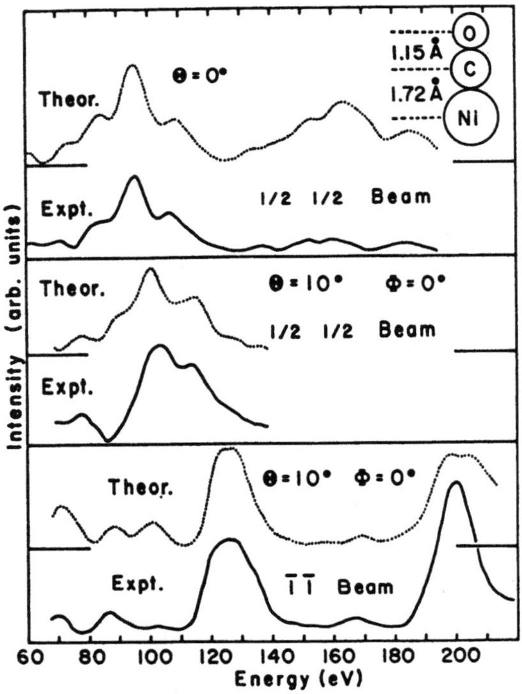

Fig. 2 Comparison of experimental and best-fit LEED I/V curves for the c(2x2)CO overlayer on Ni(100). The structural mode is indicated in the inset. After PASSLER et al.[18]. © 1979 The American Physical Society

to involve all crystal layers. However, as the electron wave is strongly damped with increasing distance from the surface, it usually suffices to perform the calculations for several layers only [11]. The calculations are mostly restricted to small unit cells owing to computing limitations. Good guides to the full theory and (working!) computer programs have been published by VAN HOVE and TONG [12]. In most cases, intensity versus voltage (energy) curves are fitted for different Bragg beams and various scattering geometries. Starting point is always a model assumption about the structure, whereby a small number of fitting parameters is left open and the degree of agreement between calculation and experiment is usually judged on the basis of different reliability factors.

LEED intensity analyses by ADAMS et al. [13] using the full dynamical theory revealed the theoretically expected oscillatory behavior [14] of the interlayer relaxation in the unreconstructed Cu(110): there is an 8.5% decrease in the lattice spacing d_1 between the first and second layers, and a 2% increase between the second and third, d_2. High-energy ion-scattering results obtained by the same group gave a 5.3% decrease of d_1 and a 3.3% increase of d_2. These numbers serve as an excellent example of the present degree of accuracy to which interlayer spacings at surfaces can be determined. More open surfaces like the sawtooth-like Fe(310) were investigated with LEED by SOKOLOV et al. [15]; they exhibit more complex relaxations,whereby atomic displacements perpendicular as well as parallel to the surface can occur; the results are in partial agreement with predictions of calculations minimizing the total energy of semi-infinite crystals with high-index surfaces [16].

Probably the most reliable LEED investigation of an adsorbate structure is that of the system Ni(100) + c(2×2)CO, for which two different groups independently measured and analyzed I/V data, and came to very similar conclusions: CO stands perpendicular to the surface with the C-atoms on top of the Ni atoms; the Ni-C bond length was determined to be 1.8 ± 0.1 Å and the C-O distance 1.1 ± 0.1 Å [17,18]. Figure 2 is reproduced to illustrate the degree of agreement between best-fit and experiment. It is worth mentioning that surface relaxations occurring at clean surfaces are usually changed, and surface reconstructions can be removed or induced upon adsorption of (sometimes very small amounts of) foreign species [19,20].

Very recently, the stringent restrictions owing to computer limitations have been relaxed by new methods which take advantage of the symmetry properties of the surface structure [21] or by using the concept of 'beam set neglect' in which third and higher-order multiple scattering is neglected [22]. In this way, more complicated structures with larger unit cells like incommensurate overlayers [21] and adsorbate structures of organic molecules [22] also seem to have been successfully tackled. As an example, Fig. 3 shows the case of a large-surface unit cell formed by as large a molecule as benzene on Rh(111).

It must be emphasized that with X-rays and electrons, light atoms in the presence of heavy ones are not easily 'seen' as their scattering power is proportional to the square of the respective electron numbers [23]. Furthermore, the relatively high energies may cause disorder in or damage of adsorbate phases; the latter problem can be circumvented by applying the video-LEED technique developed recently by MUELLER et al. [24].

4. Reflection High-Energy Electron Diffraction (RHEED)

In RHEED, electrons with energies between 1 and 10 keV are used. In this energy range, the mean free path becomes as large as 30 - 100 Å so that the surface sensitivity is lost, unless grazing incidence and emergence are used. According to the large k-vectors and large angles of incidence used, the Ewald sphere is very large compared to the reciprocal-lattice spacing. This causes the (00) and (proper-

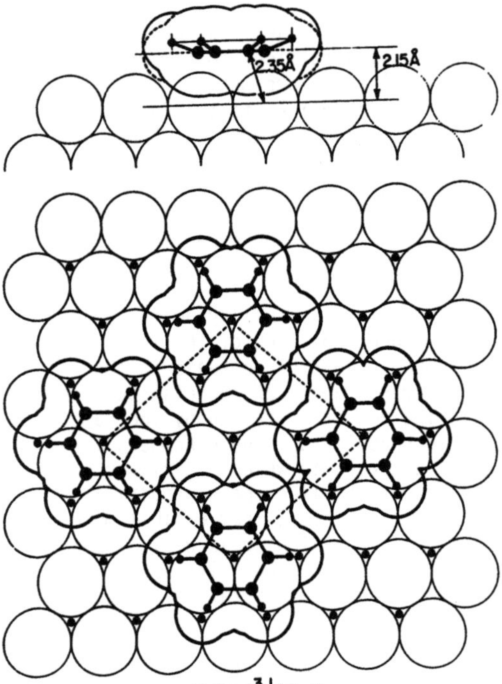

Fig. 3. Side and top views of the structure for C_6H_6 on Rh(111) as proposed by VAN HOVE et al. [22] on the basis of LEED I/V analysis. The hydrogen positions are assumed. Van der Waals' radii of 1.8 Å for C and 1.2 Å for H are indicated. The benzene on the right side shows an in-plane distortion as preferred by the authors (C-C bond lengths 1.25 and 1.6 Å). The side view shows possible CH bending away from the surface.

ly aligned) neighboring lattice rods to be touched along a great length, so that the RHEED pattern of a flat surface shows long streaks normal to the surface [2]. On the other hand, small hills on the flat surface through which the RHEED beam must pass give rise to three-dimensional diffraction, and spots are observed instead of streaks. It is clear that this method is particularly well suited to study growth mechanisms on flat surfaces.

Thus, RHEED was recently used by SEGUIN et al. [25] in a study of the growth mechanism of noble gases and molecular nitrogen on graphite. According to the relative strengths of the atom-atom interaction energies on the overlayer E_a and the adatom-substrate interaction energies E_s, one expects from theoretical considerations either layer-by-layer growth ($E_s > E_a$, 'complete wetting') or cluster growth ($E_s \lesssim E_a$, 'incomplete wetting'). For layer-by-layer condensation, the RHEED diffraction pattern is composed of parallel streaks, whereas spots appear when the adsorbate forms small bulk-like crystallites. Incomplete wetting was found to occur for He, Ne and N_2, whereas layer-by-layer growth was observed up to ten layers for Ar and Xe, in surprising contradiction to theoretical expectations.

In earlier work, MITCHELL et al. [26] applied RHEED in a study of the oxygen adsorption on and the oxidation of a Ni(110) surface. As long as pure chemisorption phases occur, the RHEED pattern exhibits streaks. The onset of oxidation is signaled by the spontaneous formation of small oxide nuclei at a critical oxygen coverage,which give rise to a pattern of broad spots. The widths of the spots [10] could be used to estimate the thickness and lateral extension of the oxide nuclei, and their lateral growth upon further oxygen uptake could be followed.

5. Photoelectron Diffraction

Photoelectron waves emitted at a particular atom in photoemission processes by ultraviolet and X-ray radiation can scatter off neighboring surface atoms in close analogy to LEED; the resulting interference pattern contains information on the distances between the scattering centers [27]. For a regular adsorbate system for example, by varying the photon energy and measuring intensity with the detector fixed in a certain escape direction, interference maxima and minima are observed related to the height of the adsorbate above the plane of substrate atoms [28]. Another possibility consists in changing the escape direction [29]. Whereas the first situation is sensitive to the distances between atomic planes, the second is more sensitive to site symmetry, particularly when the azimuthal angle is changed by rotating the surface about its normal [30].

As an example of the first case, we again discuss the system CO on Ni(100) (see Sect. 3). Figure 4 shows in the inset the intensity of the C 1s core-level photoemission normal to the surface as a function of the kinetic energy of the photoelectrons [28]. The intensity oscillations observed have no analog for the free CO molecule. They arise because of interference between the direct outgoing photo-

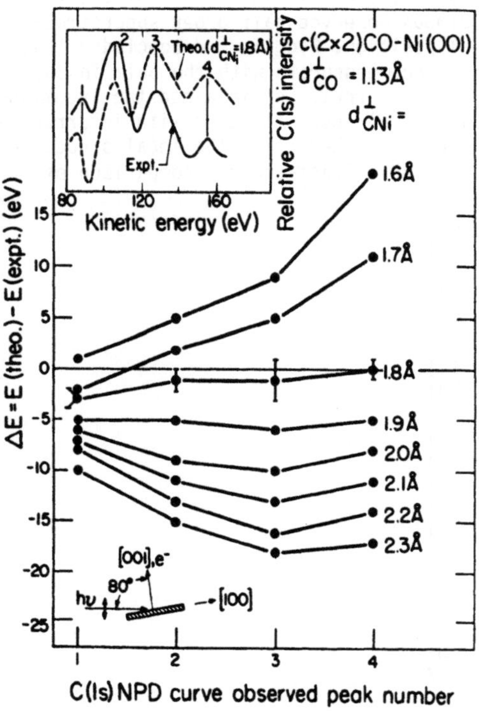

Fig. 4. Photoelectron diffraction from CO on Ni(100) is observed as modulation of the normal emission intensity against photon energy for C 1s emission. The interference maxima and minima yield a C-Ni distance of 1.8 Å in good agreement with LEED results (compare Fig. 2). After KEVAN et al. [28]. © 1981 The American Physical Society

electrons with back-reflected ones from the Ni substrate layers. The interference maxima occur, in a first approximation, when a multiple of the photoelectron wavelength matches the spacing between the C-atom and the outermost plane of Ni atoms. Owing to phase-shifts upon back reflection and multiple scattering, similar complications as in LEED theory occur and a full LEED-type calculation is necessary to obtain the accurate distances. The best-fit C-Ni distance of 1.8 Å is in good agreement with the LEED results of Refs. 17 and 18 (see Fig. 2). In order to determine the C-O distance, backscattering of photoelectrons from the O 1s level has to be investigated.

6. Glancing Incidence X-Ray Diffraction

As in the case of high-energy electrons, X-rays can be made surface-sensitive by using glancing angles of incidence. The method was developed by MARRA et al. [31] and will be especially powerful in connection with the high flux synchrotron sources. The X-ray beam, usually incident at angles less than 1°, is reflected out of the surface at small angles and diffracted parallel to the surface over large angles, so that surface Bragg reflections can be measured. The important advantage of the use of X-ray lies in the fact that the simple kinematical theory [10] can be used for data interpretation. It is clear that this method has excellent prospects for structural studies at interfaces.

Surface X-ray diffraction was applied by ROBINSON [32] in a study of the (1×2) reconstruction of the Au(110) surface. ROBINSON observed half-order superstructure peaks to be displaced from their ideal position, Fig. 5, and concluded that the surface has a long-range reconstruction incommensurate with the bulk in the direction perpendicular to the close-packed metal rows. He explained the incommensurability by an overlayer that is locally a (1×2) reconstruction with interruptions by domain walls perpendicular to the close-packed rows. The local structure was determined by a Patterson procedure [10] and a least-squares comparison of

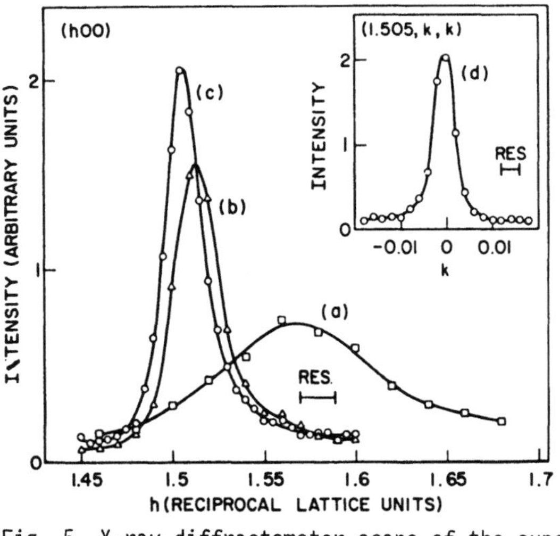

Fig. 5. X-ray diffractometer scans of the superlattice peak near (3/2,0,0) for the Au(110) (1×2) reconstructed surface. (a-c) Radial scans along the [100]-direction (perpendicular to the close-packed metal rows). (a) Sample cut at 1.5° from the ideal surface orientation. (b, c) Samples cut better than 0.1°. (d) Scan parallel to the close-packed rows of the peak in (c). The instrumental resolution is indicated. After ROBINSON [32]. © 1983 The American Physical Society

measured and calculated structure factors; in the first layer, every second surface row is missing whereby the spacing between the first two metal layers is increased; the atoms in the second layer are displaced laterally by ~0.12 Å away from the topmost atoms.

7. Atomic-Beam Diffraction

The development in the last few years of high-pressure nozzle sources yielding highly monochromatic neutral particle beams of thermal energies (20 - 300 meV) and sufficient intensity [33] has allowed particle-surface scattering experiments to be performed under ultra-high vacuum conditions. Molecular-beam scattering is presently being used in many branches of surface science,like in studies of the particle-surface interaction potential [34], of surface phonon dispersions [35], of global energy exchange [36], of surface diffusion [37], and of substrate-mediated chemical reactions [38].

Here, we concentrate on the use of He diffraction in surface-structure studies [39]. The nature of the scattering centers in the case of He scattering is revealed by a discussion of the He/surface-interaction potential: At distances not too far from the surface, the He atoms feel an attraction due to van der Waals' forces, the depth of the potential ($D \lesssim 10$ meV) never being very large compared to the incoming particle energy. Closer to the surface, the particles are repelled owing to the overlap of their electronic densities with that of the surface. This causes a steeply rising repulsive part of the potential. The classical turning points are usually farther away on top of the surface ions rather than between them, which gives rise to a periodic modulation of the repulsive part parallel to the surface. The modulation of the repulsive part of the He-surface potential follows closely the contours of constant electron density corresponding to the normal energy of the incoming particles [40], which often reflects directly the geometrical arrangement of the surface atoms. In analyzing intensities, one usually resorts to the so-called hard corrugated-wall (HCW) model [41,42], in which the repulsive part is assumed to be infinitely steep, and the lateral modulation is described by a corrugation function whose amplitude is energy-dependent; this assumption has been shown to give reliable results for the corrugation shapes and amplitudes,as long as small angles of incidence are involved [43]. With the hard corrugated-wall model, quick calculational procedures [41] have been developed; they are necessary for analyses of complicated diffraction patterns which may arise if many particles contribute to forming the corrugation in a single unit cell.

It must be emphasized that the corrugation function is a replica of the electron-density contour corresponding to the energy of the incoming He beam; thus, simple charge-surface density calculations by overlapping atomic charge densities allowed determination of adatoms with an accuracy of ± 0.1 Å in some test cases [44]. Owing to the small energies of the incoming particles, He scattering has the very important advantage of being far less destructive than any other surface method. As the surface electron density is drastically influenced by any adsorbate (including hydrogen), the method is especially valuable in studying light adsorbates on heavy substrates, problematic with almost all other methods. Last but not least, it is equally well applicable to metal, semiconductor and insulator surfaces.

As a beautiful and physically interesting example of surface-structural investigations with He diffraction, we discuss the adsorbate system hydrogen on Ni(110), where six differently ordered adsorption phases of hydrogen were identified with He scattering [45,46]. The arrangement of the adatoms in the different phases

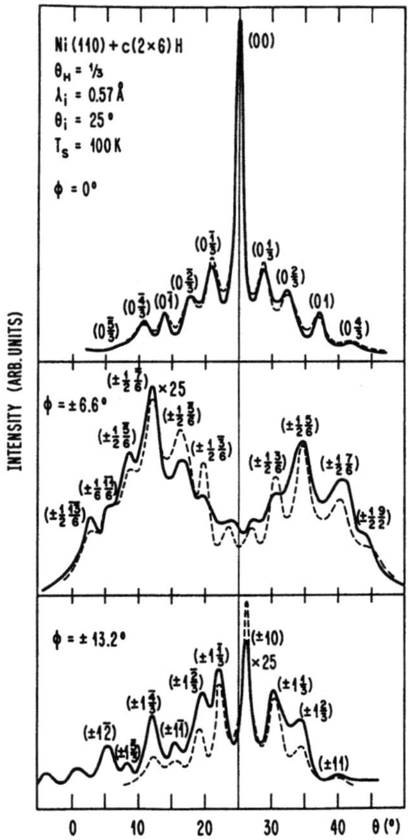

Fig. 6. In plane and out-of-plane He-diffraction traces for the c(2×6) phase of hydrogen on Ni(110) corresponding to a coverage of 1/3 monolayer. The full line corresponds to the experimental result, the broken line to the best-fit calculation. The corresponding corrugation function is shown in Fig. 7b. After [46], © 1983 The American Physical Society

gives a fascinating picture of the coverage-dependent ordering of the hydrogen atoms on this particular surface. The phases observed are: three different c(2×6) phases corresponding to coverages Θ of 1/3, 2/3 and 5/6 monolayers (ML), a c(2×4) with Θ = 1/2 ML, a (2×1) with 1 ML, and a (1×2) phase with the saturation coverage of 1.5 ML. Experimental diffraction scans for the lowest coverage c(2×6) are exhibited in Fig. 6 together with hard-wall best-fit curves. The corrugations of the clean surface and of two c(2×6) phases are shown,together with ball models of the surface structures in Figs. 7a-c. All configurations provide a direct picture of the adatom configurations,as every adsorbed hydrogen atom produces a distinct corrugation hill on the Ni(110) substrate. The hydrogens form zig-zag chains along the close-packed Ni rows even at low coverages with the adatoms in shifted two-fold coordinated sites. The exact location of the adatoms relative to the substrate is not visible direct from the corrugation functions, but has been predicted by theoretical calculations [47] and verified by comparing the observed (2×1) corrugation with charge-density calculations for different adsorption sites [48]. The lateral interaction of these zig-zag chains is so long-ranged that at 1/3 ML they form a sufficiently well-ordered c(2×6) phase with alternating zig-zag and zag-zig configurations at distances of 10.6 Å, whereby two close-packed Ni rows between the hydrogen chains remain adsorbate-free (Fig. 7b). In the next ordered c(2×4) phase, the H-chains cover half the surface,leaving each other Ni row adsorbate-free. The following ordered phase corresponding to 2/3 ML has c(2×6) periodicity; here, two zig-zag and two zag-zig chains alternate,leaving one close-packed Ni row adsorbate-free (Fig. 7c). The fourth ordered structure again has

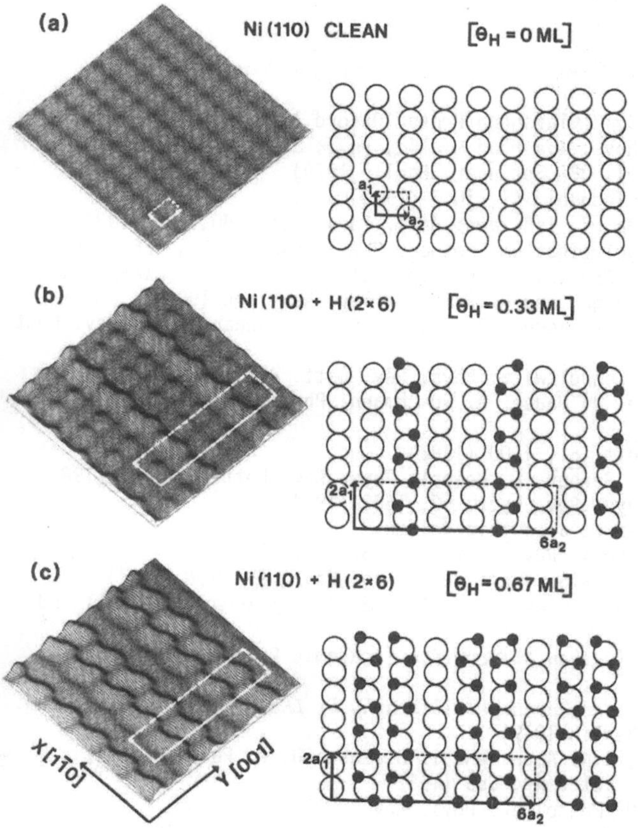

Fig. 7a-c. Corrugation functions (left side) and hard-sphere models (right side) of the clean Ni(110) surface (a) as well as two of the six ordered phases of hydrogen on Ni(110) (b) and (c). Note that the corrugations yield a direct picture of the adsorbate configurations: every pronounced hill corresponds to a H-atom. The formation of hydrogen zig-zag chains and their ordering with increasing coverage is clearly visible. The corrugations are expanded by a factor of five in the vertical direction. The maximum corrugation amplitude is ~0.27 Å for all hydrogen phases. For the clean surface, the corrugation is ~0.07 Å perpendicular and ~0.03 Å parallel to the close-packed Ni rows. After [46], © 1983 The American Physical Society

c(2×6) periodicity and corresponds to 5/6 ML; the extra 1/6 ML of H goes into the energetically obviously less-favored two-fold coordinated sites on the previously adsorbate-free Ni rows, in this way forming a distorted hexagonal overlayer of H-adatoms; this is the only phase up to 1 ML in which two different adsorption sites are occupied. It may act as a precursor to build up the (2×1) with 1 ML, where the adatoms form a slightly denser distorted hexagonal pattern with all close-packed Ni rows covered with parallel hydrogen chains; the last c(2×6) may be necessary to facilitate the phase jump of the zag-zig chains along the close-packed Ni rows in order to make all chains parallel. The transition to the (1×2) saturation phase involves an adsorbate-induced substrate reconstruction [20] with partial hydrogen accommodation in the second layer. A similar hydrogen-induced reconstruction leads to effective channels for hydrogen uptake into the bulk in the case of Pd(110) [49].

References

1. G. Ertl, J. Kueppers: *Low Energy Electrons and Surface Chemistry* (Verlag Chemie, Weinheim 1974)
2. M. Prutton: *Surface Physics* (Clarendon Press, Oxford 1975)
3. G.A. Somorjai, M.A. Van Hove: *Adsorbed Monolayers on Solid Surfaces* in *Structure and Bonding*, Vol. 38 (Springer, Berlin 1979)
4. R.M. Tromp: J. Vac. Sci. Technol. A1, 1047 (1983)
   M. Aono, Y. Hou, C. Oshima, Y. Ishizawa: Appl. Phys. Lett. 40, 178 (1982)
   E. Sailer, C. Varelas: Nucl. Instrum. Methods Phys. Research B2, 326 (1984)
5. J. Stoehr: In *Principles, Techniques and Applications of EXAFS, SEXAFS and XANES*, ed. by D. Koningsberger and R. Prins (Wiley, New York 1984)
   J. Stoehr, K. Baberschke, R. Jaeger, R. Treichler, S. Brennan: Phys. Rev. Lett. 47, 381 (1981)
   J. Stoehr, R. Jaeger, T. Kendelewitz: Phys. Rev. Lett. 49, 142 (1982)
   U. Doebler, K. Baberschke, J. Haase, A. Puschmann: Phys. Rev. Lett. 52, 1437 (1984)
6. L.D. Marks: Phys. Rev. Lett. 51, 1000 (1983)
   G. Binnig, H. Rohrer, Ch. Gerber, E. Weibel: Phys. Rev. Lett. 50, 120 (1983)
   H.W. Fink, G. Ehrlich: Phys. Rev. Lett. 52, 1532 (1984)
7. M. Henzler and P.K. Larsen et al.: this volume
8. H. Richter, U. Gerhardt: Phys. Rev. Lett. 51, 1570 (1983)
9. F. Jona, J.A. Strozier Jr., W.S. Yang: Rep. Prog. Phys. 45, 527 (1982)
10. R.W. James: *The Optical Principles of the Diffraction of X-Rays* (G. Bell and Sons Ltd., London 1950)
11. J.B. Pendry: *Low Energy Electron Diffraction* (Academic Press, London and New York 1980)
12. M.A. Van Hove, S.Y. Tong: *Surface Crystallography by LEED*, Springer Ser. Chem. Phys. Vol. 2 (Springer, Berlin 1979)
13. D.L. Adams, H.B. Nielsen, J.N. Andersen, I. Steensgard, R. Feidenhans'l, J.E. Sorensen: Phys. Rev. Lett. 49, 669 (1982)
14. U. Landmann, R.N. Hill, M. Mostoller: Phys. Rev. B 21, 448 (1980)
15. J. Sokolov, F. Jona, P.M. Marcus: IBM Research Report RC 10207 (1983)
16. R.N. Barnett, U. Landman, C.L. Cleveland: Phys. Rev. Lett. 51, 1359 (1983)
17. S. Andersson, J.B. Pendry: Phys. Rev. Lett. 43, 363 (1979)
18. M. Passler, A. Ignatiev, F. Jona, D.W. Jepsen, P.M. Marcus: Phys. Rev. Lett. 43, 360 (1979)
19. H.D. Shih, F. Jona, D.W. Jepsen, P.M. Marcus: Phys. Rev. Lett. 37, 1622 (1976)
    J.F. van der Veen, R.M. Tromp, R.G. Smeenk, F.W. Saris: Surf. Sci. 82, 468 (1979)
    T. Narusawa, W.M. Gibson, E. Toernquist, Phys. Rev. Lett. 47, 417 (1981)
    W. Ho, R.F. Willis, E.W. Plummer: Phys. Rev. Lett. 40, 1463 (1978)
20. J.Demuth: J. Colloid. Interface Sci. 58, 184 (1977)
    K. Griffiths, P.R. Norton, J.A. Davies, W.N. Unertl, T.E. Jackman, Surf. Sci. (in press)
21. W. Moritz: Habilitationsschrift, University of Munich (1983)
22. M.A. Van Hove, Rongfu Lin, G.A. Somorjai: Phys. Rev. Lett. 51, 778 (1983)
23. R.J. Behm, K. Christmann, G. Ertl: Surf. Sci. 99, 320 (1980)
24. K. Heinz, K. Mueller: In Springer Tracts in Modern Physics 91 (Springer, Berlin 1982)
25. J.L. Seguin, J. Suzanne, M. Bienfait, J.G. Dash, J.A. Venables: Phys. Rev. Lett. 51, 122 (1983)
26. D.F. Mitchell, P.B. Sewell, M. Cohen: Surf. Sci. 69, 310 (1977)
27. F.J. Himpsel: Advances in Physics 32, 1 (1983)
28. S.D. Kevan, R.F. Davis, D.M. Rosenblatt, T.G. Tobin, M.G. Mason, D.A. Shirley, C.H. Li, S.Y. Tong: Phys. Rev. Lett. 46, 1629 (1981)
29. H.H. Farrell, M.M. Traum, N.V. Smith, W.A. Roger, D.P. Woodruff, P.D. Johnson: Surf. Sci. 102, 527 (1981)

M.W. Kang, C.H. Li, S.Y. Tong: Solid State Commun. 36, 149 (1980)
30. S. Kono, S.M. Goldberg, N.F.T. Hall, C.S. Fadley: Phys. Rev. B 22, 6085 (1980)
J.G. Tobin, L.E. Klebanoff, D.H. Rosenblatt, R.F. Davis, E. Umbach, A.G. Baca, D.A. Shirley, Y. Huang, W.M. Kang, S.Y. Tong: Phys. Rev. B 26, 7076 (1982)
31. W.C. Marra, P. Eisenberger, A. Cho: J. Appl. Phys. 50, 6927 (1979)
P. Eisenberger, W.C. Marra: Phys. Rev. Lett. 46, 1081 (1981)
32. I.K. Robinson: Phys. Rev. Lett. 50, 1145 (1983)
33. J.P. Toennies, K. Winkelmann: J. Chem. Phys. 66, 3965 (1977)
34. H. Hoinkes: Rev. Mod. Phys. 52, 933 (1980)
G. Boato, P. Cantini: In *Advances in Electronics and Electron Physics*, Vol. 60 (Academic Press, New York 1983)
35. P. Toennies: In Springer Ser. Chem. Phys. 21 (Springer, Berlin 1982)
36. A.W. Kleyn, A.C. Luntz, D.J. Auerbach: Phys. Rev. Lett. 47, 1169 (1981)
37. B. Poelsema, L.K. Verheij, G. Comsa (preprint)
38. T. Engel, G. Ertl: J. Chem. Phys. 69, 1267 (1978)
39. K.H. Rieder: In Springer Ser. Chem. Phys. 21 (Springer, Berlin 1982)
T. Engel, K.H. Rieder: In Springer Tracts in Modern Physics 91 (Springer, Berlin 1982)
40. N. Esbjerg, J. Norskov: Phys. Rev. Lett. 45, 807 (1980)
J. Harris, A. Liebsch: J. Phys. C 15, 2275 (1982)
J. Harris, A. Liebsch: Phys. Rev. Lett. 49, 341 (1982)
41. U. Garibaldi, A.C. Levi, R. Spadacini, G.E. Tommei: Surf. Sci. 48, 649 (1975)
42. N. Garcia: J. Chem. Phys. 67, 897 (1977)
43. K.H. Rieder: Surf. Sci. 117, 13 (1982)
K.H. Rieder, N. Garcia: Phys. Rev. Lett. 49, 43 (1982)
44. K.H. Rieder, M. Baumberger (to be published)
45. K.H. Rieder, T. Engel: Phys. Rev. Lett. 43, 373 (1979)
K.H. Rieder, T. Engel: Phys. Rev. Lett. 45, 824 (1980)
K.H. Rieder, T. Engel: Surf. Sci. 109, 140 (1981)
46. K.H. Rieder: Phys. Rev. B 27 (RC) 7799 (1983)
47. J.P. Muscat: Surf. Sci. 110, 389 (1981)
48. D.R. Hamann: Phys. Rev. Lett. 46, 1227 (1981)
49. K.H. Rieder, M. Baumberger, W. Stocker: Phys. Rev. Lett. 51, 1799 (1983)

Defects at Surfaces

Martin Henzler

Institut für Festkörperphysik, Universität Hannover, Appelstraße 2
D-3000 Hannover, Fed. Rep. of Germany

1. Introduction

In solid state physics, two informations are most important for the description of the properties of a solid: the chemical composition and the atomic arrangement. The simplest solid is a single crystal out of one kind of atoms. Solid state physics therefore starts with the description of an ideal, that is, defect-free single crystal. Many properties may be described in this way. It is, however, necessary to proceed to real crystals by introducing more and more chemical and structural variations, since small variations may cause drastic changes, for example dislocations for mechanical strength of metals, impurities for the color of glass, or doping for the conductivity of semiconductors.

In surface physics exactly the same quantities are important. Since ultra high vacuum and Auger electron spectroscopy are standard, the chemical composition of a surface is usually well controlled quantitatively. Methods for cleaning a surface and introducing a known amount of adlayers are well established. The situation with respect to structural information is less clear. Usually, the structure is checked with a diffraction technique like Low Energy Electron Diffraction (LEED) or Reflection High Energy Electron Diffraction (RHEED) on a more qualitative basis with the guideline that a "good pattern" (= bright spots, low background, reproducibility) means a "good surface", which is meant as useful for any kind of experiment. As in bulk solid state physics, however, small amounts of defects may have drastic effects, like electronic surface states or electron traps at semiconductor surfaces (for example atomic steps at the Si/SiO_2 interface of MOS-devices). There are additional properties, specific to surfaces, which may rely heavily on defects: the crystal growth needs incomplete layers as an intermediate stage or surface adsorption, desorption and heterogeneous catalysis may be influenced or even determined by so-called active sites.

It is therefore important to know the presence and amount of all kinds of defects to understand surface properties and control surface processes or devices.

2. Systematic description of defects

An ideal crystal is strictly periodic in three dimensions with respect to composition and arrangement. Therefore an ideal surface should be strictly periodic in two dimensions. Usually, for an ideal surface, all surface atoms should be in positions given by the bulk periodicity, so that the surface periodicity is the same as that of a parallel bulk layer. Any deviation from an ideal surface may still have some periodicity, although different in dimension and absolute length. It is therefore helpful to arrange defects according to their dimension of remaining periodicity (fig. 1). Point defects may be random or correlated (for example, it may be unlikely or especially likely to find two point defects close to each other). One-dimensional defects are the border lines of limited two-dimensional entities (like terraces or superstructure domains) and two-dimensional defects have periodicities different to that of the ideal surface in absolute value (superstructure) or direction (facet). Due to their periodicity, superstructures are not always considered as defects. Three-dimensional defects are not directly surface defects, they may, however, be caused by surface processes, in any case they are important for many surface studies. All defects have characteristic diffraction patterns, which are discussed later.

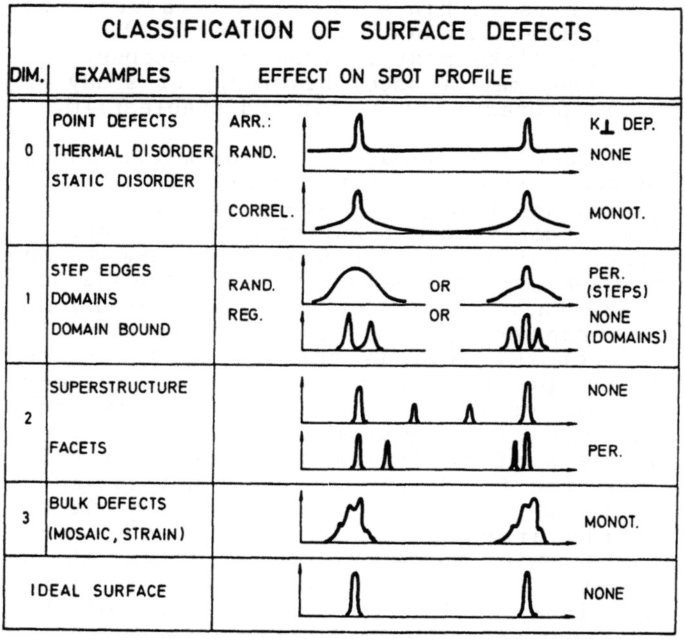

Fig. 1: Classification of surface defects with respect to their dimension (DIM). For the effect on the spot profile the arrangement (ARR) is important: random (RAND), correlated (COR) or regular (REG). The energy or $K_\perp$ dependence is shown on the right side as periodic (PER) or monotonous (MONOT).

Another important parameter for all defects is the kind of deviation from periodicity. Three kinds should be distinguished:
a) chemical composition (like periodic adlayers with or without superstructure),
b) incomplete surface layer (like random or periodic vacancies) with all atoms on regular bulk sites
c) deviation from regular bulk sites (like periodic shifting to produce superstructure or thermal disorder).

Usually, a combination of different techniques is required to characterize a defect with respect to all parameters.

3. Experimental techniques

Any technique should fulfill the following three requirements: a) atomic resolution: that means that the wavelength of the radiation used has to be smaller than atomic distances, b) surface sensitivity, so that the contribution of the bulk may be separated c) ultra high vacuum (uhv) compatibility, so that the surface is not changed during investigation.

The wave length of some radiation is shown in fig. 2 with some hints to measurement techniques. The surface sensitivity is strongly dependant on energy. For atoms and ions with low energy (a few eV) and electrons (20-200 eV) the penetration depth is within a few atomic layers, for thermal atoms the reflection occurs in front of the top layer. Many techniques (like most elctron microscopes) are not uhv compatible; they may be used, however, for study of interfaces, which are protected by a thin and stable film.

For imaging of a surface frequently electron microscopes are used (1). Whereas scanning microscopes (SEM), due to their limitated resolution (except for scanning transmission microscope STEM) and poor vacuum condition, are used mainly for study of coarse features (5-100 nm), the transmission microscopes (TEM) show atomic resolution (less than 0.2 nm). Samples have to be thinned to 5-100 nm (depending on material and electron energy). Surface sensitivity is obtained by using cross-sections through an interface structure, so that the projection of the interface is seen (2). Or special features of the interface are used for imaging (e.g. thin layer of heavy atoms, excess atoms made visible with the help of forbidden bulk spots, misfit dislocations at the interface). In some cases uhv has been obtained in the microscope, so that free surfaces could be studied with nearly atomic resolution (3). A recent development is the Scanning Tunnel Microscope (STM), which meets all requirements for surface studies (4). By scanning a piezo driven tip in a fraction of a nm in front of the surface, a tunnel current from the tip to the surface is used to probe the surface atom arrangement with atomic resolution.

Wheras the TEM provides both image and diffraction pattern, there are several techniques which are designed to provide

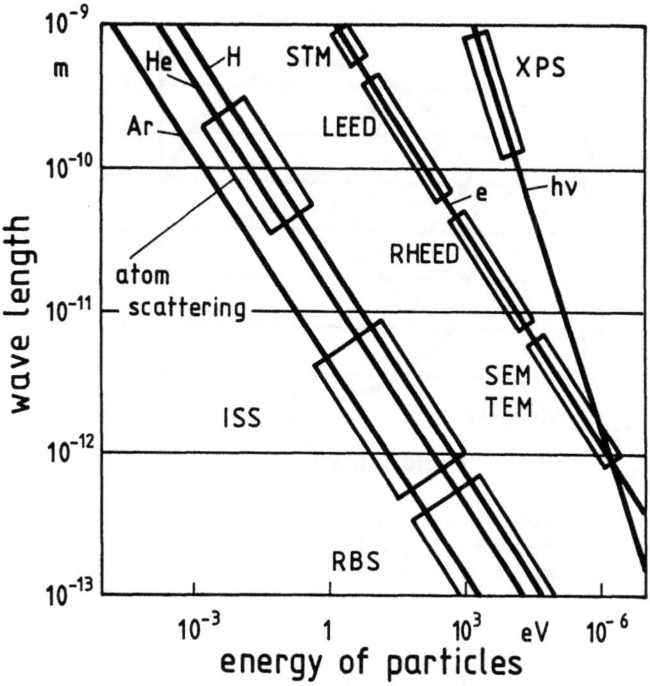

Fig. 2: Wave length of radiation used for study of structural properties of surfaces. The abbreviation for the radiation is: hv for electromagnetic radiation, e for electrons and H, He and Ar for hydrogen, helium and argon resp.. The boxes stand for typical ranges of the following techniques: XPS - x-rayPhotoemission Spectroscopy; STM = Scanning Tunnel Microscopy; LEED = Low Energy Electron Diffraction, RHEED = Reflection High Energy Electron Diffraction, SEM = Scanning Electron Microscopy, TEM = Transmission Electron Microscopy, ISS = Ion Scattering Spectroscopy, RBS = Rutherford Back Scattering.

just the diffraction pattern in the uhv and with high surface sensitivity.

The most widely used technique is Low Energy Electron Diffraction (LEED), using the energy range of minimum penetration depth (5). An arrangement is shown in Fig. 3. If high energy (RHEED) is used, gracing incidence is needed, to reduce the (projected) penetration depth for surface sensitivity and the scattering vector for reduced thermal diffuse scattering (Fig. 4). Diffraction is also possible with atom scattering, using thermal atom beam or a monochromatized beam with the help of a high-pressure nozzle beam (6).

For all diffraction experiments, an understanding is possible with the help of the reciprocal lattice. Out of the two unit vectors a_1 and a_2 within the surface a set of two reciprocal vectors a_1^* and a_2^* are formed, which fulfill the

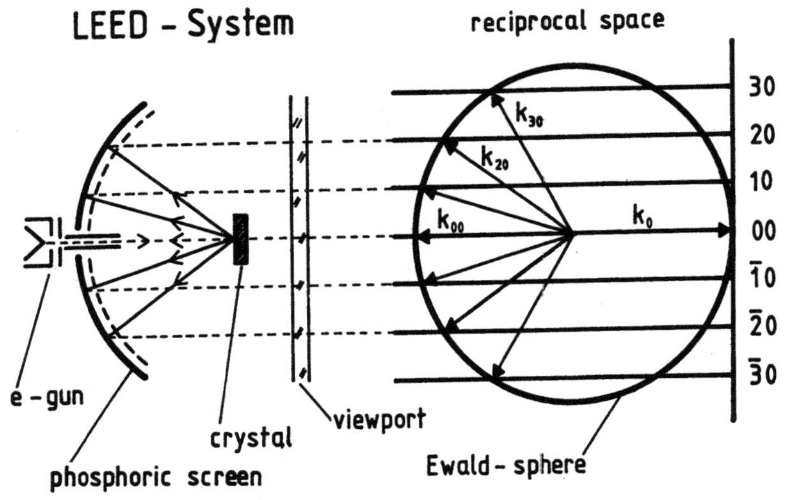

Fig. 3: Schematic arrangement of a LEED-system and the construction of the diffraction pattern in reciprocal space with the help of the Ewald sphere (for normal incidence of the electron beam).

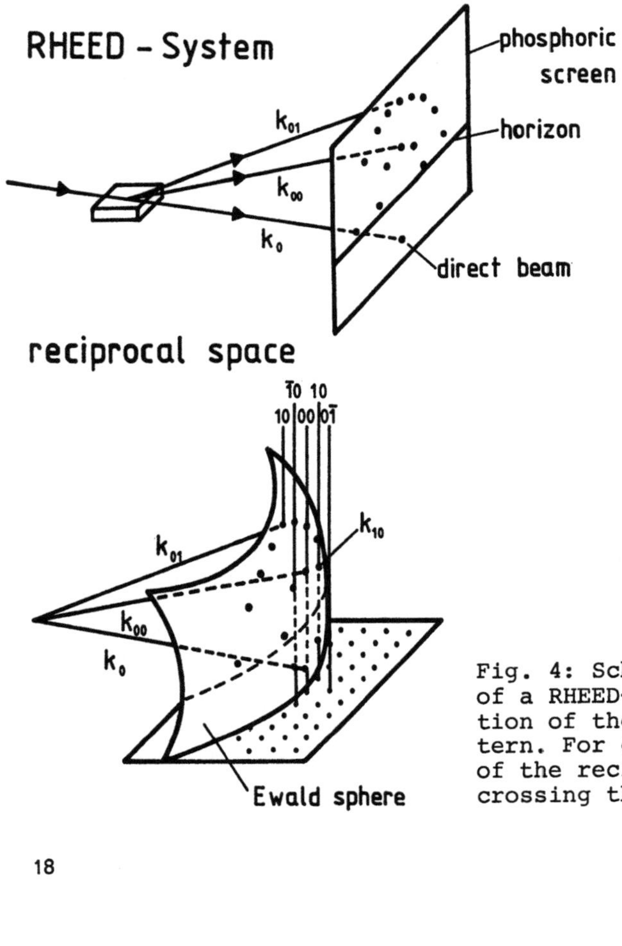

Fig. 4: Schematic arrangement of a RHEED-system and construction of the diffraction pattern. For clarity only a few of the reciprocal rods are shown crossing the Ewald sphere.

requirements $a_i\ a_j^* = 2\pi\delta_{ij}$. Since any component of a_i^* normal to the surface does not affect the requirement, all possible reciprocal vectors and their integer multiples form a set of rods normal to the surface. For any incoming waves exp ($ik_o r$) a diffracted wave exp (ikr) is found if

$$/k/ = /k_o/ \text{ (elastic scattering) and}$$

$$K = k - k_o = ha_1^* + ka_2^*$$

(meeting the Laue condition for scattering). Therefore, the Ewaldsphere is an easy way to predict or evaluate a diffraction pattern, as described in any text book on solid state physics. For LEED and RHEED the Ewald construction is demonstrated in figs. 3 and 4 resp..

For periodic structures, the rods are infinitely sharp, and therefore the diffraction spots as sharp as given by the instrument. Any new periodicity is seen as new ("extra") spots besides spots already expected from the ideal surface ("normal spots").

Since the normal spots are assigned with integer indices, the extra spots have fractional indices. In fig. 5 it is demonstrated in the first two rows, how extra periodicities (superstructures) produce sharp extra spots.

Strict periodicity means periodicity to infinity, or at least beyond that range, which is coherently covered by the

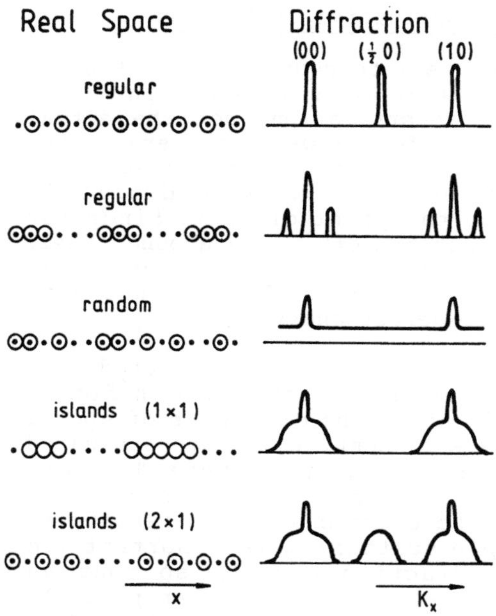

Fig. 5: Schematic representation of diffraction patterns if half of the sites in a regular array are occupied in different arrangements.

beam of the instrument (see lecture II). If the periodic structure is smaller than that range, then an essential new feature arises: the diffraction spot is broadened due to the finite number of scatterers (see textbook on physics for diffraction on a grating with a finite number of slits). The half width is now proportional to inverse of the number of scatterers. If the arrangement of scatterers is no more in periodic groups, rather at random distributed over the lattice sites, a homogenous background is produced (7). Some examples for atoms distributed over lattice sites are shown in fig. 5.

Atoms may be displaced out of regular bulk lattice sites at the surface due to the different numbers and arrangements of neighbors. This may result in a homogenous displacement (e.g. a reduced top layer distance) with no effect on background and spot shape (only spot intensity is altered) or in a periodic displacement producing sharp extra spots as those in fig. 5. If a non-periodic displacement occurs (due to thermal vibration or static disorder after bombardment or due to disordered low-temperature adsorption), the same background as with random adsorption on lattice sites is observed. Now, however, the background increases and the spot intensity decreases with increasing scattering vector (Debye-Waller-factor, see fig. 7). In fig. 1 for each defect is shown in principle which features are expected and which features are important to distinguish the defects qualitatively (fig. 8).

4. Schematic description of quantitative evaluation

A diffraction pattern is the absolute square of the Fourier transform of the surface atom arrangement. The back transformation produces the autocorrelation, not the actual surface. The information is therefore always an average over a given area, which is both an advantage (high precision, easy averaging) and a disadvantage (more difficult identification) depending on the point of view. Since for microscopy techniques it is just the other way around, their respective merits are supplementary to each other as summarized in fig. 6. It is therefore helpful to use both microscopy and diffraction for defect analysis.

Since diffraction patterns are not so easily evaluated as a direct image, first the information is summarized in fig. 7. The easiest information is that of the existence of spots. It shows that periodicities are present, which are evaluated with diffraction angles or with the Ewald sphere. The next information is the integral intensity within the spots. Due to the strong scattering of the electrons (which results in surface sensitivity) multiple scattering is important (in contrary to most x-ray diffraction cases), so that an extensive calculation is required, which is so far possible only for periodic arrangements with small unit meshes. Then by fitting of calculated and measured intensity-versus-voltage curves (IV-curves), the atom position within a unit mesh is determined. A third information is the spot profile

Comparison of Diffraction and Imaging	LEED, RHEED atom diffraction	STM, TEM SEM
Qualitative evaluation	−	+
Quantitative evaluation	+	−
Advantages	high precision good averages	easy identification single or polycrystals
Disadvantages	only single crystals only averages	averages troublesome

Fig. 6: Comparison of diffraction (e.g. LEED) and imaging methods (e.g. STM) with respect to defect detection on surfaces.

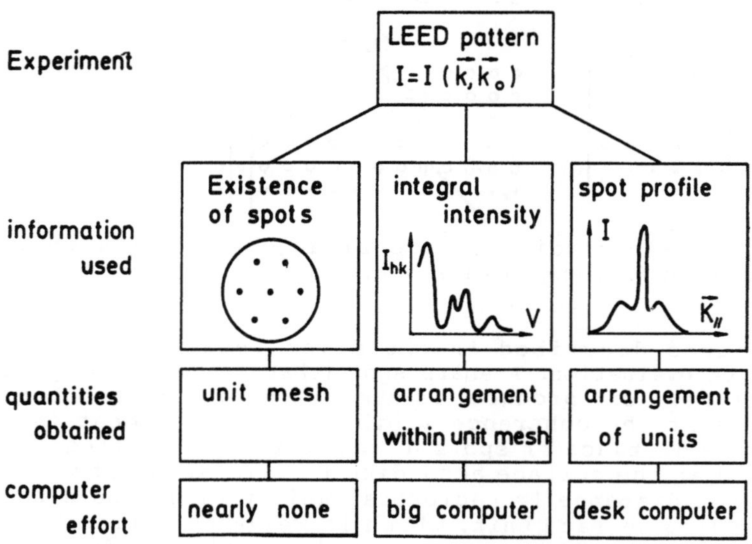

Fig. 7: Schematic representation of the information in a diffraction pattern.

(splitting or broadening) which yields information on the arrangement of units, not the content of a unit. The content of a unit is only available if both intensity and profile are evaluated simultaneously, which so far is not possible. Since defects are large, frequently non-periodic units, the spot profile is the main information used.

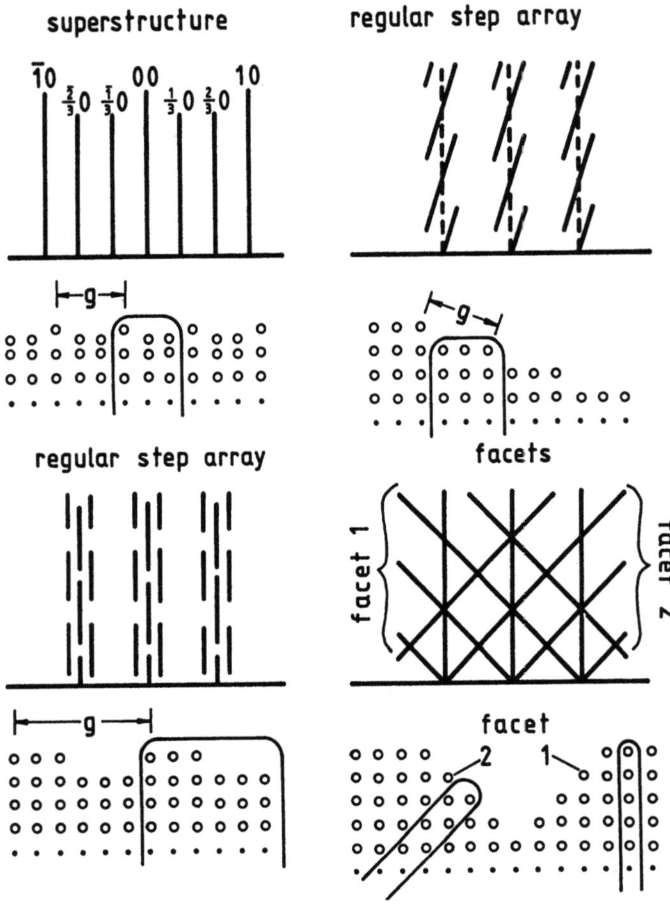

Fig. 8: Some periodic surface arrangements with their representation in reciprocal space.

The features described in the first lecture enable a qualitative evaluation. A sharp spot (that means as sharp as the instrument can produce) is due to a periodic structure over at least the coherence width of the instrument (see below). The distance of spots d^* corresponds to a frequent distance d at the surface according to $d^*/a^*=a/d$ with a^* a normal spot distance (= reciprocal unit vector) and a the corresponding surface unit, which is well-known from bulk structure. The different two-dimensional defects distinguished from the energy-dependance of the spot position as seen from the reciprocal lattice (fig. 8). Each surface is described by a unit,and its periodic repetition by a vector g, as marked in fig. 8. Therefore,its reciprocal lattice (that is its Fourier transform) is given by the product of the Fourier transform of the (finite) unit and the Fourier transform of the (infinite) array of identical units (for a more extensive description, see ref. 9). It is therefore necessary to reconstruct the reciprocal space out

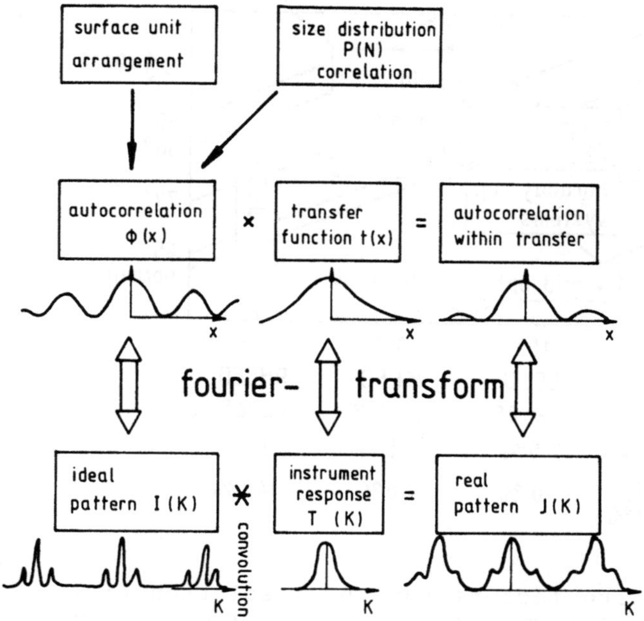

Fig. 9: The quantities which are used in computation and evaluation of a spot profile in a one-dimensional representation: The product of the autocorrelation Φ (x) with the transfer function t(X) (its halfwidth is the transferwidth) yields the autocorrelation within transfer width. From all three functions the counterpart in reciprocal space is obtained by Fourier transform. For evaluation of a measured profile,the scheme may be used in reverse order.

of the observed diffraction pattern, so that the new periodicities and their inclination to the low index face (or average face) are derived.

For diffuse spots the evaluation procedure is the same: the reciprocal space shows especially, if several layers are involved (stepped surface) or if displacements are involved (monotonous decrease of intensity with increasing scattering vector). The half-width (instead of the spot distance) is now an average size of a periodic unit-like terrace or domain (instead of a frequent distance).

For a quantitative analysis of a diffraction profile, a full calculation is needed. The necessary steps are shown in fig. 9. First, a model surface is constructed. Usually a probability distribution for occurrence of a given size (terrace, domain) is chosen. Additionally, assumptions on the correlation are needed. It makes a big difference if the sizes are mixed at random or if, as the opposite correlations, all units of the same size are side by side. Although the probability distribution may be the same for the two cases, the diffraction patterns (and probably some other properties too) are quite different. From that distribution the

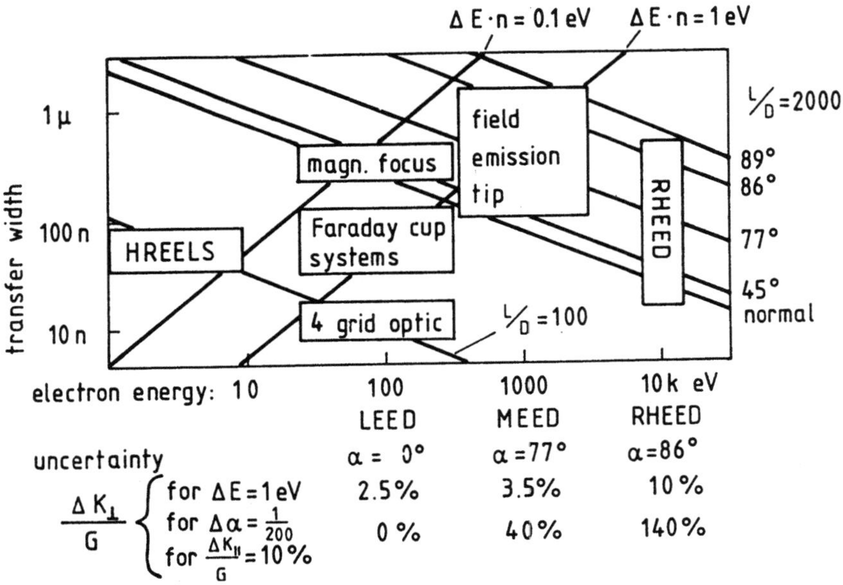

Fig. 10: The limiting factors for the resolution of an electron diffraction system: width D of a recorded spot due detector and source diameter and lense aberration for different distances L from sample to detector. Two typical values for L/D are shown: 100 for usual 4-grid-systems and 2000 for faraday cup systems. The influence of the angle α of the outgoing beam is shown for RHEED application. A second limiting factor is the energy-spread of the electron source, which is only important for non-specular beams (order n, lattice constant a),. A third limiting quantity is the inherent uncertainty in $K_{\perp}$, which is important for gracing incidence. Faraday cup systems are reported in (19,15) magnetic focusing in (17), a field emission tip in (15, 16). RHEED covers a wide range of transfer width with maximum in the plain of incidence and minimum normal to it (20, 24). HREELS stands for High Resolution Electron Energy Loss Spectrometer (40), which usually is not used for diffraction. MEED stands for Medium Energy Electron Diffraction. A review is given in (15).

autocorrelation is calculated, that is the probability of finding another surface atom in the distance $\underline{r}$ (which has a component parallel and one normal to the surface). By Fourier transform, the diffraction pattern is obtained, which is idealized with respect to the instrument and to the scattering (kinematic approximation). Those two additional contributions to the real diffraction pattern deserve a special explanation.

5) The kinematic approximation and inclusion of dynamic effects

The kinematic approximation is mostly used for calculation of the spot profiles. It does not mean, however, that only surface atoms in a single scattering approximation are used. The crystal may be described by units, consisting of one or several surface atoms and all underlying atoms (columns). For a periodic surface, all columns are identical. Therefore the intensity I is exactly given by I = F x G with F the structure factor giving the intensity coming from a single unit, including all multiple scattering contributions within the column or via neighboring columns. The lattice factor is just

$$G = \sum_{n,m} \exp(i\underline{K}(\underline{r}_n - \underline{r}_m)),$$

containing only the scattering vector $\underline{K}$ and the position vectors $\underline{r}_n$ of the units and describes the spot positions. If a surface is disordered, not all columns are exactly the same. For example in fig. 10 a cross-section through a stepped surface shows 14 columns. If all columns are identical and the structure factor F varies only slowly with $\underline{K}$, then the profile of a spot is given by the lattice factor G and therefore the terrace width distribution derived as given in fig. 9. The kinematic approximation, therefore, only neglects differences of the units according to the individual neighborhood. So column 6 in fig. 10 is taken identical to column 5 or 7, which has to be modified, if multiple scattering is taken explicitly into account. An improvement has been presented for overlayer structures, distinguishing two different columns for clean substrate sites and overlayer covered sites. It is demonstrated that the overlayer atomic scattering factor, the overlayer distance and the damping strongly influence the spot shape as a function of energy (13). Dynamic effects are included in (10). Here, the scattering factor of a column depends explicitly on the surrounding. If in fig. 10 only neighboring columns are considered for multiple scattering, at least five different scattering factors F_n (as a function of energy and angle) have to be calculated separately: the columns 2-6 are different, whereas all others agree with one of those (except translation). Then an autocorrelation has to be constructed using a size distribution and arrangement taking into account the probabilities for possible configurations, as described in (10). An actual calculation for the disordered Au(110) 2x1 shows that only the intensity, not the spot profile, however, is different for the kinematic and dynamic calculation (10). The validity range of the kinematic calculation may be checked by measuring the spot profile for different spots and energies. If a strict periodicity with the vectors of reciprocal space is observed, the kinematic approximation with identical structure factors F (varying only slowly with $\underline{K}$) is justified. So far, no experimental results are reported which require dynamic calculations with respect to disorder.

Dynamic effects will be important, as soon as overlayers with atoms different from the bulk are studied. If the overlayer has a superstructure, then only the overlayer columns are contributing to the extra spot. Their shape is therefore completely described within kinematic approximation given above (11,12). For overlayers without superstructure, the normal spots have to be used for analysis. Here, at least two different columns have to be used, which have to be calculated within some approximations (13) or according to the full formalism used for intensity computation.

A second problem is the comparison of one-dimensional model calculations with diffraction patterns of surfaces with two-dimensional disorder. There are only few examples of one-dimensional disorder (for example Au(110) 2x1 in ref. 10). Experimentally, mostly a one-dimensional spot profile with a point detector or slit detector is recorded and evaluated; that means instead of $I\ (K_x,\ K_y)$ the function

$$I_1\ (K_x, 0) \text{ or } I_2\ (K_x) = \int I\ (K_x, K_y)\ dK_y$$

The two functions are quite different, as demonstrated in (14). The terrace width distribution $P_1(N)$ and $P_2(N)$ as derived from the functions I_1 and I_2 by fitting have a different meaning: $P_1(N)$ is the width distribution derived by projecting the whole surface onto the x-axis and treating that composite arrangement as system for (coherent) diffraction. $P_2(N)$ is obtained by treating the atomic rows in x-direction independently, and summing the probabilities of all rows (incoherent superposition). It has therefore to be specified which distribution is meant.

6) Instrumental limitations

As shown in fig. 9 the observable spot profile is derived from the ideal profile by convolution with the instrument response function T(K) (11,12). This convolution implies that the different electron waves producing the profile from an ideal surface (T(K)) are independant, and are therefore superpored incoherently. By Fourier transform of the convolution, a product of Fourier transforms of the two functions is obtained:

$$F(I) = F(G * T\ (K_{\|})) = F(G) \times F(T(K)) = P(X) \times t(X)$$

(F( ): Fourier transform, *: convolution, P(X): the pair correlation function of units, and t(X): the transfer function). The instrument therefore limits the range of detectable distances, since the pair correlation P(X) is reduced by t(X) for distances larger than a given length (transfer width or coherence width) due to the shape of t(X) (which in many cases is approximated by a Gaussian function). It is therefore important to have a sharp focus (= narrow instrument response function) to detect large defect distances (due to a large transfer width). With high precision measurements, even distances larger than the transfer width may be detected with the help of deconvolution. Another important effect of the instrument function is, that qualitative features like splitting or central spike may be hidden in a broad profile due to a bad instrument response.

The instrument response function T(K) should be a δ - function as close as possible. Finite spot size in the recording of a profile may be caused by finite electron source, imperfect lenses for imaging and finite size detector. It is convenient to describe this limiting size not on the screen, rather on the crystal surface, by the Fourier transform to the transfer function (fig. 9), yielding the transfer width. The transfer width may be calculated with the spot diameter d, the distance L from screen to crystal, the angle α of the diffracted beam to the crystal normal and the wave length λ to $t = \lambda * L/(d * \sin\alpha)$, if the profile is measured in the plane of incidence. In the direction normal to that plane α has to be taken as Zero. Fig. 10 shows some results with L/d = 100 (for the usual 3 or 4 grid LEED optics) and L/d = 2000(for high resolution faraday cup systems) with different angles for LEED and RHEED.

The second limitation is the broadening of a diffracted beam due to the energy-spread of the electron source in the direction to the specular beam only for non-specular beams. This broadening converted into a transfer width yields $t = 2 * E * a/(\Delta E * n)$ (with ΔE the energy-spread, n the order of the nonspecular beam corresponding to the surface periodicity a), which is also shown in fig. 10.

A third limitation is the uncertainty in $K_\perp$ due to the above limitations and a variation of $K_\perp$ during scanning in $K_\parallel$ direction, which is especially important for gracing incidence. An example is given in fig. 10. It is of course possible to do precision measurements both with high and low energy electrons, both regimes having their special advantages. In any case, high resolution provides not only an increased range of detectibility and higher accuracy, it may also reveal new qualitative features, like a central spike on top of a broad shoulder, not detectable with low resolution instruments, as demonstrated in (8). The detectable range for distances, sizes and correlations is a few atomic distances up to the transfer width, or a multiple of it, depending on the accuracy of data collection (9,12,15).

The usual instruments are sketched already in fig. 3 and 4. Their transfer width is shown in fig. 10. New developments use better electron sources (like field emission tips (15), (16)) or better focussing (magnetic focussing (17) or increased distances (18) (19)). Also, the principles of data collection are different. Either the data are taken from the phosphonic screen with a spot photometer (or closed TV-system) or the Faraday eagle and the diffracted electron beam are displaced versus each other (by either moving the electron beam or the Faraday cage or by turning the crystal). An extensive description is given in (15).

7) Examples

Some examples will show what kind of information is available and what has been concluded from measurements.

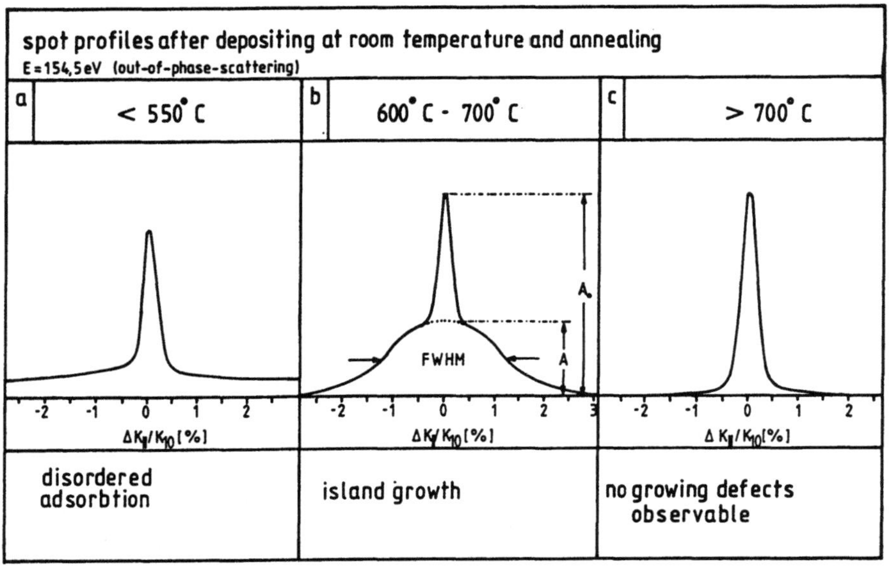

Fig. 11: Schematic spot profiles seen in epitaxy depending on substrate temperature.

In epitaxy LEED or RHEED can reveal in situ the growth of two- dimensional islands. The kind of information is seen in fig. 11. For sufficiently high temperatures the number of islands is so small and the size of the islands so large that no change of the diffraction spots is observed at all. That means that the average distance and average size of the islands is always much larger than the transfer-width of the system. By decreasing the substrate temperature, a typical shoulder is observed with each spot, if the energy of the electron beam is chosen for out-of-phase interference of neighboring terraces. This shoulder does not change its shape with varying energy and varying coverage, only its amplitude, as demonstrated in fig. 6. The formula for the amplitude ratio A/A_0 has been derived independently both by P. Cohen (20) and coworkers and T.M. Lu (21), so that both the dependence with coverage and with energy is described as observed in the experiment (19, 22, 23). In RHEED, instead of the spot profile, just the peak intensity has been observed, and found to be oscillatory with coverage (20,24). The damping of the oscillatory behaviour with increasing coverage may be due to inhomogeneity at the (large) area diffracted by the electron beam, or due to starting of a new layer before the former layer is completed. It has been reported that some spots show maximum intensity for incomplete layers and minimum intensity for complete layers, which has been interpreted as due to dynamic effects of diffraction (24). A more simple explanation calls for a measurement outside the central spike, which is especially likely when the central spike is below the horizon and therefore invisible, and only the broad shoulder detectable at all. It is surprising that the ratio A/A_0 is described by the formula on fig. 12 with a very high accuracy, which is

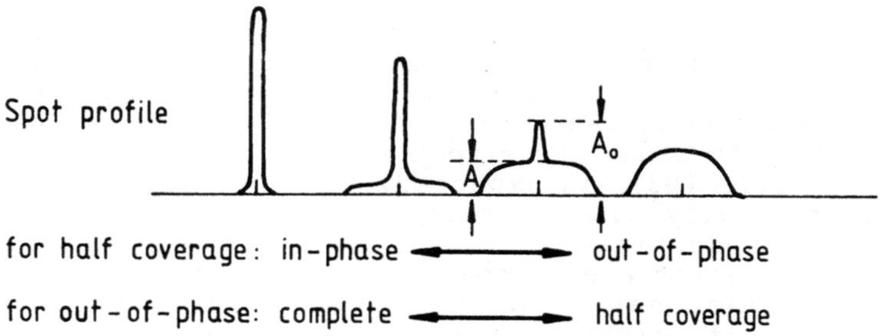

Fig. 12: Spot profiles during epitaxial growth for varying electron energy and varying coverage. The ratio A/A_o is given by

$$A/A_o = \frac{G(K_{\parallel})\,(1-(1-2\Theta)^2)\,\sin^2\left(\frac{K_{\perp} d}{2}\right)}{G(K_{\parallel})\,(1-(1-2\Theta)^2)\sin^2\left(\frac{K_{\perp} d}{2}\right) + G_o(K_{\parallel})\,\left[1-(1-(1-2\Theta)^2)\,\sin^2\left(\frac{K_{\perp} d}{2}\right)\right]}$$

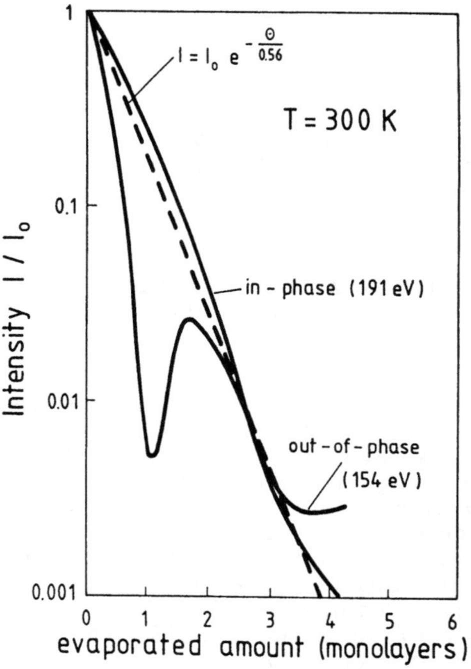

Fig. 13: Intensity of the 00-beam of a Si(111) surface after evaporation of the given amount of silicon at room temperature. The spot profile always has been given by the instrumental response function T(K).

demonstrated in (23). It shows that the kinematic approximations work for amplitude ratios to a much higher degree than expected. It is a support also for that part of evaluation which is not so easily checked, like influence of dynamic effects on spot profile. In a more recent study, the profile has been measured for increasing substrate temperatures, so that the density of islands (as derived from the half-width of the shoulder) is seen as a function of substrate temperature (23). As expected, the island density, and therefore the nucleation probability, decreases with increasing temperature due to the larger size for the critical nucleus and the increased surface mobility. In another experiment, the epitaxy was tried at room temperature with the intention to study the effect of an amorphous layer on the spot profile (23). As expected, the spot intensity decreases exponentially with increasing coverage (fig. 13). As a surprise, the decrease is faster than linear, which requires that each evaporated atom is not just in a random position, it rather has to disturb the exact lattice position of a lot of neighboring atoms. A second surprise has been that for out-of-phase voltage, the intensity decreases by another factor of 50 close to half coverage (one monolayer is a half double layer, since a double layer is required for a complete layer in a 111 plane of the diamond lattice). This observation requires that all adsorbed atoms are very close to a regular position, at least with respect to the vertical distance. If they form minimum size islands (e.g. a pyramid out of 4 atoms) in a random distribution, then the observed decrease may be explained. This is consistent with a recent calculation of that arrangement (25).

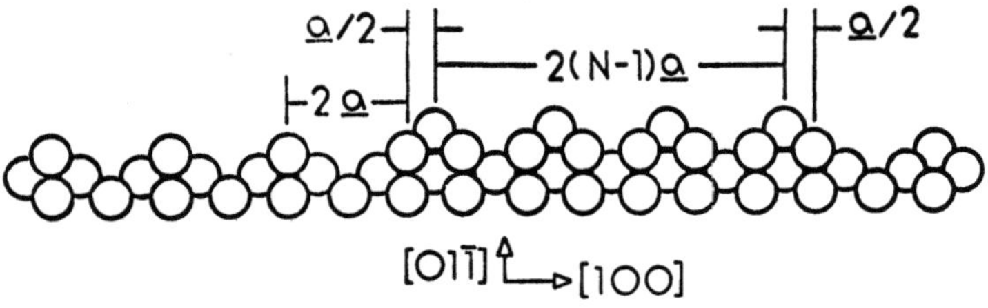

Fig. 14: Cross-section through a Au(110) surface with superstructure and atomic step, causing broadening of the spots.

The clean Au(110) surface shows a 2x1 superstructure, which has been investigated with a lot of different techniques. With LEED the superstructure has been identified (10, 26). Spot profile analysis and intensity measurements have been interpreted as missing row model and domain disorder, including an atomic step going from one domain to the next one (fig. 14) (10, 26). Also microscopy with STM (4) and high-resolution electron microscopy (27) are in support of this model. A new diffraction technique using glancing incidence x-ray diffraction (28) detected a shift in the center position of the broadening extra spot. The broadening is due to finite size domains, and the shift of spot position due to a shift of neighboring domains both in lateral and normal direction (see fig. 14) (28, 29).

The Si/SiO_2 interface has been studied with different techniques: As shown in fig. 15 electron microscopy has been used to image thin cross-section of the interface for a direct evaluation of the atomic steps, as seen in the photograph fig. 16 (2, 30). Although, due to the projection along the interface, it is difficult to locate individual

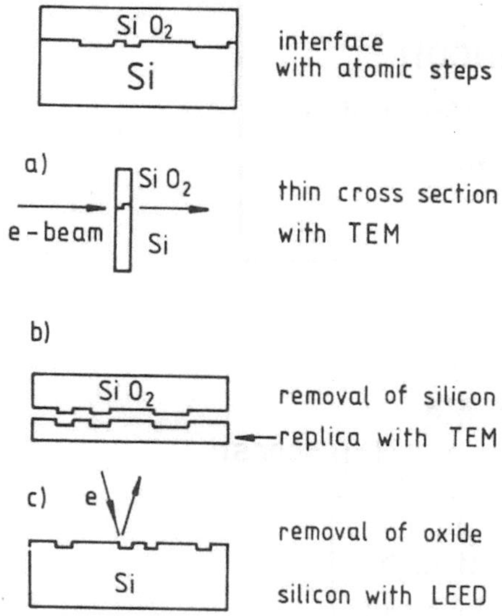

Fig. 15: Three techniques to study the Si/SiO_2 interface for atomic steps (see (30-34)).

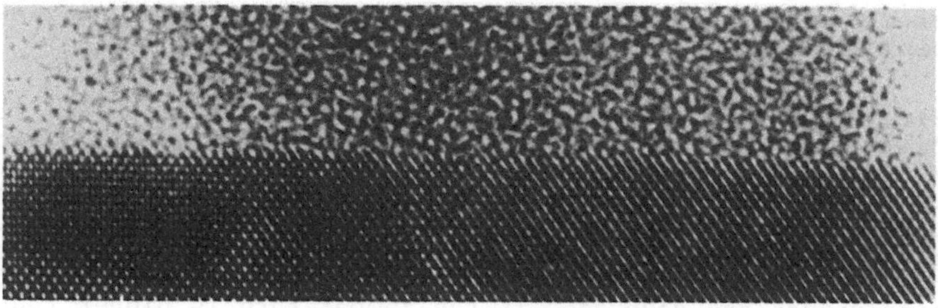

Fig. 16: Micrograph of a cross-section of a Si/SiO_2 interface with lattice resolution in the silicon lattice (from (30)).

steps, some average data have been derived and correlated with measurement of the electron mobility (30). In a different approach (see fig. 15), the silicon has been dissolved chemically, and a replica of the bare oxide (made with oblique metal evaporation and carbon) has been imaged with TEM (31). Here also, some correlation between step density and mobility has been found. The most extensive study has been made with LEED after dissolving the oxide with HF and transferring the bare silicon into uhv under a cover of methanol to avoid oxidation (32, 33). Here correlations of the atomic roughness are measured by LEED and both the oxidation parameters and the electrical properties are measured by Hall mobility and capacitance. A summary of the results is given in fig. 17. Recently, more details have been resolved with a high resolution LEED system , showing that the step distribution is not just a geometrical distribution, it rather consists of two distributions, one with a small average terrace-width (10-20 atomic distances) and one with a large average width (more than 150 distances) (34). The reason for this inhomogeneity is not known, some possibilities are discussed in (34).

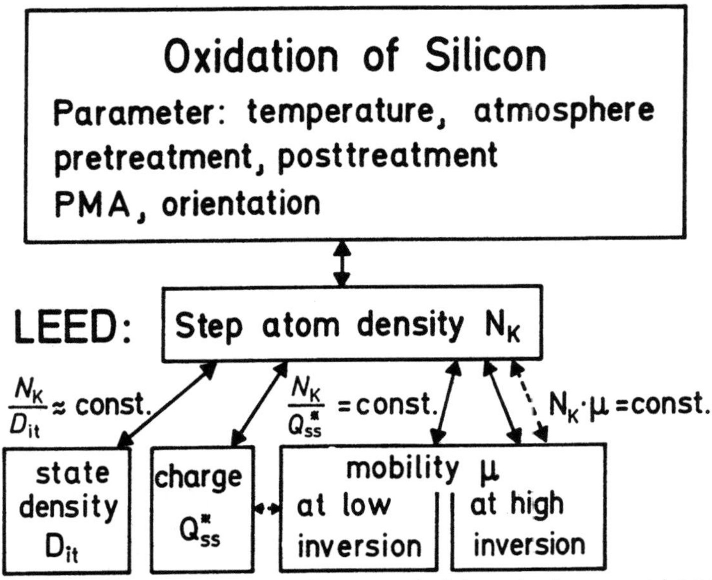

Fig. 17: Summary of correlation between atomic steps and process parameters of oxidation and electronic device parameters.

Finally, an example of a surface phase-transition will be discussed. During the adsorption of oxygen on W(112) and W(110), a superstructure 2x1 is formed by the saturation coverage of a half monolayer (35-37). If disorder is present, out-of-phase domains broaden the half-order spot so that peak intensity or half-width of the extra spots are used to derive average domain size as a function of temperature and coverage. If the temperature is higher than a coverage dependant limit, a completely disordered lattice is formed with a well-defined phase-transition during cooling. Also during adsorption at sufficiently low temperature (T 225 K), a disordered phase is formed. By down-quenching the lattice gas or "up-quenching" the low temperature disordered phase, a growth of the superstructure domains with time at constant temperature has been derived from the peak intensity of the extra spots (35-37). The first increase has been described by a square root dependance on time $t^{1/2}$, which agrees with theoretical predictions (38,39).

8) Conclusion

Defects at surfaces may be studied both with microscopy and diffraction. Whereas microscopy provides easy identification of defects, diffraction is especially suited for quantitative analysis. Already, general shape (like broadening, splitting), peak intensity and half-width provide qualitative evaluation and quantitative determination of average values. With the new high resolution instruments, more details will be available on the size distributions and kinetic data over a broad range of sizes and distances (from a few to a few thousand atomic distances). In this way, it will be possible to study and control reactions and other processes which involve order and disorder, so that defects at surfaces are no more unknown species, but rather part of surface physics.

References

1) J.M. Cowley, "Diffraction physics", North Holland Pub.Co., Amsterdam 1981
2) O.L. Krivanek and J.H. Mazur: Appl.Phys.Lett. 37, 392 (1980)
3) N. Osakabe, Y. Tanishiro, K. Yagi, G. Honja, K. Takayanaji: Surf.Sci. 97, 393 (1980); 102, 424 (1980); 104, 527 (1981)
4) G. Binnig, H. Rohrer, C. Gerber and E. Weibel: Rhys.Rev.Lett. 49, 57 (1982); Phys.Blätt. 39, 16 and 176 (1983); Surf.Sci. 126, 236 (1983)
5) G. Ertl, J. Küppers "Low Energy Electrons and Surface Chemistry" Verlag Chemie, Weinheim (1974)
6) T. Engel, K.H. Rieder: Springer Tracts in: Modern Physics, Springer, Berlin, 1982, p 55
K.H. Rieder: this volume
7) M. Henzler in Electron Spectroscopy for Surface Analysis, ed. H. Ibach, Topics Current Phys. 4, (Springer Verlag Berlin, Heidelberg, New York 1977)
8) M. Henzler: Appl.Phys. A34 (1984)
9) M. Henzler: Appl.Surf.Sci. 11/12, 450 (1982)
10) H. Jagodzinski, W. Moritz, D. Wolf: Surf.Sci. 77, 233, 249, 265, 283 (1978)
11) T.M. Lu, M.G. Lagally, Surf.Sci. 99, 695 (1980) and 120, 47 (1982)
12) M.G. Lagally, Appl.Surf.Sci. 13, 260 (1982) and "Diffraction Techniques" in "Methods of Experimental Physics: Surfaces", Academy Press (1984)
13) D. Saloner, M.G. Lagally, I.V.S.T. A2, 935 (1984)
14) M. Henzler, P. Marienhoff, J.V.S.T. B2 (1984)
15) J.A. Martin, M.G. Lagally, J.V.S.T. A1, 1210 (1983) and Rev.Sci.Instr. 54, 1273 (1983)
16) E. Williams, R.L. Park, Bull.Am.Phys.Soc. (1983)
17) F.W. Wulfert, PhD-Thesis, Hannover 1982
18) E.G.McRae, priv.comm.
19) K.D. Gronwald, M. Henzler, Surf.Sci. 117, 180 (1982)
20) C.S. Lent, P.I. Cohen, Surf.Sci. in press
21) J.M. Pimbley, T.M. Lu, J.Appl.Phys. 55, 182 (1984)
22) P.O. Hahn, J. Clabes, M. Henzler, J.Appl.Phys. 51, 2079 (1980)
23) M. Henzler, K. Rudnik to be published
24) P.J. Dobson, J.H. Neave, B.A. Joyce, Surf.Sci. 119, L 339 (1982)
P.J. Dobson, N.G. Norton, J.H. Neave, B.A. Joyce, Vacuum 33, 593 (1983)
25) J.M. Pimbley, T.M. Lu private comm.
26) W. Moritz, D. Wolf, Surf.Scoi. 88, L 29 (1979)
27) L.D. Marks, D.L. Smith, Nature 303, 316 (1983) and Surf.Sci. 139, 281 (1984)

28) I.K. Robinson, Phys.Rev.Lett. 50, 1145 (1983)
I.K. Robinson, Y. Kuk, L.C. Feldman, Phys.Rev. B (1984)
29) T.M. Lu, P. Fenter, priv. comm.
30) S.M. Goodnick, R.G. Gann, J.R. Sites, D.K. Ferry, C.W. Wilmsen, J.V.S.T. B1, 803 (1983)
31) T. Sugano, J.J. Chen, T. Hamano, Surf.Sci. 98, 154 (1980)
32) P.O. Hahn, M. Henzler, J.Appl.Phys. 52, 4122 (1981) and 54 6492 (1983) and J.V.S.T. A2, 574 (1984)
33) P.O. Hahn, S. Yokohama, M. Henzler, Surf.Sci. (1984) in press
34) M. Henzler, P. Marienhoff, J.V.S.T. B2 (1984)
35) M.G. Lagally, G.-C. Wang, T.M. Lu in "Ordering in two dimensions" ed. S.K. Sinha (North Holland, New York 1980) p 113 andin"Chemistry and Physics of Solid Surfaces" ed. R. Vanselow (CRC Press, Boca Raton, Florida 1979), Vol. 2, p. 153
36) G.-C. Wang, T.M. Lu, Phys.Rev.Lett. 50, 2014 (1983)
37) J.H. Perepezko, J.T. McKinney, P.K. Wu, M.G. Lagally, Phys.Rev.Lett. 50, (1983)
38) I.M. Lifshitz, Zk.Eksp.Teor.Fiz. 42, 1354 (1962) (Sov.Phys. JETP 15, 939 (1962))
39) S.M. Allen, J.W. Cahn, Acta Metall. 27, 1085 (1979)
40) H. Froitzheim, in "Electron Spectroscopy for Surface Analysis" ed. H. Ibach, Top.CurrentPhysics 4, (Springer Berlin 1977)

Metal-Metal Superlattices

Charles M. Falco

Departments of Physics and of Optical Sciences and in the Arizona Research Laboratories, University of Arizona, Tuscon, AZ 85721, USA

Wayne R. Bennett and Ahmed Boufelfel

Department of Physics, University of Arizona, Tucson, AZ 85721, USA

An introduction to metal-metal superlattices is presented, along with a brief review of some of the materials which have been made, and phenomena which have been observed so far. We describe in detail the results of a number of structural characterization techniques applied to these materials, including Bragg θ-2θ x-ray diffraction, transmission and reflection Laue diffraction, wide film Debye-Scherrer diffraction, Rutherford Backscattering Spectroscopy (RBS) and Auger Electron Spectroscopy (AES). Particular emphasis is placed on recent results obtained with Ta/Mo superlattices produced by sputtering techniques.

1. Introduction to Metal - Metal Superlattices

Recent advances in vacuum and vapor deposition technology have made possible the sequential deposition of elements, alloys and compounds under conditions where contamination during the deposition process is small [1]. For many combinations of materials,this can be done with a regularity of a few Angstroms. This present paper is concerned with the structural properties of metal-metal superlattices produced by sputtering techniques. Our emphasis is on recent results obtained with Ta/Mo superlattices. We begin with an introduction of some useful definitions, and a brief review of some of the materials which have been made, and phenomena which have been observed, so far.

Figure 1 shows an idealized picture of a metal-metal superlattice. This ideal specimen consists of alternating layers of two different materials with an integer number of atoms making up the thickness at every point in each layer. By this we mean that each layer of a particular material is, say, 6 atoms thick; not 5 atoms thick in some places and 7 in other places located a microscopic distance away, giving an average thickness of 6. The interfaces are perfectly sharp (no interdiffusion), the material is a single crystal within each layer (typically ~1 cm^2 in area), and long-range structural coherence is maintained perpendicular to the layers over the entire thickness of the sample (typically ~1 micrometer). All artificially layered metal-metal samples produced to date deviate in various ways from this ideal material, as described below.

Metals which are artificially layered on a length scale from ten's to thousand's of Angstroms have degrees of structural perfection which fall roughly into three classes, as shown schematically in Fig.2. Also indicated in this figure are how the the two layer thicknesses, d_1 and d_2, as well as the superlattice wavelength Λ are defined. For many purposes the ideal single

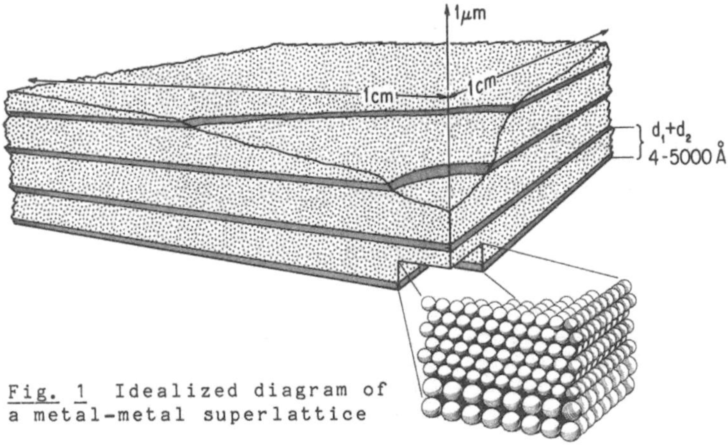

Fig. 1 Idealized diagram of a metal-metal superlattice

crystal material depicted in Fig. 1, and found in some semiconductors such as GaAs/GaAlAs [1], is not necessary in order for an artificially layered metal to exhibit interesting or useful behavior. Consequently, the preparation and study of layered metals which fall into each class of this scheme is worthwhile.

In increasing order of structural perfection, the first class of layered metals depicted in Fig. 2 are "multilayer" films which may have sharply defined boundaries at the interfaces between two components, and have regular spacing from layer to layer. However, either, or both, layers may be polycrystalline or amorphous, so that long-range structural coherence may not exist within the plane of the layers, and does not exist perpendicular to the layers. There is no superlattice in this class of materials due to the absence of long-range perpendicular structural coherence. However, such materials may be used to study phenomena such as superconductivity, where the relevant length scale at which layering begins to alter the physical properties is quite large. Typical superconducting coherence lengths are in the range 50 - 10,000 Å. The ability to alter the layer thicknesses in a highly controlled way has enabled studies of dimensional crossover effects [2], and of thickness effects on superconducting critical temperatures [3].

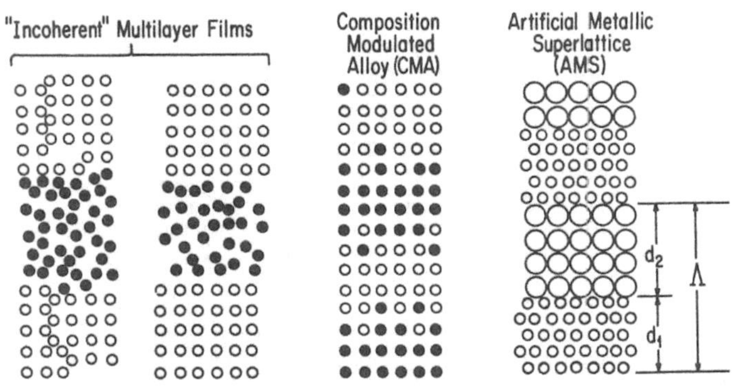

Fig. 2 Schematic representation of the various degrees of structural coherence in the three classes of metal-metal superlattices

"Composition Modulated Alloys (CMA's)" have a single lattice constant, and have lattice sites which may be occupied by either of the two materials, since they are mutually soluble in each other. Even though well-defined layers may be put down in the deposition process, the composition gradient at the interfaces strongly encourages interdiffusion, so that the boundaries between layers are generally not sharp, and the amplitude of composition modulation in the center of a layer can be in the range from slightly greater than 0% to 100%. A superlattice does exist in this class of materials, although the absence of sharply defined interfaces can preclude the possibility of observing many phenomena of interest. For example, it would not be possible with a CMA to study the properties of the interface layer of a magnetic material in contact with a non-magnetic material, i.e., a layer of magnetic atoms surrounded on all sides but one by other magnetic atoms. Ni/Cu has been the most heavily studied CMA [4-14].

As discussed above, and shown in Fig. 1, an ideal "Artificial Metallic Superlattice (AMS)" would have sharp interfaces between the two components, with long range structural coherence maintained across many layers of the material, as well as in the plane of the layers. No such ideal material has yet been produced, although some metal-metal superlattices, such as Nb/Cu [15-18], Nb/Al [19] and Nb/Ta [20-22] have certain of the properties of an ideal AMS. As mentioned above in the discussion of multilayer films, the ability to alter layer thicknesses has already enabled these AMS's to be used for studies of dimensional crossover effects [23], of thickness and proximity effects on superconducting critical temperatures and energy gaps [18], and of tunneling density of states [22]. The rest of this paper will be concerned with our progress to date to grow, and structurally characterize, materials which are in this class.

It has been found that most metals grown by vapor deposition techniques tend to grow with the highest density crystallographic planes in the plane of the substrate [25]. For BCC metals these are the (110) planes and for FCC metals the (111). However, by suitable control over the deposition process, it is possible to grow metals with other crystallographic orientations. For metals of different symmetries, perfect lattice matching at the interfaces is impossible. However, epitaxial growth in this case may be most likely to occur when the in-plane densities of the two elements are nearly equal. Figure 3 shows the in-plane densities of various crystallographic orientations for a number of common metals of different crystalline symmetry [26].

Artificial metal-metal layering can influence a number of physical properties. Among these are electrical transport, superconductivity, magnetism and elastic properties. For details, see the many articles cited in Ref. [1]

2. Preparation of Metallic Superlattices by Sputtering

A schematic diagram of our sputtering system is shown in Fig. 4. The base pressure of this diffusion-pumped system is $<7 \times 10^{-8}$ torr. The primary impurity limiting the base pressure is water vapor, with the total nitrogen and oxygen partial pressures typically being 2-5 times smaller. It has been shown that such a system is capable of depositing many metals with purity equivalent to the starting target material [27]. During sputtering (typical pressure $\sim 5 \times 10^{-3}$ torr)

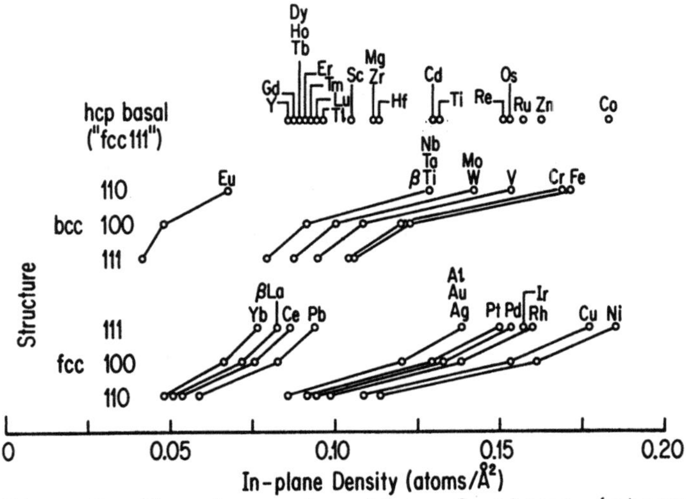

<u>Fig. 3</u> In-plane density of atoms (atoms/$Å^2$) of various crystallographic orientations for a number of common metals which can be deposited by sputtering

the Ar sputtering gas pressure is held constant to ± 0.2% using gas flow controllers. Since the sputtering rate is only weakly pressure dependant in this range, this reduces pressure induced sputtering rate variations to a negligible level. In the usual mode of operation, a microprocessor is programmed to operate a servo controlled motor to move the substrates alternately over the deposition sources at known rotation rates, or to hold them in place above each source for determined lengths of time. Thus, variations in layer thickness due to rotation speed or to timing errors are negligible.

There are two types of variation in superlattice wavelength which are possible for our sputtered samples. The first is the variation due to fluctuations or drift in the sputtering rate during the deposition. At a given point in the x-y plane (defined by the plane of the substrate), this causes Λ to fluctuate or vary as a function of z. The second is caused by the spatial variation in sputtering rate due to a finite size target. This causes a variation in Λ as a function of x and/or y. For the results reported here, two of the three magnetically-confined-plasma sputtering guns were used to deposit the metal layers. Feedback control of the power supplies is achieved at 40 Hz using an empirical relationship relating sputtering rate to the target voltage and current. This system is capable of regulating the sputtering rates (typically 20 - 100 Å/sec.) to ± 0.1%. This is the dominant source of error in depositing these materials, so that it should be possible with this system to hold the superlattice wavelength constant at any given point in the x-y plane of the sample to ± 0.1% as the superlattice stack is built up . Experimentally, we find ± 0.3% as an upper limit on the wavelength variation in our samples.

The above discussion refers to the control over Λ as a function of z which is possible at any given point in the x-y plane of the sample. However, because of the finite size of any sputtering

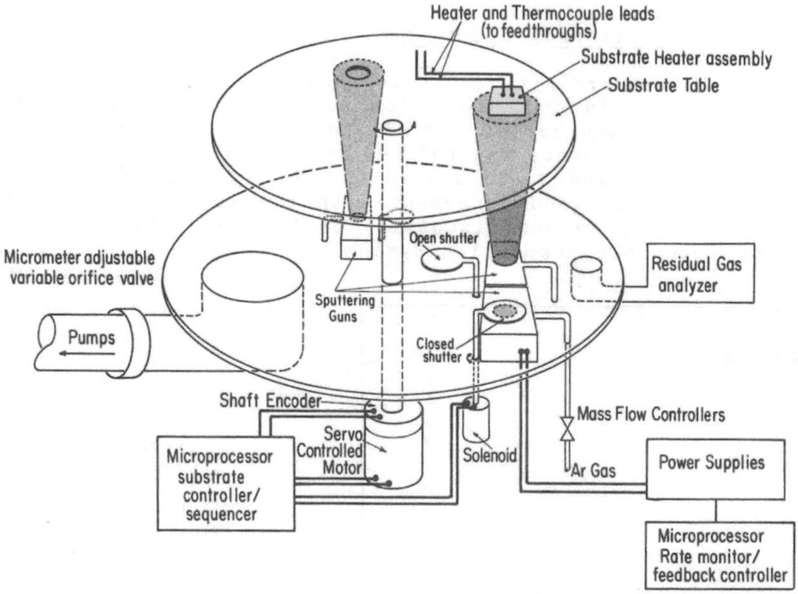

Fig. 4 Schematic diagram of sputtering system used for depositing metal-metal superlattices

target, absolute uniformity of layer thicknesses to arbitrary accuracy over the entire plane of the sample is impossible to achieve unless a planetary system is used for holding the substrates. Unfortunately, use of a substrate heater, which is essential for producing epitaxial growth with many materials, would be difficult with a planetary. Instead of trying to reduce this intrinsic spatial variation of sputtering rate from a finite radius target, we have taken advantage of it to produce a continuous "wavelength spread" of samples during each deposition [28]. The advantage of this is that in each deposition we are able to produce under absolutely identical sputtering conditions a set of samples which have a small, continuous variation in superlattice wavelength. Typically we have a total variation of ~15% over a length ~8 cm. Thus, we can look for small superlattice wavelength-induced effects on physical or structural properties, which might be masked by normal variations in the deposition process for samples produced in different depositions.

By altering the size and shape of chimneys used to confine the flux of sputtered particles, it is possible to produce wavelength spreads along the length of our samples of any desired magnitude [28]. As discussed below, samples of wavelength Λ ~ 40-60 Å are typically the most suitable to use for quantitative θ-2θ x-ray diffraction measurements. For samples in this wavelength range, a variation of 15% corresponds to changing the number of atoms, n, making up the thickness of each layer by approximately 2, e.g., from n=14 to n=16. Thus, with each deposition it is possible to produce samples with Λ crossing several values of n in order to look for n-dependant effects on the structural, electronic, or magnetic properties.

3. Characterization

3.1 Rutherford Backscattering Spectroscopy (RBS)

Figure 5 shows a Rutherford Backscattering Spectrum (RBS) of a Ta/Mo superlattice obtained in collaboration with Prof. John Leavitt. Although the absolute accuracy of the RBS technique is limited to a few percent, such measurements can be used to determine the relative layer thicknesses of Ta and Mo to a precision of approximately 0.5%. These RBS results are in good agreement with the layer thicknesses expected based on the previous discussion of the regulation of the sputtering rates with our deposition system.

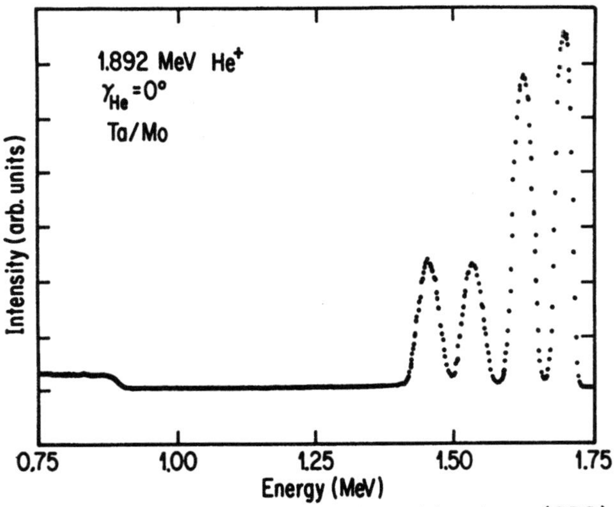

Fig. 5 Rutherford backscattering (RBS) spectrum from a Ta/Mo superlattice of individual layer thickness ~300 Å. The integrated areas under each peak can be used to determine the relative layer thicknesses to a precision of approximately 0.5%

Preliminary analysis of a number of RBS measurements on Ta/Mo layers indicates that the sticking coefficient of, for example, the first Ta atomic monolayer on a fresh Mo surface is different than for subsequent Ta monolayers [29]. That is, the Ta-Mo sticking coefficient is different than the Ta-Ta sticking coefficient. The same is true for Mo deposited on Ta. This has significant consequences for the growth of metal-metal superlattices with a precisely determined number of atomic layers within each superlattice layer. For example, for fixed sputtering rates it means that it is not possible to rotate the substrates above the sputtering targets twice as fast to produce superlattices with exactly half of the wavelength. While this inverse proportionality between substrate rotation rate and superlattice wavelength holds to some degree of accuracy (which can be quite high), the differences in sticking coefficents cause increasing error as the wavelength is decreased. RBS is a very useful tool for determining the relative sticking coefficients of materials of interest for metal-metal, as well as other types of superlattices.

3.2 X-Ray Diffraction

To determine the crystal structure of these materials, we have used a variety of x-ray diffraction techniques including Bragg θ-2θ scans with the sample held in various orientations, transmission and reflection Laue diffraction of samples still attached to the substrates as well as removed, and wide film Debye-Scherrer diffraction [30] patterns at various grazing incidence angles. These various techniques provide complementary information, as described below.

Superlattice lines observed with standard Bragg θ-2θ diffraction as shown in Fig. 6 indicate that Ta/Mo samples preferentially grow with the (110) crystal planes in the plane of the sample, and that long range structural coherence exists perpendicular to the layers. The intrinsic linewidths of each of these superlattice lines can be obtained by deconvoluting them from each other and from the instrument resolution function. These linewidths, along with the Scherrer equation [31] give a value of ~300 Å (i.e. ~ 9 layers) as a typical perpendicular structural coherence length for Ta/Mo superlattices grown on single crystal sapphire substrates at room

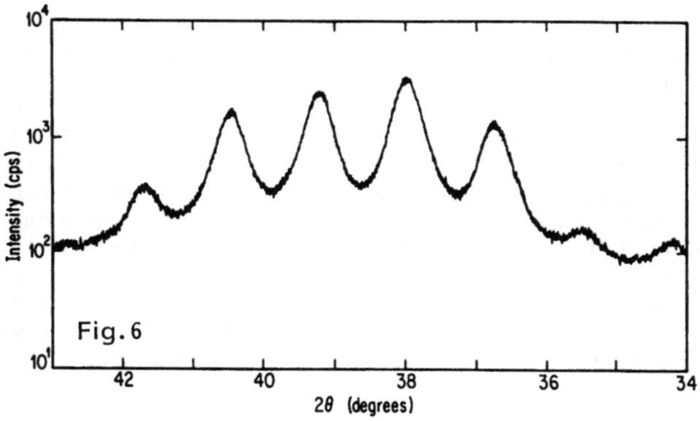

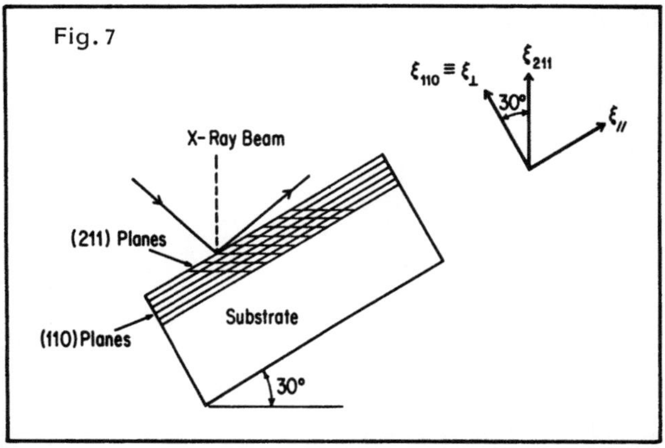

Fig. 6 X-ray diffraction from a Ta/Mo superlattice of 75 Å. wavelength, showing superlattice lines near the positions of the bulk (110) values (38.6° and 40.5° respectively for Ta and Mo)

Fig. 7 Schematic diagram showing how the in-plane structural coherence length ξ can be measured using diffraction measurements from crystal planes other than those in the plane of the sample

temperature. Bragg diffraction measurements on different crystal planes can also be made with the sample surface rotated to some other fixed angle. This is shown in Fig. 7. For example, the (211) planes in a metal with BCC structure (such as Ta and Mo) are at an angle of 30° with respect to the (110) planes. Consequently, the sample can be mounted in a standard diffractometer at 30° with respect to the substrate normal, and θ-2θ scans made through the (211) reflections. Knowing the structural coherence length perpendicular to the substrate and at some other angle relative to the substrate normal is sufficient to obtain the in-plane structural coherence length. A typical value for the Ta/Mo samples discussed above is >300 Å.

Transmission and reflection Laue diffraction using polychromatic Mo radiation is useful for determining the symmetry of growth of the material. The primary difficulty with Laue diffraction is the small diffraction intensity from the thin film as compared to the diffraction intensity from the substrate. This can be overcome in some cases by removing the sample from the substrate. However, strong adhesion often means that the sample is damaged in the removal process. For materials which grow best on substrates such as single crystal sapphire, removal by dissolving the substrate is not possible.

Wide film Debye-Scherrer x-ray diffraction ("Read camera" [30]) measurements have been found to be very useful for studying the crystalline structure of our metal-metal superlattices. The Read camera uses a glancing angle incident x-ray beam at some fixed angle with respect to the substrate (typically 5° - 50°, depending on what features are being studied), and the diffracted beams then go out in various directions determined by the crystal structure. Unlike the case of the θ-2θ diffractometer, the scattered wavevector is no longer constrained to be perpendicular to the substrate. Thus one photograph contains information about the crystal structure perfection both in- and out- of the plane of the substrate. Figure 8 is a Read camera diffraction pattern from a polycrystalline Mo

Fig. 8. Wide field Debye-Scherrer diffraction pattern from a Mo thin film showing a complete Debye arc at $2\theta = 40.5°$ characteristic of preferential, randomly oriented, polycrystalline (110) growth in the plane of the substrate

film grown on a room temperature single crystal sapphire substrate showing the expected complete Debye ring at $2\theta = 40.5^{\circ}$ due to diffraction from randomly oriented (110) planes in the Mo. Sharp spots from the single crystal sapphire substrate are clearly visible.

In contrast to the polycrystalline Mo film shown in Fig. 8, Fig. 9 shows diffraction patterns from a Ta/Mo superlattice sample of wavelength Λ = 70 Å grown under identical conditions as for the Mo film. The incident x-ray beam is at a 28° grazing angle for the top photograph in Fig. 9 and 43° for the bottom. Both photographs were made from the same point on the same sample, with only the angle of the incident x-ray beam changed. It is necessary to use several angles to completely cover the regions of interest. Unlike the case of the Mo film shown in Fig. 8, the structure of these superlattice

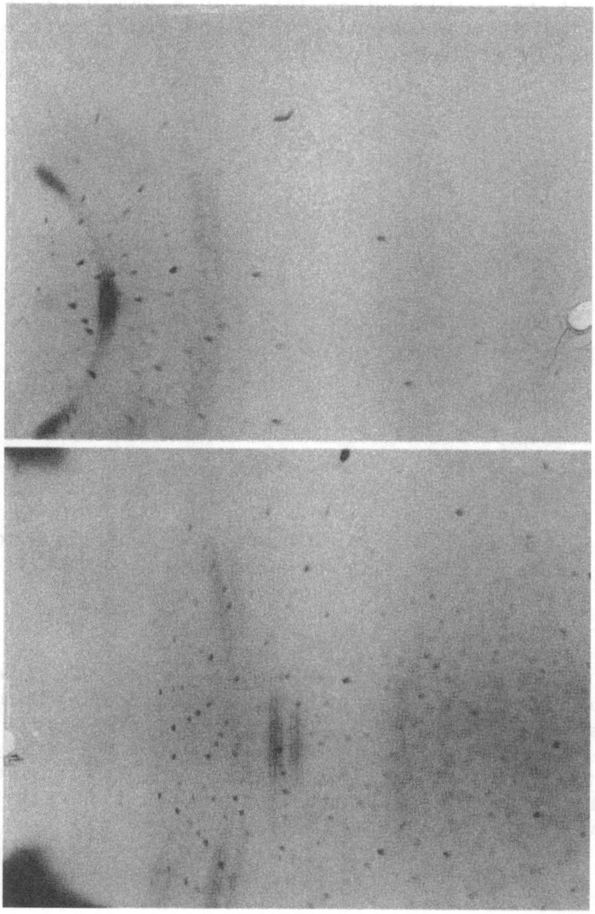

Fig. 9 Wide field Debye-Scherrer diffraction patterns from the same spot on a Ta/Mo superlattice, with the incident beam at a 28° grazing angle for the top photograph and 43° for the bottom

films is much more like single crystalline. The angular width of the superlattice "spots" indicates a mosaic spread of only about ± 7°. The superlattice lines are clearly apparent. In addition to the (110) "spot" at 2Θ = 40° which one would have seen in a standard diffractometer, one sees diffraction from the (011) and (01$\bar{1}$) planes oriented at 60° with respect to the (110) spot. Although these spots all have the same inter-planar spacing, and thus occur at the same 2Θ value, the Read camera shows that they have the proper crystallographic orientation with respect to each other, as expected of a sample which is close to single crystal in quality. At higher angles one sees the (200), (211), etc. planes, all of which lie at the correct "single crystal" values with respect to each other.

Low angle measurements of Bragg diffraction directly from the superlattice can be useful for determining the variation in wavelength throughout the sample. Figure 10 shows the calculated low angle reflectivity for Cu K_α radiation from a "perfect" Ta/Mo superlattice, i.e., one with perfectly sharp interfaces and with no variation in wavelength. The first Bragg peak from the superlattice at an angle Θ = 0.8° is clearly visible.

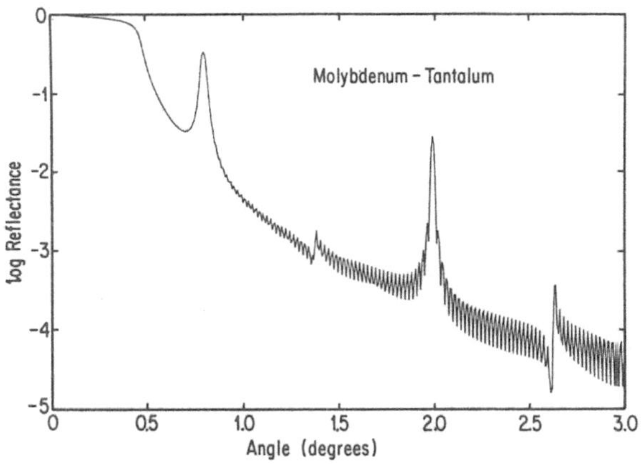

Fig. 10 Calculated low angle Bragg diffraction from an ideal Ta/Mo superlattice with no variation of wavelength throughout the sample. Note that the horizontal axis is Θ rather than 2Θ

To see the effect of small variations in the superlattice wavelength we have calculated a series of low angle diffraction patterns with wavelengths fluctuating by ± 1% throughout the sample. It can be seen from Fig. 11 that the sub-structure between the Bragg peaks is washed out by this variation. Consequently, low angle diffraction seems a promising way to check on the sample uniformity. Experiments to examine this behavior are presently underway.

3.3 Auger Electron Spectroscopy (AES)

Auger Electron Spectroscopy (AES) can be very useful for the study of surfaces, since the electron escape depth can be less than 10 Å [32]. By monitoring certain characteristic peaks such as carbon and oxygen while ion milling the sample, it is possible to differentiate

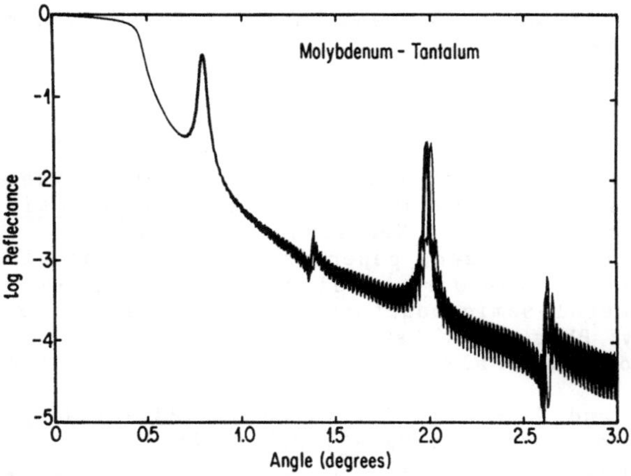

Fig. 11 Calculated low angle Bragg diffraction from a set of Ta/Mo superlattices differing in wavelength by ± 1%. Note that the horizontal axis is Θ rather than 2Θ. The sub-structure between Bragg peaks can be seen to be washed out by this variation.

between contamination in the bulk of the sample caused by impurities incorporated during the deposition process, and contamination of the top ~50 Å, caused by oxidation or contamination after the sample was made [33]. For many purposes, contamination of the latter type is inconsequential, since it only alters the properties of a small fraction of the sample (which is typically between several thousand Angstroms and several micrometers thick).

Ion mill AES is much less useful for studying the actual composition profile of metal-metal superlattices [33]. The process of ion milling the sample to expose subsequent layers and interfaces, also causes atomic mixing and preferential sputtering which rapidly degrades the apparent composition profile [34]. The combination of a finite escape depth for the Auger electrons, along with local degradation of the material caused by the ion milling process, limits the depth resolution of this technique to ~40 Å for determining the composition profile of the superlattice [33,34].

4. Summary

It is possible to use sputtering techniques to produce certain metal-metal superlattices with long range structural coherence perpendicular to the layers as well as within the layers. Control of the superlattice wavelength throughout a sample to within 0.3% is easily possible. The structural properties of Ta/Mo superlattices produced by these sputtering techniques have been described. The usefulness of a variety of techniques for characterizing metal-metal superlattices have been discussed, including Bragg Θ-2Θ x-ray diffraction, transmission and reflection Laue diffraction, wide film Debye-Scherrer diffraction, Rutherford Backscattering Spectroscopy and Auger Electron Spectroscopy.

Leavitt. We thank Felix Fernandez for generating Figs. 10 and 11. It is a pleasure to acknowledge Royal W. Stark for many helpful discussions and support. This work was supported by the Director, Office of Energy Research, Office of Basic Energy Sciences,

Materials Sciences Division of the U. S. Department of Energy under Contract No. DE-AC02-83ER45025.

References

1 For a recent review of both metal and semiconductor superlattices see the various chapters in L. Chang and B.C. Giessen: Synthetic Modulated Structures (Academic Press, New York, in press)

2. Although the discussion in the present paper is limited to metal-metal superlattices, we note the work on dimensional crossover effects observed using metal-semiconductors by S. Ruggiero, T.W. Barbee, and M.R. Beasley: Phys. Rev. Lett. 45, 1299 (1980)

3 J.Q. Zheng, J.B. Ketterson, C.M. Falco and I.K. Schuller: Physica 108B, 945 (1981)

4 A.A. Hirsch: N. Friedman and Z. Eliezer: Physica 30, 2314 (1964)

5 A.A. Hirsch: IEEE Trans. on Magnetics MAG-1, 254 (1965)

6 B.J. Thaler, J.B. Ketterson and J.E. Hilliard: Phys. Rev. Lett. 41, 336 (1978)

7 R.M. White and C. Herring: Phys. Rev. B22, 1465 (1980)

8 T. Jahlborg and A.J. Freeman: Phys. Rev. Lett. 45, 653 (1980)

9 E.M. Gyorgy, J.F. Dillon, Jr., D.B. McWhan, L.W. Rupp, Jr., L.R. Testardi and P.J. Flanders: Phys. Rev. Lett. 45, 57 (1980)

10 E.M. Gyorgy, D.B. McWhan, J.F. Dillon, Jr., L.R. Walker and J.V. Waszczak: Phys. Rev. B25, 6539 (1982)

11 J.Q. Zheng, C.M. Falco, J.B. Ketterson and I.K. Schuller: Appl. Phys. Lett. 38, 424 (1981)

12 J.Q. Zheng, J.B. Ketterson, C.M. Falco and I.K. Schuller: J. Appl. Phys. 53, 3150 (1982)

13 G.P. Felcher, J.W. Cable, J.Q. Zheng, J.B. Ketterson and J.E. Hilliard: J. of Magnetism and Mag. Materials 21, L198 (1980)

14 W.S. Zhou, H.K. Wong, J.R. Owers-Bradley and W.P. Halperin: Physica 108B, 953 (1981)

15 I.K. Schuller and C.M. Falco: in Inhomogeneous Superconductors-1979, ed. by D.U. Gubser, T.L. Francavilla, J.R. Leibowitz and S.A. Wolf (American Institute of Physics, New York, 1980) p.197

16 I.K. Schuller: Phys. Rev. Lett. 44, 1597 (1980)

17 W.P. Lowe, T.W. Barbee, Jr. ,T.H. Geballe and D.B. McWhan: Phys. Rev. B24, 6193 (1981)

18 C.M. Falco and I.K. Schuller: in Superconductivity in d- and f-Band Metals, ed. by W. Buckel and W. Weber (Kernsforschungs-anslage Karlsruhe, Germany, 1982) p. 283

19 J. Geerk, M. Gurvitch, D.B. McWhan and J.M. Rowell: Physica 109 &110 B&C, 1775 (1982)

20 S.M. Durbin, J.E. Cunningham, M.F. Mochel and C.P. Flynn: J. Phys. F - Metal Phys. 11, L223 (1981)

21 S.M. Durbin, J.E. Cunningham and C.P. Flynn: J. Phys. F - Metal Phys. 12, L75 (1982)

22 G. Hertel, D.B. McWhan and J.M. Rowell: in Superconductivity in d- and f-Band Metals, ed. by W. Buckel and W. Weber (Kerns-forschungsanslage Karlsruhe, Germany, 1982) p. 299

23 I. Banerjee, Q.S. Yang, C.M. Falco and I.K. Schuller: Phys. Rev. B28, 5037 (1983)

24 I. Banerjee, Q.S. Yang, C.M. Falco and I.K. Schuller: Solid State Commun. 41, 805 (1982)

25 See, for example, several chapters in Epitaxial Growth, ed. by J.W. Matthews (Academic Press, New York, 1975)

26 We would like to thank Dr. John Rowell for pointing out the usefulness of such a plot for the bcc (110)-fcc(111) orientation.

27 C.M. Falco: J. Appl. Phys. (in press)

28 W.R. Bennett, A. Boufelfel and C.M. Falco: to be published
29 W.R. Bennett, J.A. Leavitt and C.M. Falco: to be published
30 M.H. Read and C. Altman, Appl. Phys. Lett. 7, 51 (1965)
31 See, for example, H.P. Klug and L.E. Alexander: X-Ray Diffraction Procedures, 2nd edition (Wiley, New York, 1974) Chapter 9
32 J.M. McDavid and S.C. Fain, Jr.: Surface Science 52, 161 (1975)
33 I.K. Schuller and C.M. Falco: Surface Science 113, 443 (1982)
34 R.Ludeke, L. Esaki and L.L. Chang: Appl. Phys. Lett. 24, 417 (1974)

Advances in Semiconductor Superlattices, Quantum Wells and Heterostructures

Leo Esaki

IBM Thomas J. Watson Research Center, Yorktown Heights
Yorktown, NY 10598, USA

In 1969, research on synthesized semiconductor superlattices was initiated with a proposal for a one-dimensional periodic structure consisting of alternating ultra-thin layers by Esaki and Tsu.[1,2] Two types of superlattices: doping and compositional, were envisioned. The idea of the superlattice occurred to us while investigating the possible observation of resonant tunneling through double and multiple potential-barriers.[3] In general, if characteristic dimensions such as superlattice periods and widths of potential wells in semiconductor nanostructures are reduced to less than the electron mean free path, the entire electron system will enter a quantum regime of reduced dimensionality with the presence of nearly-ideal hetero-interfaces. Our effort for the semiconductor superlattice[4] is viewed as a search for novel phenomena in such a regime with engineered structures.

It was recognized at the beginning that, while the structure was undoubtedly of considerable interest, the engineering of such a crystal consisting of ultra-thin layers would be a formidable task. Nevertheless, the proposal of a semiconductor superlattice inspired a number of material scientists.[5-8] Indeed, steady improvements in thin-film growth techniques such as MBE or MOCVD during the last decade have made possible high-quality heterostructures having designed potential profiles and impurity distributions with a dimensional control close to interatomic spacing and with virtually defect-free interfaces in a lattice-matched case such as GaAs-$Ga_{1-x}Al_xAs$. Such great precision, indeed, has allowed access to a new quantum regime.

In this review, I will present the evolution of semiconductor superlattices and quantum wells, surveying significant milestones chosen from the research occurring over the past fifteen years.

1) Superlattice Band Model and Experiment (1969-1972)

A synthesized superlattice possesses unusual electronic properties of quasi-two-dimensional character. The introduction of the superlattice potential perturbs the host bandstructure in such a manner as to give rise to narrow subbands separated by forbidden regions, analogous to the Kronig-Penney band model.[9] The electron dynamics in the superlattice direction, with a simplified path integration method,[10] was analyzed for conduction electrons in narrow subbands. This calculation predicted an unusual current-voltage characteristic including a differential negative resistance. In 1972, Esaki et al.[11] found that a

MBE-grown GaAs-GaAlAs superlattice[12] exhibited a negative resistance in its transport properties, which was, for the first time, interpreted in terms of the superlattice effect.

It is worthwhile mentioning here that, in 1974, Gnutzmann and Clauseker[13] pointed out an interesting possibility; namely, the occurrence of a direct-gap superlattice made of indirect-gap host materials, because of Brillouin-zone folding which was later reexamined by Madhukar.[14] The idea suggests the synthesis of new optical materials.

2) Multibarrier Tunneling and Quantum Wells: Theory and Experiment (1973-1974)

In 1973, Tsu and Esaki[15] computed the resonant transmission coefficient T*T for double, triple, and quintuple barrier structures from the tunneling point of view, leading to the derivation of the current-voltage characteristics. The superlattice band model previously presented assumed an infinite periodic structure, whereas, in reality, not only a finite number of periods is prepared with alternating epitaxy, but also the electron mean free path is limited. Thus, this multibarrier tunneling model provided useful insight into the transport mechanism. In early 1974, Chang, Esaki and Tsu[16] observed resonant tunneling in double-barriers, and subsequently, Esaki and Chang[17] measured quantum transport properties for a superlattice having a tight-binding potential. The resonance is achieved at such applied voltages as to align the Fermi level of the electrode with the bound states in the well. Their energies can be obtained from resonant voltages. A number of double barriers were MBE-grown with four different well widths, 35, 40, 50 and 65Å, for which measured values for the bound states were found to agree with the calculation.

These experiments, probably constitute the first clear observation of man-made bound states in both single and multiple potential wells. Such achievement can be viewed as a laboratory practice of elementary one-dimensional quantum physics described in textbooks (Do-It-Yourself Quantum Mechanics!). Sollner et al.[18] recently demonstrated dramatically-improved I-V characteristics in resonant double-barrier tunneling. Such an improvement clearly endorses the evolution of MBE for the last decade.

3) Optical Absorption for Quantum Wells and Superlattices (1974-1975)

Dingle et al.[19,20] observed pronounced structure in the optical absorption spectrum, representing bound states in isolated[19] and double quantum wells.[20] For the former, GaAs well widths in the range between 70Å and 500Å were prepared. The GaAs wells were separated by $Ga_{1-x}Al_xAs$ barriers which were normally thicker than 250Å. In low-tempereature measurements for such structures, several exciton peaks, associated with different bound-electron and bound-hole states, were resolved. For the latter study, a series of structures, with GaAs well widths in the range between 50Å and 200Å and $Ga_{1-x}Al_xAs$($0.19< x < 0.27$) barrier widths between 12Å and 18Å, were grown by MBE on the GaAs substrates. The spectra at low temperatures clearly indicated the evolution of resonantly split discrete states into the lowest subband of a superlattice.

4) Raman Scattering (1976-1980)

Manuel et al.[21] reported the observation of enhancement in the Raman cross section for photon energies near electronic resonance in GaAs-$Ga_{1-x}Al_xAs$ superlattices of a variety of configurations. The results elucidated the two-dimensionality in the electronic structure. Later, however, the significance of resonant inelastic light scattering as a spectroscopic tool was pointed out by Burstein et al.,[22] claiming that the method yields separate spectra of single particle and collective excitations which will lead to the determination of electronic energy levels in quantum wells as well as Coulomb interactions. Subsequently, Abstreiter et al.[23] and Pinczuk et al.[24] observed light scattering by intersubband single particle excitations between discrete energy levels, of two-dimensional electrons in GaAs-$Ga_{1-x}Al_xAs$ heterojunctions and quantum wells.

Meanwhile, Colvard et al.[25] reported the observation of Raman scattering from folded acoustic longitudinal phonons in a GaAs(13.6Å)-AlAs(11.4Å) superlattice. Before this observation, Narayanamurti et al.[26] showed selective transmission of high-frequency phonons due to narrow band reflection determined by the superlattice period.

5) Modulation Doping (1978) and Subsequent Developments

In the original article,[1] the spatial separation of carriers and their parent impurities was proposed in order to reduce impurity scattering by means of concentrated doping in the regions of the potential hills. In 1978, Dingle et al.[27] successfully implemented such a concept in modulation-doped GaAs-GaAlAs superlattices, achieving electron mobilities which far exceed the Brooks-Herring predictions. Soon after, Störmer et al.[28] reported a two-dimensional electron gas at modulation-doped GaAs-GaAlAs heterostructures. Such heterostructures were used to fabricate a new high-speed field-effect-transistor[29,30] called MODFET. Hall mobilities in the dark at 4.2K for such confined electrons recently exceeded 1000000 $cm^2/V{\bullet}sec$.[31,32] Subsequently, a similar technique provided a two-dimensional hole gas at the hetero-interface.[33] Such a hole gas not only revealed characteristic quantum effects[34] but also was found useful for p-channel MODFETs.[35] More recently, Wang et al.[36] reported a hole mobility as high as 97000$cm^2/V{\bullet}sec$ at 4.2K and deduced a valence-band offset of 210$\pm$30meV for $Ga_{0.5}Al_{0.5}As$-GaAs heterojunctions, corresponding to $\Delta E_c/\Delta E_g = 0.62 \pm 0.05$.

6) Quantized Hall Effect (1980-1981) and Discovery of Fractional Filling (1982)

In 1980, v. Klitzing et al.[37] demonstrated the interesting proposition that quantized Hall resistance in Si MOSFETs can be used for precision determination of the fine structure constant α. Subsequently, Tsui and Gossard[38] found the modulation-doped GaAs-GaAlAs heterostructures quite desirable for the same purpose, primarily because of their high electron mobilities, which led to the determination of α with a great accuracy.[39] The quantized Hall effect in the two-dimensional electron or hole[34,40] system is observable at sufficiently high magnetic fields and low temperatures; in such a range of magnetic fields as to locate the Fermi level in the localized states between the extended states, the magneto-resistance ρ_{xx} vanishes and the Hall resistance ρ_{xy} goes through plateaus. This surprising result can be understood by the argument that the localized states do not take part in quantum transport.[41]

Recently, Tsui, Störmer and Gossard[42] discovered a striking phonomenon: the existence of an anomalous quantized Hall effect, a Hall plateau in ρ_{xy} and a dip in ρ_{xx}, at a fractional filling factor of 1/3 in the extreme quantum limit at temperatures lower than 4.2K. This discovery has spurred a large number of experimental and theoretical studies. Laughlin,[43] as an explanation of such a fractional filling, presented variational wave functions which describe the condensation of a two-dimensional electron gas into a new state of matter, an incompressible quantum fluid; the elementary excitations of such quantum fluid are fractionally charged.

Mendez et al.[44] performed magnetotransport measurements at 0.51K (later 68mK)[45] and up to 28 T for a dilute two-dimensional electron gas with a concentration of $6x10^{10}cm^{-2}$ in a GaAs-GaAlAs heterojunction. The magnetoresistance indicated a substantial deviation from linearity above 18 T and exhibited no additional features for filling factors beyond 1/5, which suggested a transition to a crystalline state. Meantime, Lam and Girvin[46] calculated the critical Landau-level filling factor for transition from Laughlin's liquid state to a Wigner crystal in comparing the energies of these states. The result appears to be consistent with the experimental observation.

7) Variety of Heterojunctions and Superlattices and Relevant Topics

A major portion of the studies reported up to now have been carried out with the GaAs-$Ga_{1-x}Al_xAs$ system. However, a variety of other systems, notably InAs-GaSb(-AlSb), InAlAs-InGaAs,[47] InP-lattice matched alloys,[48,49,50,51] Ge-GaAs,[52,53] CdTe-HgTe,[54,55,56] PbTe-PbSnTe,[57,58] and lattice-mismatched pairs of III-V compounds[59,60] (strained-layer superlattices) have also been seriously explored from scientific as well as technical aspects.

Semiconductor hetero-interfaces exhibit the abrupt discontinuity in the local band structure, usually associated with a gradual band bending in its neighborhood which reflects space-charge. Hetero-interfaces may be classified into four kinds: type I, type II-staggered, type II-misaligned, and type III. The conduction band discontinuity ΔE_c is equal to the difference in the electron affinities of the two semiconductors. Type I is applied to the GaAs-AlAs, GaSb-AlSb, GaAs-GaP systems, etc., where their energy-gap difference $\Delta E_g = \Delta E_c + \Delta E_v$. On the other hand, type IIs are applied to pairs of InAs-GaSb, $(InAs)_{1-x}(GaAs)_x$-$(GaSb)_{1-y}(GaAs)_y$,[61] InP-$Al_{0.48}In_{0.52}As$,[51] etc., where electrons and holes are confined in the different semiconductors at their heterojunctions and superlattices. Particularly, in type II-misaligned, the top of the valence band in GaSb is located above the bottom of the conduction band in InAs by the amount of E_s, differing from type II-staggered. In this classification, we add a unique member of the family, HgTe-CdTe,[56] as type III, where HgTe is a zero-gap semiconductor due to the inversion of the relative positions of Γ_6 and Γ_8 edges. The Γ_8 light-hole band in CdTe becomes the conduction band in HgTe.

The bandedge discontinuities at the hetero-interfaces obviously command all properties of quantum wells and superlattices, and thus constitute the most relevant parameters for device design[62] as well. Recently, considerable efforts have been made to understand the electronic structure at interfaces[63] or heterojunctions.[64,65,66] Even in an ideal situation, the discontinuity provides formidable tasks in theoretical handling: Propagation and evanescent Bloch waves should be matched across the interface, satisfying continuity conditions on the envelope wave functions.[67,68] The fundamental understanding as well as the experimental determination for such parameters as ΔE_c and ΔE_v, however, are still not satisfactory, even in the most-studied GaAs-$Ga_{1-x}Al_xAs$ system. Values of $\Delta E_c/\Delta E_g$, 85%, and $\Delta E_v/\Delta E_g$, 15%,[20] determined in 1975 apparently are to be revised to about 60% and 40%, respectively.[36,69,70,71,72]

Another breed called "n-i-p-i," the outgrowth of a doping superlattice, was pursued by Döhler[73] and Ploog et al..[74] Döhler et al.[75] observed in GaAs doping superlattices that the photon energies in luminescence were varied by the laser excitation intensity. Superlattice structures were also made of amorphous semiconductors. Abeles and Tiedje[76] pioneered the development of such structures consisting of alternating layers of hydrogenated amorphous silicon, germanium, silicon nitride, and silicon carbide. Esaki et al.[77] proposed the introduction of a third constituent such as AlSb in the InAs-GaSb system. Such a triple-constituent system leads to a new concept of man-made polytype superlattices, which offers an additional degree of freedom.

In heterostructures, it is certainly desirable to select a pair of materials closely lattice-matched in order to minimize defect formation or stress. However, heterostructures lattice-mismatched to a limited extent can be grown with essentially no misfit dislocations, if the layers are sufficiently thin, because the mismatch is accommodated by uniform lattice strain.[78] On the basis of such premise, Osbourn[59] and his co-workers[60] prepared a number of strained-layer superlattices from lattice-mismatch pairs, claiming their relatively high-quality superlattices to be suitable for opto-devices. It is certainly true that, without the requirement of lattice matching, the number of available pairs for superlattice formation is greatly inflated.

Voisin et al.[79] made optical absorption measurements on GaSb-AlSb superlattices, of which the spectrum exhibits the two-dimensional density of states and pronounced free exciton peaks. In comparison with the reported optical absorption spectrum[20] for the GaAs-GaAlAs case, it was noticed that the first two exciton peaks appear to be exchanged. Using an effective mass theory, such observation was interpreted as the occurrence of the reversal of the heavy- and light-hole bands due to the strain effect induced by a lattice mismatch equal to 0.65%.

8) InAs-GaSb Superlattices (1977-1982) and Quantum Wells (1982-)

In 1977, while searching for a new type of superlattice, InAs-GaSb was selected, because of its extraordinary bandedge relationship at the interface, called type II-misaligned. It was observed that, in the study of $(InAs)_{1-x}(GaAs)_x$-$(GaSb)_{1-y}(GaAs)_y$ p-n heterojunctions,[61] the rectifying characteristic changes to nonrectification as both x and y approach zero, implying the change-over from the "staggered" heterojunction to the "misaligned" one. At the misaligned-bandgap interface, electrons which "flood" from the GaSb valence band to the InAs conduction band, leaving holes behind, produce a dipole layer consisting of two-dimensional electron and hole gases.

First, Sai-Halasz et al.[80] made a one-dimensional calculation and, subsequently, the LCAO band calculation[80] was performed for InAs-GaSb superlattices. The calculated subband structure is strongly dependent upon the period: The energy gap decreases with increase in the period, becoming zero at 170Å, corresponding to a semiconductor-to-semimetal transition. In those calculations, the misaligned magnitude, E_s, was set at 0.15eV: a value which had been derived from analysis of optical absorption.[82] Recently, Altarelli[83] performed self-consistent electronic structure calculations in the envelope-function approximation with a three-band k•p formalism for this superlattice.

The electron concentration in superlattices was measured as a function of InAs layer thickness;[84] it exhibited a sudden increase of an order-of-magnitude in the neighborhood of 100Å Such increase indicates the onset of electron transfer from GaSb to InAs which is in good agreement with theoretical prediction. Far-infrared magneto-absorption experiments[85,86] were performed at 1.6K for semimetallic superlattices which confirmed their negative energy-gap.

MBE-grown GaSb-InAs-GaSb quantum wells have been investigated, where the unique bandedge relationship allows the coexistence of electrons and holes across the two interfaces. Before an experimental approach, Bastard et al.[87] had performed self-consistent calculations for the electronic properties for such quantum wells, predicting the existence of a semiconductor-to-semimetal transition as a result of electron transfer from GaSb when the InAs quantum well thickness reaches a threshold, somewhat similar to the mechanism in the InAs-GaSb superlattices. Such a transition was confirmed experimentally by Mendez et al.[88] The threshold thickness was found to be about 60Å. The electron mobility in InAs quantum wells of 150-200Å ranges between 1.5 and $2.0 \times 10^5 cm^2/V \cdot sec$ at 4.2K. Such values of the mobility were compared with theoretical calculations.[89]

Measurements of magnetoresistance ρ_{xx} and the Hall resistance ρ_{xy} caused a few surprises: No holes are evident in ρ_{xy} whereas the increase of ρ_{xx} is proportional to B^2 at low fields, suggesting a two-carrier conduction mechanism; there is a nearly 20% decrease of electrons at high fields; the most of all is the advent of anomalous peaks adjacent to ordinary peaks in Shubnikov-de Haas oscillations. Washburn et al.[90] observed that such anomalous peaks weaken with decreasing temperature and finally vanish at 19mK.

Such phenomena appear to be too unusual to be interpreted in terms of the known effects at this time. Nevertheless, we can present the following considerations: First of all, the electron number, n, in the InAs layer, may be balanced not by the hole number, p, alone, but rather by $p + N_D$, where N_D is the concentration of ionized donor states in the neighborhood of the interfaces. The InAs well is also susceptible to unwanted modulation-doping from GaSb. It seems possible that p is a minor fraction of n, which means a rather large deviation from the theoretical calculations.[87] Secondly, although these holes are contributing to conduction at low magnetic fields, they will form electron-hole bound states, possibly induced by a certain magnetic field strength, 5T or so. The reduction of n may be indicative of the formation of such two-particle states which manifest themselves as anomalous peaks, but cease to contribute to conduction at very low temperatures. Along with this line, the binding energy for such an excitonic state is derived to be 6.4meV from the difference of the positions between the normal and anomalous peaks. The obtained value is about three times the effective Rydberg of InAs (1.9meV), comparable to the hydrogenic binding energy[91] in the two-dimensional system, but larger than the previously calculated exciton energy[92] at no magnetic field condition. Limited experiments have also been carried out for AlSb-InAs-AlSb quantum wells.[93]

We have witnessed the remarkable development of an interdisciplinary nature on this subject. Such development can be considered to be a *renaissance* in semiconductor research. Indeed, a variety of *engineered* structures exhibited extraordinary transport and optical properties; some of them, such as ultrahigh carrier mobilities, semimetallic coexistence of electrons and holes, etc., may not even exist in any *natural* crystal. Thus, this new degree of freedom offered in semiconductor research *through advanced material engineering* has inspired many ingeneous experiments, resulting in observations of not only predicted effects but also totally unknown phenomena such as fractional quantization which require novel interpretations. Activities in this new frontier of semiconductor physics, in turn, give immeasurable stimulus to device physics, leading to novel devices such as MODFETs,[29,30] MQW-lasers,[94] advanced APDs[95,96] and real-space electron-transfer devices,[97] or provoking new ideas[98,99] for applications.

This article concludes with some comments on the prospective research;

1) The route to one dimension is still wide open. A number of attempts such as GaAs quantum well wires[100] or pinched accumulation layers in Si MOSFET,[101] have already been made. A true one-dimensional electron gas, however, can be created with a combination of a superlattice potential and the surface inversion.

2) Our early efforts focussed on transport properties in the direction of the one-dimensional periodic potential including resonant tunneling. Such efforts apparently demand the stringent requirements for lateral uniformity as well as the defect density in grown wafers. With improved techniques there will be more promising studies in this direction.

3) The electronic properties of two-dimensional holes associated with the complexity of the valence-band structure at the hetero-interfaces deserve further scrutiny.

4) The discovery of the fractionally quantized Hall effect is probably one of the most significant events in this field. More efforts will be needed for in-depth understanding of the quantum fluid, including a transition to a Wigner crystal.[44,45,46]

5) In two dimensions, there is not true metallic conduction:[102] all of the two-dimensional electron gas systems exhibit a logarithmic increase of the resistance as the temperature decreases,[103] which is characteristic of the weak localization. Thus, localization studies are pertinent to the physics of reduced dimensionality.

6) Still required are innovations in theory and experiment for accurate determination of the bandedge offsets at the hetero-interfaces.

7) It is worth pursuing multilayer structures which exhibit electric field-induced effects in either optical[104] or transport properties. Such structures clearly possess device potentials.

Hopefully, this presentation, which cannot possibly cover every landmark, provides some flavor of the excitement in this field.

REFERENCES

1. L. Esaki and R. Tsu, IBM Research Note RC-2418 (1969). (This paper was a supplement to a proposal later awarded as Contract #DAHCO4-69-C-0069).
2. L. Esaki and R. Tsu, IBM J. Res. Develop. 14, 61 (1970).
3. D. Bohm, *Quantum Theory* (Prentice Hall, Englewood Cliffs, N.J. 1951), p. 283.
4. L. Esaki, *Les Prix Nobel en 1973,* Imprimerie Royale P. A. Norstedt & Söner, Stockholm 1974, p. 66; Science 183, 1149 (1974).
5. A. E. Blakeslee and C. F. Aliotta, IBM J. Res. Develop. 14, 686 (1970).
6. L. Esaki, L. L. Chang, and R. Tsu, in Proceedings 12th International Conference on Low Temperature Physics, Kyoto, Japan, September 1970. (Keigaku Publishing Co., Tokyo, Japan), p. 551.
7. A. Y. Cho, Appl. Phys. Lett. 19, 467 (1971).
8. J. M. Woodall, J. Cryst. Growth 12, 32 (1972).
9. R. de L. Kronig, and W. J. Penney, Proc. Roy. Soc. A130, 499 (1930).
10. R. G. Chambers, Proc. Phys. Soc. (London) A65, 458 (1952).
11. L. Esaki, L. L. Chang, W. E, Howard, and V. L. Rideout, Proceedings of the 11th International Conference on the Physics of Semiconductors, Warsaw, Poland, 1972, edited by the Polish Academy of Sciences (PWN-Polish Scientific Publishers, Warsaw, Poland, 1972), p. 431.

12. L. L. Chang, L. Esaki, W. E. Howard, and R. Ludeke, J. Vac. Sci. Technol. 10, 11 (1973); L. L. Chang, L. Esaki, W. E. Howard, R. Ludeke, and G. Schul, J. Vac. Sci. Technol. 10, 655 (1973).
13. U. Gnutzmann and K. Clauseker, Appl. Phys. 3, 9 (1974).
14. A. Madhukar, J. Vac. Sci. Technol. 20, 149 (1982).
15. R. Tsu and L. Esaki, Appl. Phys. Lett. 22, 562 (1973).
16. L. L. Chang, L. Esaki, and R. Tsu, Appl. Phys. Lett. 24, 593 (1974).
17. L. Esaki and L. L. Chang, Phys. Rev. Lett. 33, 495 (1974).
18. T. C. L. G. Sollner, W. E. Goodhue, P. E. Tannenwald, C. D. Parker, and D. D. Peck, Appl. Phys. Lett. 43, 588 (1983).
19. R. Dingle, W. Wiegmann, and C. H. Henry, Phys. Rev. Lett. 33, 827 (1974).
20. R. Dingle, A. C. Gossard, and W. Wiegmann, Phys. Rev. Lett. 34, 1327 (1975).
21. P. Manuel, G. A. Sai-Halasz, L. L. Chang, C.-A. Chang, and L. Esaki, Phys. Rev. Lett. 25, 1701 (1976).
22. E. Burstein, A. Pinczuk, and S. Buchner, Physics of Semiconductors 1978, Institute of Physica Conference Series 43, London 1979, p. 1231.
23. G. Abstreiter and K. Ploog, Phys. Rev. Lett. 42, 1308 (1979).
24. A. Pinczuk, H. L. Störmer, R. Dingle, J. M. Worlock, W. Wiegmann, and A. C. Gossard, Solid State Commun. 32, 1001 (1979).
25. C. Colvard, R. Merlin, and M. V. Klein, and A. C. Gossard, Phys. Rev. Lett. 45, 298 (1980).
26. V. Narayanamurti, H. L. Störmer, M. A. Chin, A. C. Gossard and W. Wiegmann, Phys. Rev. Lett. 43, 2012 (1979).
27. R. Dingle, H. L. Störmer, A. C. Gossard, and W. Wiegmann, Appl. Phys. Lett. 33, 665 (1978).
28. H. L. Störmer, R. Dingle, A. C. Gossard, W. Wiegmann, and M. D. Sturge, Solid State Commun. 29, 705 (1979).
29. T. Mimura, S. Hiyamizu, T. Fujii and K. Nanbu, Jpn. J. Appl. Phys. 19, L225 (1980).
30. D. Delagebeaudeuf, P. Delescluse, P. Etienne, M. Laviron, J. Chaplart, and N. T. Linh, Electron. Lett. 16, 667 (1980).
31. M. Heiblum, E. E. Mendez, and F. Stern, Appl. Phys. Lett. 44, 1064 (1984).
32. E. E. Mendez, P. J. Price, and M. Heiblum, to appear in Appl. Phys. Lett.
33. H. L. Störmer and W. T. Tsand, Appl. Phys. Lett. 36, 685 (1980).
34. H. L. Störmer, Z. Schlesinger, A. Chang, D. C. Tsui, A. C. Gossard, and W. Wiegman, Phys. Rev. Lett. 51, 126 (1983).
35. H. L. Störmer, K. Baldwin, A. C. Gossard, and W. Wiegmann, Appl. Phys. Lett. 44, 1062 (1984).
36. W. I. Wang, E. E. Mendez, and F. Stern, Appl. Phys. Lett. 45, 639 (1984).
37. K. v. Klitzing, G. Dorda, and M. Pepper, Phys. Rev. Lett. 45, 494 (1980).
38. D. C. Tsui and A. C. Gossard, Appl. Phys. Lett. 38, 550 (1981).
39. D. C. Tsui, A. C. Gossard, B. F. Field, M. E. Cage, and R. F. Dziuba, Phys. Rev. Lett. 48, 3 (1982).
40. E. E. Mendez, W. I. Wang, L. L. Chang, and L. Esaki, Phys. Rev. B 28, 4886 (1983).

41. T. Ando and Y. Uemura, J. Phys. Soc. Jpn. 36, 959 (1974).
42. D. C. Tsui, H. L. Störmer, and A. C. Gossard, Phys. Rev. Lett. 48, 1559 (1982).
43. R. B. Laughlin, Phys. Rev. Lett. 50, 1395 (1983).
44. E. E. Mendez, M. Heiblum, L. L. Chang, and L. Esaki, Phys. Rev. B, 28, 4886 (1983).
45. E. E. Mendez, L. L. Chang, M. Heiblum, L. Esaki, M. Naughton, K. Martin, and J. Brooks, to appear in Phys. Rev. B.
46. Pui K. Lam and S. M. Girvin, Phys. Rev.B, 30, 473 (1984).
47. R. People, K. W. Wecht, K. Alavi, and A. Y. Cho, Appl. Phys. Lett. 43, 118 (1983).
48. K. Y. Cheng, A. Y. Cho, and W. R. Wagner, Appl. Phys. Lett. 39, 607 (1981).
49. M. Razeghi and J. P. Duchemin, J. Vac. Sci. Technol. B, 1, 262 (1983).
50. M. Voos, J. Vac. Sci. Technol. B, 1, 404 (1983).
51. E. J. Caine, S. Subbanna, H. Kröemer, J. L. Merz, and A. Y. Cho, to appear in Appl. Phys. Lett.
52. P. M. Petroff, A. C. Gossard, A. Savage, and W. Wiegmann, J. Cryst. Growth 46, 172 (1979).
53. C-A. Chang, A. Segmüller, L. L. Chang, and L. Esaki, Appl. Phys. Lett. 38, 912 (1981).
54. J. N. Schulman and T. C. McGill, Appl. Phys. Lett. 34, 883 (1979).
55. G. Bastard, Phys. Rev. B, 25, 7584 (1982).
56. Y. Guldner, G. Bastard, J. P. Vieren, M. Voos, J. P. Faurie and A. Million, Phys. Rev. Lett. 51, 907 (1983).
57. H. Kinoshita and H. Fujiyasu, J. Appl. Phys. 51, 5845 (1980).
58. E. F. Fantner and G. Bauer, *Two-Dimensional Systems, Heterostructures, and Superlattices,* ed. G. Bauer, F. Kuchar and H. Heinrich, Springer Series in Solid-State Science 53, Springer-Verlag, 1984, p.207.
59. G. C. Osbourn, J. Appl. Phys. 53, 1586 (1982).
60. G. C. Osbourn, R. M. Biefeld and P. L. Gourley, Appl. Phys. Lett. 41, 172 (1982).
61. H. Sakaki, L. L. Chang, R. Ludeke, C.-A. Chang, G. A. Sai-Halasz, and L. Esaki, Appl. Phys. Lett. 31, 211 (1977).
62. H. Kröemer, Surf. Sci. 132, 543 (1983).
63. M. L. Cohen, Advances in Electronics and Electron Physics, Vol. 51, 1, Academic Press, 1980.
64. W. Harrison, J. Vac. Sci. Technol. 14, 1016 (1977).
65. W. R. Frensley and H. Kröemer, Phys. Rev. B16, 2642 (1977).
66. J. Tersoff, to appear in Phys. Rev. B.
67. G. Bastard, Phys. Rev. B24, 5693 (1981).
68. S. R. White and L. J. Sham, Phys. Rev. Lett. 47, 879 (1981).
69. H. Kröemer, Wu-Yi Chen, J. S. Harris, Jr., and D. D. Edwall, Appl. Phys. Lett. 36, 295 (1980).
70. R. C. Miller, A. C. Gossard, D. A. Kleinman and O. Munteanu, Phys. Rev. B 29, 3740 (1984).
71. R. C. Miller, D. A. Kleinman, and A. C. Gossard, Phys. Rev. B29, 7085 (1984).

72. T. W. Hickmott, P. M. Solomon, R. Fischer, and H. Morkoc, to appear in J. Appl. Phys.
73. G. H. Döhler, Phys. Status Solidi (b)52, 79 and 533 (1972).
74. K. Ploog, A. Fischer, G. H. Döhler, and H. Künzel, in *Gallium Arsenide and Related Compounds 1980*, Institute of Physics Conference Series No. 56, edited by H. W. Thim (Institute of Physics, London, 1981), p. 721.
75. G. H. Döhler, H. Künzel, D. Olego, K. Ploog, P. Ruden, H. J. Stolz, and G. Abstreiter, Phys. Rev. Lett. 47, 864 (1981).
76. A. Abeles and T. Tiedje, Phys. Rev. Lett. 51, 2003 (1983).
77. L. Esaki, L. L. Chang, and E. E. Mendez, Jpn. J. Appl. Phys. 20, L529 (1981).
78. J. H. van der Merwe, J. Appl. Phys. 34, 117 (1963).
79. P. Voisin, C. Delalande, M. Voos, L. L. Chang, A. Segmüller, C. A. Chang, and L. Esaki, to appear in Phys. Rev. B.
80. G. A. Sai-Halasz, R. Tsu, and L. Esaki, Appl. Phys. Lett. 30, 651 (1977).
81. G. A. Sai-Halasz, L. Esaki, and W. A. Harrison, Phys. Rev.B18, 2812 (1978). br;.un4 82.G. A. Sai-Halasz, L. L. Chang, J-M Welter, C.-A. Chang, and L. Esaki, Solid State Commun. 25, 935 (1978).
83. M. Altarelli, Phys. Rev. B28, 842 (1983).
84. L. L. Chang, N. J. Kawai, G. A. Sai-Halasz, R. Ludeke, and L. Esaki, Appl. Phys. Lett. 35, 939 (1979).
85. Y. Guldner, J. P. Vieren, P. Voisin, M. Voos, L. L. Chang, and L. Esaki, Phys. Rev. Lett. 45, 1719 (1980).
86. J. C. Maan, Y. Guldner, J. P. Vieren, P. Voisin, M. Voos, L. L. Chan g, and L. Esaki, Solid State Commun. 39, 683 (1981).
87. G. Bastard, E. E. Mendez, L. L. Chang, L. Esaki, J. Vac. Sci. Technol. 21, 531 (1982).
88. E. E. Mendez, L. L. Chang, C-A. Chang, L. F. Alexander, and L. Esaki, Surf. Sci. 142, 215 (1984).
89. E. E. Mendez, G. Bastard, L. L. Chang, C-A Chang, and L. Esaki, Bull. Am. Phys. Soc. 29, 471 (1984).
90. W. Washburn, R. A. Webb, E. E. Mendez, L. L. Chang, and L. Esaki, to appear in Phys. Rev. B.
91. G. Bastard, Phys. Rev. B 24, 4714 (1981).
92. G. Bastard, E. E. Mendez, L. L. Chang, and L. Esaki, Phys. Rev. B 26, 1974 (1982).
93. C-A. Chang, E. E. Mendez, L. L. Chang, and L. Esaki, Surf. Sci. 142, 598 (1984).
94. W. T. Tsang, Appl. Phys. Lett. 39, 786 (1981).
95. F. Capasso, J. Vac. Sci. Technol. B1, 457 (1983).
96. T. Tanoue and H. Sakaki, Appl. Phys. Lett. 41, 67 (1982).
97. K. Hess, M. Morkoc, H. Shichijo, and B. G. Streetman, Appl. Phys. Lett. 35, 469 (1979).
98. J. J. Quinn, U. Strom, and L. L. Chang, Solid State Commun. 45, 111 (1983).
99. T. Nakagawa, N. J. Kawai, K. Ohta, and M. Kawashima, Electronics Lett. 19, 822 (1983). Phys. Rev. B 29, 3752 (1984).
100. P. M. Petroff, A. C. Gossard, R. A. Logan and W. Wiegmann, Appl. Phys. Lett. 41, 635 (1982).

101. A. B. Fowler, A. Hartstein, and R. A. Webb, Phys. Rev. Lett. 48, 196 (1982).
102. E. Abrahams, P. W. Anderson, D. C. Licciardello, and T. V. Ramakrishnan, Phys. Rev. Lett. 42, 673 (1979).
103. S. Washburn, R. A. Webb, E. E. Mendex, L. L. Chang, and L. Esaki, Phys. Rev. B29, 3752 (1984).
104. E. E. Mendez, G. Bastard, L. L. Chang, and L. Esaki, Phys. Rev. B26, 7101 (1982).

Contributions

Electronic Surface Transitions in Si(111)2×1 Studied by Polarized Light

A. Cricenti, F. Ciccacci*, S. Selci*, P. Chiaradia, and G. Chiarotti*

Istituto di Struttura della Materia del CNR, Via E. Fermi 44
I-00044 Frascati, Italy

In the last few years, great attention has been given to the study of electron surface states on clean semiconducturs, Si(111)2×1 in particular receiving most of the interest. Despite the great amount of experimental and theoretical work, some controversies still remain to be explained (for example, dynamical LEED data are in disagreement with all the reconstruction models [1]). In order to discriminate between the different models for the reconstruction, we performed measurements of differential reflectivity by means of polarized light. Theoretical calculations [2] based on the selection rules for the optical transitions have been reported in the range 0.3 - 4.0 eV, where two peaks in ΔR/R with unpolarized light occur [3]. Figure 1 shows the spectral dependence of ΔR/R for a single domain Si(111)2×1 surface with electrical field of light along the [0$\bar{1}$1] and [$\bar{2}$11] directions, which in our case, corresponds to the ΓJ and ΓJ' lines of the surface Brillouin zone, respectively.

Figure 2 shows the dependence upon polarization of the peaks at 0.45 eV and at 2.4 eV: the dashed and dotted lines are a $\cos^2(\theta)$ and const. + $\cos^2(\theta)$ fit, respectively (θ being the azimuthal angle of polarization).

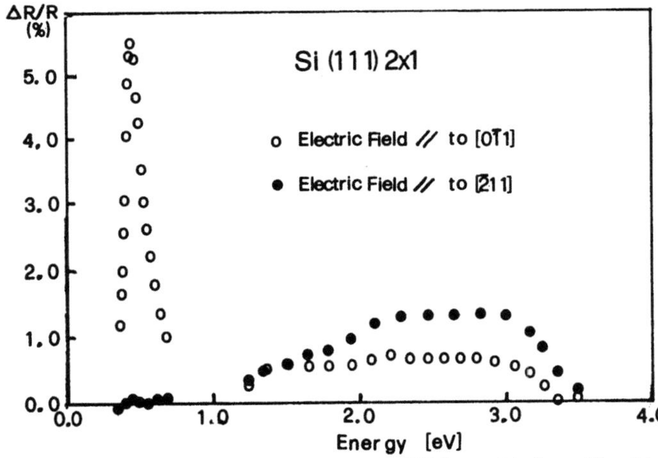

Figure 1: Spectral dependence of differential reflectivity of Si(111)2×1 for light polarizations along the [$\bar{2}$11] and [0$\bar{1}$1] directions

*Dipartimento di Fisica, II Università di Roma, Tor Vergata
I-Roma, Italy

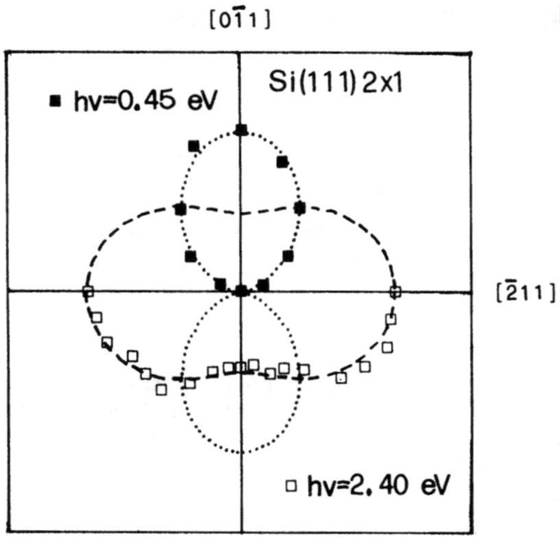

Figure 2: Dependence of the peak intensities upon various polarizations

The peak at 0.45 eV is strongly anisotropic: the transition only occurs for light polarized along the [0$\bar{1}$1] direction. Contrarily, the broad structure around 2.5 eV has a much smaller anisotropy with opposite polarization dependence (maximum is along the [$\bar{2}$11] direction) [4]. The experimental results are in favour of Pandey's model [5] of the reconstructed Si(111)2×1 surface that predicts quasi-unidimensional chains along the [0$\bar{1}$1] direction. In this model [2], the results at about 2.5 eV would correspond to transitions perpendicular to the chains or around the J' point of the surface Brillouin zone.

References

1. H. Liu, M.R. Cook, F. Jona and P.M. Marcus: Phys. Rev. B 28, 6137 (1983)
2. R. Del Sole and A. Selloni: Solid State Commun. 50, 825 (1984)
3. S. Nannarone, P. Chiaradia, F. Ciccacci, R. Memeo, P. Sassaroli, S. Selci and G. Chiarotti: Solid State Commun. 33, 593 (1980)
4. Further explanations are given in P. Chiaradia, A. Cricenti, S. Selci and G. Chiarotti: Phys. Rev. Lett. 52, 1145 (1984); S. Selci, F. Ciccacci, A. Cricenti, P. Chiaradia and G. Chiarotti: to be published
5. K.C. Pandey: Phys. Rev. Lett. 47, 1913 (1981)

Short Range Bonding Interaction at Metal-Metal Interfaces

Wayne R. Bennett and John A. Leavitt

Department of Physics, University of Arizona, Tucson, AZ 85721, USA

Charles M. Falco

Departments of Physics and of Optical Sciences and in the Arizona Research Laboratories, University of Arizona, Tucson, AZ 85721, USA

Using the techniques discussed by Falco, Bennett, and Boufelfel (in this volume), freshly deposited Mo surfaces were exposed to a beam of sputtered Ta atoms for varying, controlled lengths of time. The resultant number of deposited Ta atoms/cm^2 (from here on referred to as Ta coverage) was subsequently measured using Rutherford Backscattering (RBS).

Since the Ta flux is held constant (to within ±0.3%) any change in the growth rate of the sample is due to a change in the net sticking coefficient of the sputtered Ta. Consequently, we are able to investigate the dependence of this net Ta sticking coefficient on the Ta coverage.

As shown in Fig. 1, each of our samples was designed to have three Ta depositions (corresponding to three distinct Ta beam-exposure times) each separated from the others by ~360 Å of Mo. This configuration has the following two advantages, i) the different Ta layers are made in the same run under identical conditions, and ii) all layers are simultaneously analyzed with the same Rutherford $^4He^+$ beam as shown in Fig. 2. This has the effect of improving the relative precision with which the layer thicknesses can be determined (typically we achieve ~1%).

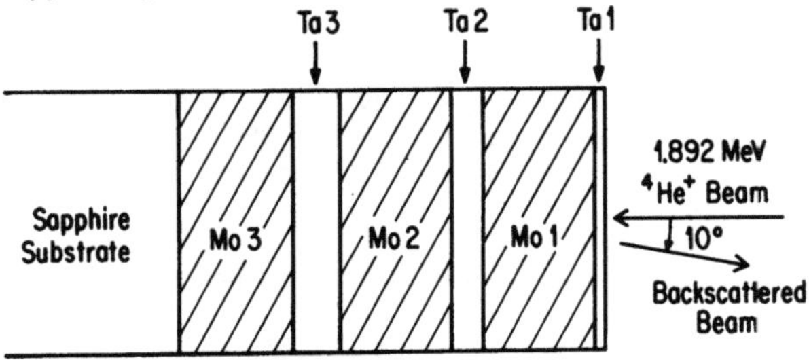

Fig. 1. Schematic diagram of a sample (not to scale)

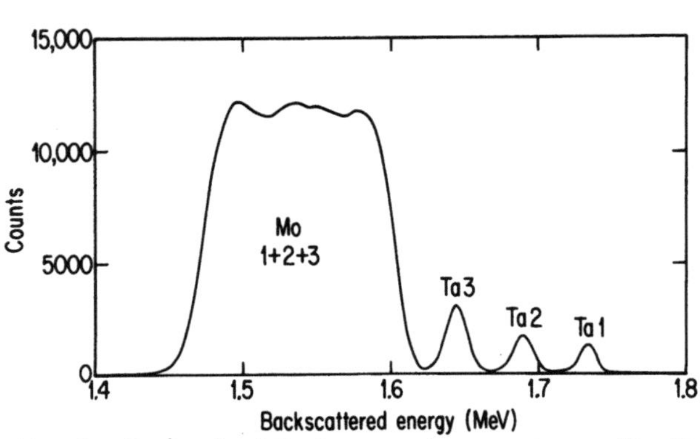

Fig. 2. Rutherford Backscattering spectrum. The Ta coverage from each layer can be determined from the corresponding peak integration

Our measurements show that the Ta growth rate, and therefore the sticking coefficient, decreases with increasing Ta coverage until, beyond a coverage equivalent to approximately 3-4 monolayers, it remains constant.

This variation in growth rate has important consequences for the deposition of superlattices of chosen superlattice wavelength.

A more detailed account of the variation of the sticking coefficient with coverage will be published elsewhere.

Acknowledgements

J. A. Leavitt acknowledges the assistance of P. Stoss and D. B. Cooper, and financial support for the IBM/UA Ion Beam Analysis Facility from IBM General Products Division, Tucson.

This work was supported by the Director, Office of Energy Research Office of Basic Energy Sciences, Materials Sciences Division of the U. S. Department of Energy under contract number DE-AC02-83ER45025.

Part II

Lattice Dynamics

Phonons at Interfaces and Superlattices

G. Benedek

Gruppo Nazionale di Struttura della Materia del Consiglio Nazionale delle Ricerche, Dipartimento di Fisica dell'Università, Via Celoria 16
I-20133 Milano, Italy

V.R. Velasco

Instituto de Fisica del Estado Solido, Consejo Superior de Investigaciones Cientificas, Serrano 123, E-28006 Madrid, Spain

1. Introduction

Since the discovery of surface waves in semi-infinite isotropic elastic media by Lord RAYLEIGH |1| until recently the dynamics of surfaces was in most cases confined to the realm of the theory of elasticity. In the meantime the theory of surface waves has been applied to the study of anisotropic media |2| and it has evolved from its original geophysical and seismological grounds |3| to technical applications in delay lines, signal-processing, opto-acoustic, and acousto-electronic devices, etc. |4|.

On the other hand the surface lattice dynamics in the dispersive range, where the microscopic effects and the discrete character of the lattice become important, was centered mainly on theoretical studies devised for ideal, unrelaxed surfaces |5,6|. These simple models were the main source of information, as long as no spectroscopy of surface phonons was available. Actually, the determination of the dispersion curves of surface phonons was considered much more difficult than for bulk phonons, since the standard probes of lattice vibrations, such as thermal neutrons, do not work so well for surfaces due to their large penetration depth and consequent lack of sensitivity.

The turning point in surface phonon spectroscopy was achieved less than four years ago by means of high-resolution measurements of atom beam scattering. Thanks to the development of intense, highly monochromatic helium beam sources |7|, the dispersion relations of Rayleigh modes in different crystal surfaces were observed for the first time |8-12|, thus confirming earlier theoretical predictions |13-15|.

Another important breakthrough has been made recently by means of electron energy-loss spectroscopy (EELS). The dispersion curves of Rayleigh waves on a clean Ni(001) surface and of the adsorbate vibrational modes for a c(2x2) oxygen overlayer on Ni(001) have been measured with fair resolution |16,17|.

These techniques cannot be used for solid-solid interfaces, in thin films and superlattices. For these systems the optical techniques, such as Brillouin and Raman scattering, and infrared absorption, are more suitable. In the particular case of superlattices, a fairly vast amount of experimental information is emerging from Raman scattering experiments |18-21|.

The theoretical study of the surface lattice dynamics has been approached by means of several methods. One method is based on the construction of a trial solution for the semi-infinite lattice |22| in the same spirit as in the elastic wave theory |2|. This method was later extended to crystals with long-range forces |23| and used for the study of metal surfaces |24|.

Another method consists in the direct calculation of the eigenvalues and eigenvectors of a slab formed by a small number of atomic layers |25,26|. This method yields all the variety of acoustic and optical surface modes all over the surface Brillouin zone, provided their penetration depth is much smaller than the slab thickness. It has been shown that ten to twenty atomic layers normally suffice to yield all microscopic surface modes with excellent precision |26|. This method also shows in a clear way the effects of surface relaxation and of adsorbed layers |25|. As a consequence, the slab method is of practical use for several purposes. Nevertheless, this method has two important shortcomings. The first is characterized by the poor statistics in the calculation of the surface phonon densities, due to the limitations in the number of atomic layers imposed by the computer capability. The surface phonon densities are of great importance, as they enter several surface response functions. The second shortcoming of the slab method lies in the difficult identification of the surface resonances within the bulk phonon spectrum.

The Green function (GF) method in surface dynamics considers the surface as a perturbation modifying the bulk phonon spectrum. It was introduced about forty years ago by LIFSHITZ and ROZENZWEIG |27|. The main advantage of the GF method is that it enables one to work in the perturbation subspace rather than in the large space of the crystal coordinates. Moreover, the GF method provides the most natural way for determining surface resonant modes and for the calculation of the surface phonon densities. The main drawback of the GF method is represented by the computation difficulties arising from the singular character of the surface -projected Green functions, which may partly counterbalance the great advantage of handling small matrices.

The GF method has been used in the study of the intrinsic unrelaxed surface |28|, but it applies as well to more complex perturbations like the change in surface force constants |29-31|. Stepped surfaces and reconstruction have been succesfully studied by the method of GF generating coefficients |32|. The GF method has also been employed in the problem of the 7x7 reconstruction of the Si(111) surface and related phonon instabilities |33|.

The surface GF matching method, originally developed for the study of the electronic surface states |34|, appears to be very convenient for the dynamics of solid interfaces |35-36|.

The results obtained by the above surface phonon spectroscopic techniques are putting in evidence new aspects of surface dynamics that were unexpected on the basis of the underlying bulk dynamics. The existence of the surface implies a lowering of the translational and rotational symmetries that relax certain selection rules valid in the three-dimensional cyclic lattice. Many microscopic mechanisms which are forbidden or inhibited in the bulk for symmetry reasons, are promoted by the symmetry reduction and manifest themselves at the surface. As an example we quote the surface phonon anomalies induced by electron-phonon coupling, which may eventually show up even in crystals with regular bulk phonon dispersion. It is in this light that surface dynamics is coming out from the applicative areas and rapidly gaining positions among the fundaments of condensed matter physics.

The recent progress in the theoretical and experimental determination of the surface phonon dispersion curves has been reviewed in several recent articles |37-38|, which the reader is referred to. Also the GF method for ionic crystals with long-range forces has been illustrated in several papers and lecture notes |28,39,40|, and are here considered as a basis for the discussion of interfaces and superlattice dynamics given in Sec. 2 and 3, respectively. Actually the GF method has been already used in a variety of problems involving the dynamics of interfaces |41-43| and superlattices |44-46|.

2. Interface Lattice Dynamics

The interface between two different solids has long been used for practical applications, like metal-semiconductor junctions (diodes) or junctions between two different semiconductors (heterojunctions). The last fifteen years have seen a growing interest in the understanding of the microscopic properties of interfaces. Lamellar solids formed by intercalated layers of two different crystals are extensively studied in view of many technical applications, particularly in electronics |47|.

Lattice vibrations localized at solid interfaces play a role in numerous interface phenomena such as growth, structure and stability of epitaxial systems |48|, the inelastic electron tunneling in superconducting junctions |49|, and also affect the optical properties of semiconductor superlattices |18-21|, and lamellar eutectics |50|.

The existence of acoustic waves propagating along the interface of two semi-infinite isotropic elastic media with sagittal polarization was predicted a long time ago by Stoneley (Stoneley waves) |51|. In the elastic limit, however, their existence is severely constrained by conditions among the elastic constants and densities of the two media |52|. The same is true for interfaces between anisotropic crystals |41,53,54|, so that Stoneley waves are predicted only in a small number of the possible real interface systems.

Interface dynamics affects, in principle, several integral response functions, such as the excess of specific heat of composite materials and the thermal conductivity across interfaces. These quantities are expected to exhibit small anomalies at finite temperatures associated with spectral features of interface modes. However, only penetrating probes such as light scattering, neutron scattering or infrared spectroscopy can give some information on interface phonon densities, but to our knowledge the infrared reflectivity studies on LiF/NaF lamellar eutectics |50| are the only measurements where something ascribable to interface modes is observed.

Brillouin scattering experiments in supported films of variable thickness have been performed in recent years |55,56|. These experiments have revealed a number of Sezawa waves of the film |57| as well as its Rayleigh wave (which can be regarded as the Sezawa wave of zeroth order), but no Stoneley wave has been identified.

Several authors have approached the theoretical study of interface lattice dynamics on the basis of simple approximations such as the Born-von Kàrmàn model and identical lattice constants for the adjacent crystals (coherent interface approximation). |41,42,58| A variety of short-wave interface phonons was found even in the absence of Stoneley waves |41|. However, interesting dynamical effects can be expected to arise from the different symmetries occurring at a real interface in analogy with what happens in commensurate superlattices |59|.

Some of these effects were put in evidence in a previous work of ours |43|. We have studied the lattice dynamics of an interface with squared symmetry between two ionic crystals with commensurate lattice constants r_1 and r_2 by means of a GF method |43|. In that work, the dispersion curves and the projected phonon densities were calculated for the LiF(001)/KF(001) interface, where the lattice constants r_1 = 2.013 Å and r_2 = 2.674 Å are approximately in the ratio 3/4. The model for the unrelaxed interface is illustrated in fig. 1. It consists of a short-range coupling between the first layers of two semi-infinite ionic lattices with equal orientation.

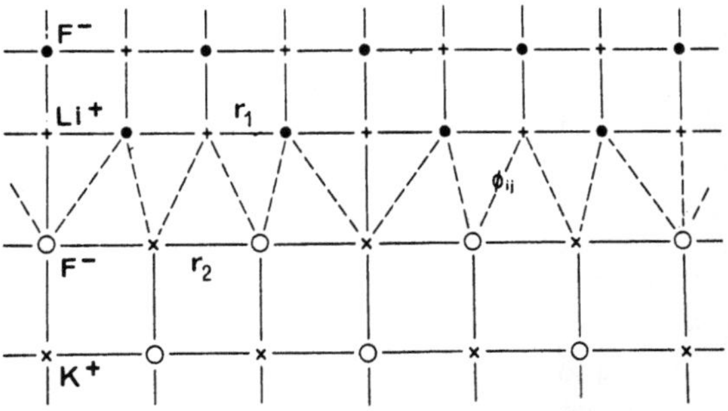

Fig. 1 Model of commensurate interface formed by two unrelaxed ionic surfaces with equal orientation. The illustrated case is that of LiF(001)/KF(001), where the interionic distances are in the ratio $r_1/r_2 \simeq 3/4$. The matrix ϕ_{ij} represents the coupling force constants between ions of the two topmost layers.

Actually, this model applies also when the two adjacent surface lattices, originally incommensurate (I), lock in to a commensurate (C) phase, i.e., $n_1r_1 = n_2r_2$ with integer n_1, n_2, through the formation of some surface dislocation network. This situation is somewhat restrictive, since a weak interface pinning potential combined with thermal fluctuations |60| favour I phases (epitaxial interfaces). As long as phonons are concerned, these phases can be approximated by an interface model of suitably large prime values of n_1 and n_2 (low commensurability), the basic difference in the dynamical behaviour between I and C phases being in the spectrum of low-lying excitations |60|.

In the case of low commensurability, the interface supercell contains several molecules of the two crystals. Thus, the phonon branches of the two lattices have to be folded many times in the interface-reduced Brillouin zone (RBZ), defined as the highest common factor of the two surface Brillouin zones. The interface modes exhibit many branches coming from the folding procedure itself, with opening of gaps at the center and the boundaries of the RBZ, like in the case of superlattices |59|.

For many purposes, such gaps can be explicitly ignored, as done in ref. 43. This amounts to neglect umklapp processes in the phonon-interface scattering which would admix any phonon wave $|\mathbf{K}\rangle$ of parallel wavevector $\mathbf{K}$, with the waves $|\mathbf{K} + \mathbf{G}_1\rangle$ and $|\mathbf{K} + \mathbf{G}_2\rangle$, $\mathbf{G}_1$ and $\mathbf{G}_2$ being the surface reciprocal lattice vectors of the two crystals. Within this approximation, $\mathbf{K}$ is a good quantum number all over the reciprocal space. This plane-wave approximation avoids unneccesary complications, reduces the computational task and provides a readable representation of the interface phonon branches in the extended Brillouin zone (EBZ), defined as the lowest common multiple of the original surface Brillouin zones.

The dynamics of each semi-infinite lattice in the absence of interface coupling is given through its surface GF matrix $\mathfrak{g}_{ii}$ (i = 1,2) |28|. For the ideal (unrelaxed) surface of each semi-infinite lattice an effective surface perturbation matrix can be expressed in terms of inverse bulk GF, so that no adjustable parameter is required to describe the dynamical perturbation of the intrinsic surface |28|. The relationship connecting the perturbation matrix to the inverse GF is given by the translational and rotational invariance conditions. The effective perturbation obtained in this way is so rapidly decaying inside the crystal, that it can be restricted to the topmost layer even in ionic crystals with long-range forces, with excellent approximation |28|.

The interface coupling is described by the nondiagonal blocks of the force constant matrix ϕ, where for the interionic potential a Born-Mayer potential was employed |43|.

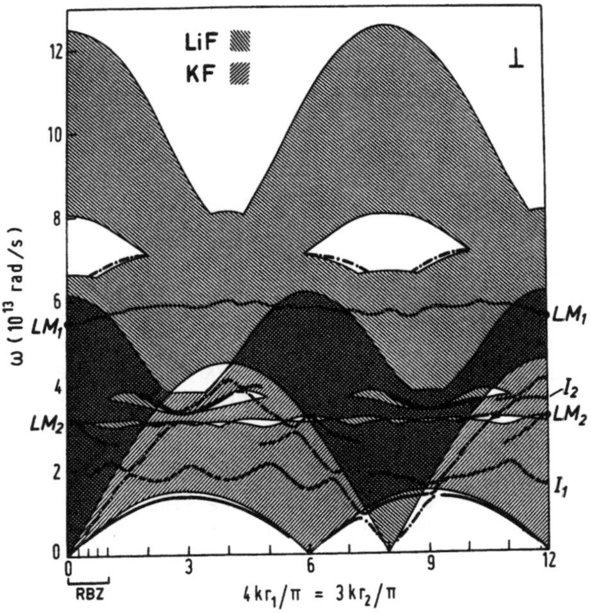

Fig. 2 Phonon dispersion curves along the <100> direction of the LiF(001)/KF(001) interface for sagittal polarization. Full lines represent interface modes originated from KF local modes. Broken lines represent resonant modes originated from KF modes. Dotted-dashed lines are local modes from LiF. Dotted lines are resonant modes from LiF.

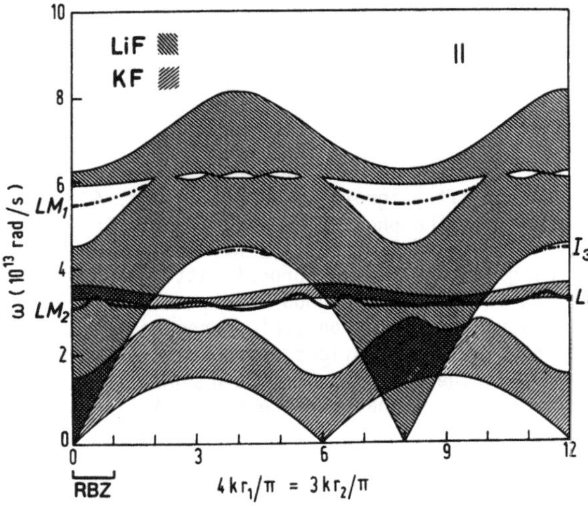

Fig. 3 Shear-horizontal mode dispersion curves along the <100> direction for the LiF(001)/KF(001) interface. The conventions are the same as in Fig. 2.

The dispersion curve and the phonon densities of states for LiF(001)/KF(001) interface have been calculated along the symmetry direction <100> in the framework of the breathing shell model |57|, using room-temperature input data.

In figs. 2 and 3 we show the calculated dispersion curves in the EBZ representation for sagittal polarization (⊥) and shear horizontal polarization (||), respectively. They are superimposed on the bulk phonon bands of LiF and KF. The dispersion curves lying outside the bands of both crystals correspond to interface localized modes, whereas all the other branches are resonances. Most of the curves correspond to existing surface modes of the two crystals, such as the optical modes LM_1 and LM_2 (Lucas modes), only slightly perturbed by the interface coupling. However, few of them are new modes induced by the interface.

In figs 2 and 3 full lines represent interface modes originated from KF local modes. The lower one in fig. 2 is the Stoneley wave, coming from KF Rayleigh wave; the higher one comes from a surface optical mode of KF at the zone boundary. The Stoneley wave curve becomes tangent to the acoustic band edge for $\mathbf{K} \to 0$, since in the elastic limit no Stoneley wave is predicted for this interface. Broken lines show the KF sagittal quasi-transverse optical mode S_2 and two strongly resonant modes I_1 and I_2 which are peeled off from the upper edges of the TA and TO KF bands, respectively, because of the interaction with LiF. The resonances I_1 and I_2 are typical interface modes. Here the contact with a lighter crystal (LiF) stiffens the frequencies of the heavier crystal (KF), making the modes rise above the KF transverse bands. The Stoneley wave is always above the KF Rayleigh wave dispersion curve due to the same reason.

In fig. 3 we can see the appearance of a local mode (I_3) from the bottom of the LiF acoustic branch. This is a new interface mode generated by a mechanism opposite to that giving I_1 and I_2. However, no interface mode is found about the edges of the KF bands.

We show in fig. 4 the perturbed interface-projected phonon densities for sagittal and shear-horizontal polarizations at the RBZ edge (reduced wavevector = 1). Most of the peaks correspond to the local and resonant modes quoted above. Other important features, however, come from the band edges. Local modes should be represented by delta functions, but they have a finite width corresponding to the frequency mesh used in the calculation. |43| The folding of the EBZ into RBZ may give, for some of the modes, a multiple structure with more sharp peaks.

As can be seen in fig. 4 at the RBZ edge, the Stoneley wave is resolved from the continuous bulk distribution, but its intensity is much lower than the resonant Rayleigh wave (RW_1) of LiF. On the other hand, the pure interface resonances I_1, I_2 and I_3 yield intense peaks in the phonon density.

We show the comparison of the perturbed densities projected onto LiF and KF layers with the unperturbed ones in Figs. 5 and 6, respectively, for reduced wavector equal to 0.5 and for sagittal polarization only. The unperturbed densities exhibit the repetitive peaks of RW_1 (fig. 5) and RW_2 (fig. 6) due to the folding. The interface perturbation introduces strong modifications in the projected spectra. The intensity of the Stoneley wave is much smaller than the original RW_2 intensity, in agreement with the fact that its frequency is much closer to the band edge. The sagittal KF perturbed density is dominated by the peaks of the interface modes I_1 and I_2.

From the study of the phonon densities it appears that the strong resonances I_1, I_2 and I_3, rather than Stoneley waves, are the important features of interface dynamics.

3. Dynamics of Superlattices

Also the dynamics of layered media in the continuum limit has been studied in the early days of geophysics |3| and revived later in connection with technical applications of superlattices. |47| Recent structural and spectroscopic studies of superlattices |18-21| and metal-metal heterostructures |62,63| have stimulated new investigations on the dynamics of periodically layered media. A systematic study of the transverse and sagittal elastic waves propagating in infinite and semiinfinite superlattices has been undertaken in the continuum limit by means

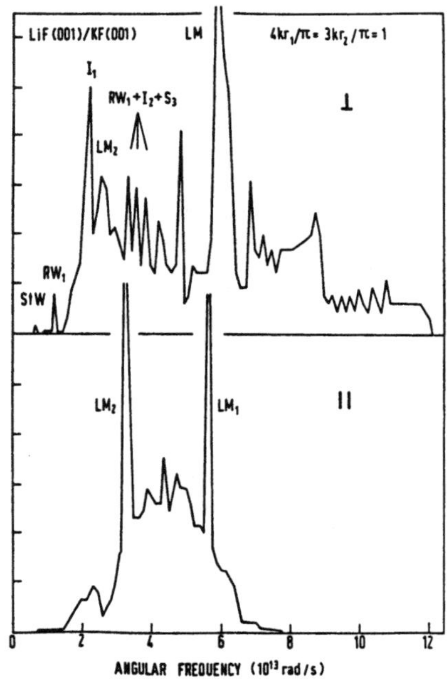

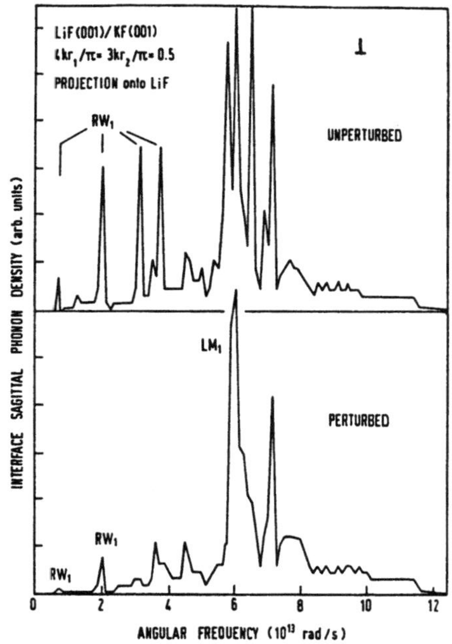

Fig. 4 Perturbed interface-projected phonon densities for sagittal and shear-horizontal polarizations at the edge of the RBZ.

Fig. 5 Perturbed and unperturbed phonon densities for sagittal polarization projected on the topmost LiF layer for reduced wavevector equal to 0.5.

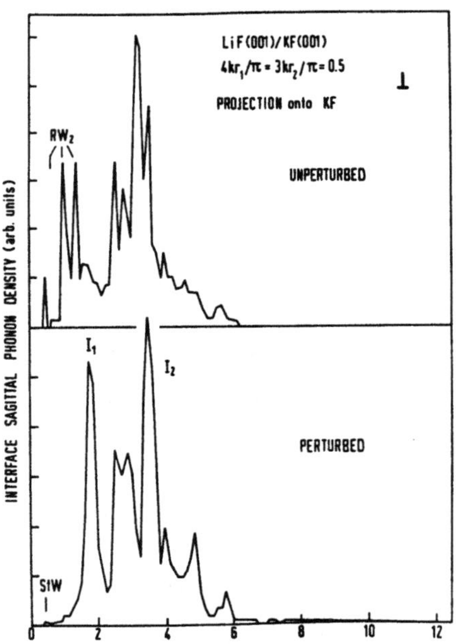

Fig. 6 Perturbed and unperturbed phonon densities for sagittal polarization projected onto the topmost KF layer for reduced wavevector equal to 0.5.

of powerful techniques such as the transfer-matrix and the Green's function methods |44,45|. A lattice theory for ideal superlattices (no mismatch in parallel periodicity) has been approached by DOBRZYNSKI et al. |46|, who have also calculated the phonons localized at a free surface parallel to the layers, and by DJAFARY-ROUHANI et al. |64| who have considered the case of a free surface cut normal to the superlattice interfaces. These authors have used a GF method and obtained analytical solutions for a simple cubic superlattice formed by a small number of alternating layers.

Here we outline the dynamics of compositional as well as doping superlattices as applications of the surface GF technique in the most general lattice case.

3.1 Compositional superlattices

We consider a superlattice as a periodic repetition along the x_3 direction of a unit cell of thickness a+b (fig.7), where a and b are the thicknesses of the two sublattices. The two sublattices contain N_a and N_b layers, whose positions are given by the coordinates κ_3 and λ_3, respectively. Thus the interlayer distances are $r_a=a/N_a$ and $r_b=b/N_b$. The unit cell is also repeated periodically in two independent directions parallel to the interface.

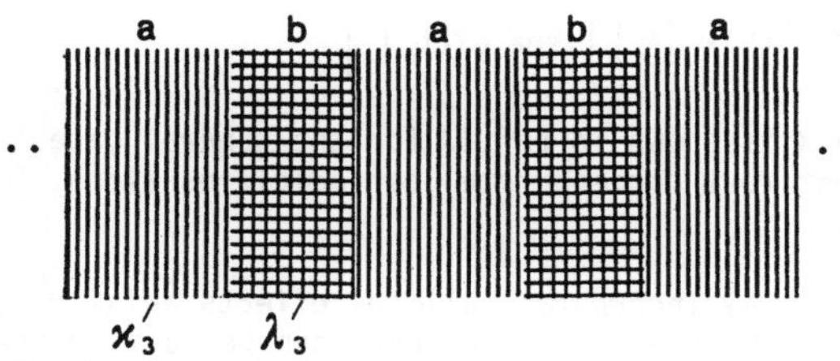

Fig.7 Compositional superlattice formed by a periodic repetition along the x_3 direction of a unit cell of thickness a+b.

The modes of each unperturbed sublattice (unperturbed means that each sublattice belongs to its own infinite lattice with cyclic boundary conditions) are classified by the parallel wavevector **K**, the normal wavevector k_3+K_3 where $|k_3|\leq\pi/(a+b)$ and K_3 is any superlattice reciprocal vector, namely

$$K_3 = 2\pi\times\text{integer}/(a+b). \tag{1}$$

The unperturbed Green functions for the two sublattices are

$$g_{a\,\alpha\beta}(\kappa_3\kappa_3';\mathbf{K},k_3;\omega^2) = \frac{1}{n_a}\sum_{K_3j}\frac{w^*_{a\alpha}(\mathbf{q}j)\,w_{a\beta}(\mathbf{q}j)\,e^{iK_3(\kappa_3-\kappa_3')}}{\omega_a^2(\mathbf{q}j)-\omega^2-2i\omega0^+} \tag{2a}$$

$$g_{b\,\alpha\beta}(\lambda_3\lambda_3';\mathbf{K},k_3;\omega^2) = \frac{1}{n_b}\sum_{K_3j}\frac{w^*_{b\alpha}(\mathbf{q}j)\,w_{b\beta}(\mathbf{q}j)\,e^{iK_3(\lambda_3-\lambda_3')}}{\omega_b^2(\mathbf{q}j)-\omega^2-2i\omega0^+} \tag{2b}$$

where

$$\mathbf{q} \equiv (\mathbf{K},\ k_3 + K_3) \tag{3}$$

is a three-dimensional wavevector, $\omega_i(\mathbf{q}j)$ and $\mathbf{w}_i(\mathbf{q}j)$ are angular frequencies and eigenvectors of sublattice i(=a,b) for the j-th branch. The numbers n_a and n_b are the maximum integers in eq.(1) for the respective summations (2a) and (2b). They are given by

$$n_a = \frac{a+b}{r_a} = N_a + N_b \frac{r_b}{r_a} , \qquad (4)$$

$$n_b = \frac{a+b}{r_b} = N_b + N_a \frac{r_a}{r_b} , \qquad (5)$$

provided that N_a, N_b, r_a and r_b are approximated so as to give integer values of n_a and n_b. (If this is not possible, n_a and n_b are the smallest integer multiples of $(a+b)/r_a$ and $(a+b)/r_b$ which are integer numbers).

When both sublattices have a symmetry plane parallel to the interface, the problem is reduced through the symmetry transformations discussed for the free-surface case in a semiinfinite lattice |28|, to that of a single interface discussed in the previous section. Thus we have to introduce the interface coupling matrix ϕ_{ab}, as in fig.1; ϕ_{ab} fulfills translational and rotational invariance conditions. In particular, these invariance conditions are needed for the determination of the diagonal blocks ϕ_{aa} and ϕ_{bb}. |43| Thus, the frequencies of the interface modes and their dispersion are obtained from the roots of the equation

$$\text{Re det} \left[1 + \begin{vmatrix} \tilde{g}_a(\mathbf{K},k_3;\omega^2) & 0 \\ 0 & \tilde{g}_b(\mathbf{K},k_3;\omega^2) \end{vmatrix} \begin{vmatrix} \phi_{aa} & \phi_{ab} \\ \phi_{ba} & \phi_{bb} \end{vmatrix} \right] = 0. \qquad (6)$$

The way the elements of ϕ depend upon **K** is shown in ref.43 for the general case where the two sublattices have different (but commensurate) periodicities along the interface.

It is important to remark that $\phi_{ab}(\mathbf{K})$ may incorporate all the effects arising from the electron-phonon coupling at the interface. Such effects can be particularly strong when the superlattice absolute gap is so small to be comparable with the phonon frequencies. This may easily happen in type II superlattices. In this case, the perturbation matrix ϕ contains an important dynamical contribution, reflecting the response of the electron system to the modulation of the interatomic distances at the interface. This can be easily understood in a dielectric formulation of the dynamical matrix . |65| The contribution of the electronic response to ϕ_{ab} would introduce a more complex K-dependence than that arising from short-range repulsive interactions, and possibly also a dependence on k_3 due to the nature of the electron-phonon interaction which is not strictly short-range.

3.2 Doping superlattices

Doping superlattices, besides their important electronic properties, |66| have an interesting dynamical structure, as they combine localized vibrational features with folding effects arising from the periodic distortion of impurities. Furthermore, doping superlattices, like type II compositional superlattices, may have a narrow gap or even semimetal character, with an enhanced electron-phonon interaction in the periodical narrow-gap regions. We consider a monoatomic lattice, with a concentration of substitutional impurities oscillating along the x_3 direction (fig.8). The impurities are assumed to be randomly distributed within any volume element, small compared to the period Nr_0 of the oscillation (r_0=interatomic distance). The average concentration p_0 is small enough to give negligible interactions among impurities. The oscillating concentration is described by a probability p_ℓ of having an impurity at a lattice site in the ℓ-th layer.

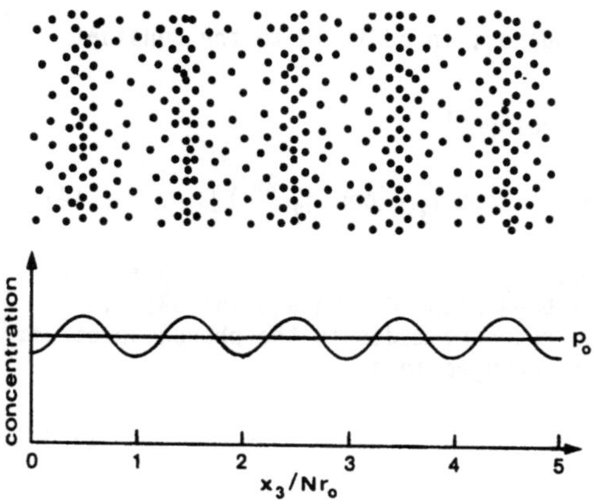

Fig. 8 Oscillating concentration of substitutional impurities in a doping superlattice.

The impurity perturbation $\Lambda_{\ell\ell'}$ of the dynamical matrix consists in a local change of mass ΔM and nearest-neighbour force constant $\Delta\phi$. It is written as: |67-69|

$$\Lambda_{\ell\ell'} = \delta_{\ell\ell'}\left[-(\varepsilon\omega^2 + \lambda)p_\ell - \tfrac{1}{2}\lambda(p_{\ell+1} + p_{\ell-1})\right] + \tfrac{1}{2}(\delta_{\ell,\ell'+1} + \delta_{\ell,\ell'-1})\,\lambda\,(p_\ell + p_{\ell'}) , \tag{7}$$

where

$$\varepsilon \equiv -\Delta M/M, \qquad \lambda \equiv -\Delta\phi/M, \tag{8}$$

and M is the atomic mass of the unperturbed lattice. Here ε and λ have to be taken as matrices in the subspace of the monoatomic cell; the generalization to polyatomic lattices is straightforward). The special form of $\Lambda_{\ell\ell'}$, eq.(7), is obtained from the requirement that the force-constant perturbation obeys the translational invariance condition

$$\sum_{\ell' \neq \ell} [\Lambda_{\ell\ell'}]_{\omega=0} = \sum_{\ell \neq \ell'} [\Lambda_{\ell\ell'}]_{\omega=0} = 0 . \tag{9}$$

Consider then the Fourier transform of $\Lambda_{\ell\ell'}$,

$$\Lambda_{q_3q_3'} = \frac{1}{N_3} \sum_{\ell\ell'} e^{iq_3\ell}\Lambda_{\ell\ell'}\, e^{-iq_3'\ell'} , \tag{10}$$

where N_3 is the number of layers of the entire superlattice, $q_3=k_3+K_3$ and $q_3'= k_3' + K_3'$. Again K_3 and K_3' are superlattice reciprocal vectors, and k_3, k_3' are restricted to the superlattice Brillouin zone. Due to the superlattice periodicity we have diagonality with respect to k_3, and we shall deal with the matrix on the reciprocal lattice

$$\Lambda_{K_3K_3'}(k_3) = \frac{1}{N_3}\sum_{\ell\ell'} e^{ik_3(\ell-\ell')}\, e^{iK_3\ell}\Lambda_{\ell\ell'}e^{-iK_3'\ell'} . \tag{11}$$

After some algebra, this becomes

$$\Lambda_{K_3K_3'}(k_3) = -\delta_{K_3K_3'}\, p_0\, \varepsilon\omega^2 - \sum_G \delta_{K_3\,K_3'-G}\, p_G\, \lambda\,(1 + \cos G - \cos q_3 - \cos q_3') , \tag{12}$$

where

$$p_G = \frac{1}{N_3}\sum_\ell e^{iG\ell} p_\ell . \tag{13}$$

Using a single cosinusoidal term of amplitude p_1, in addition to the constant term, namely $G = 0, \pm 2\pi/N$, we have

$$\Lambda_{K_3K_3'}(k_3) = -\delta_{K_3K_3'} p_0 [\, \varepsilon\omega^2 + 2\lambda(1 - \cos q_3)]$$
$$-\delta_{K_3\, K_3' \pm 2\pi/N}\, p_1 \lambda [\, 1 + \cos\frac{2\pi}{N} - \cos q_3 - \cos(q_3 \mp \frac{2\pi}{N})\,]. \qquad (14)$$

The dynamical effects of the first term, representing a uniform random distribution of non-interacting impurities, have been studied long ago |67-69|. They consist in the appearance of localized and resonant modes in the phonon spectrum; the bulk phonon squared frequencies $\omega^2_{\mathbf{q}j}$ are changed into |70|

$$\tilde{\omega}^2_{\mathbf{q}j} = \omega^2_{\mathbf{q}j} + p_0\, T_{\mathbf{q}j}(\tilde{\omega}^2_{\mathbf{q}j}) , \qquad (15)$$

where $T_{\mathbf{q}j}(\tilde{\omega}^2_{\mathbf{q}j})$ is the $(\mathbf{q}j)$ diagonal element of the transition matrix for the phonon scattering from an isolated impurity, evaluated at the renormalized frequency itself. $T_{\mathbf{q}j}$ is a complex-valued correction, whose imaginary part gives the inverse lifetime of bulk phonons due to the scattering processes. By using the renormalized frequencies as an unperturbed basis, one has to deal only with the term proportional to p_1 in the superlattice perturbation, and look for the roots of the determinantal equation

$$\mathrm{Re}\,\det [\, \delta_{K_3K_3'} I + g(\mathbf{K}, k_3+K_3; \omega^2)\cdot \Lambda_{K_3K_3'}(k_3)\,] = 0 , \qquad (16)$$

where

$$g(\mathbf{K}, k_3+K_3; \omega^2) = \sum_j \mathbf{w}^*(\mathbf{q}j)\mathbf{w}(\mathbf{q}j)/(\tilde{\omega}^2_{\mathbf{q}j} - \omega^2 - 2i\omega 0^+) , \qquad (17)$$

and $\mathbf{q}$ is given by eq. (3). Due to the lateral interactions of the impurity atoms with their neighbors, the perturbation matrix, eq. (14), has also an implicit dependence on the parallel wavevector $\mathbf{K}$. Needless to say that the response of the electron systems would introduce into the perturbation matrix a new specific dependence on $\mathbf{K}$ and k_3, as for the case of compositional superlattices.

The solutions of eq. (16), of the form

$$\omega^2 = \omega^2_\nu(\mathbf{K}, k_3), \qquad (18)$$

give the dispersion relations (labelled by ν) of those special modes induced by the modulation of the impurity concentration. We note that the matrix in eq. (16) is N x N, which makes the evaluation of the determinant quite lengthy in practical cases. However, if the modulation amplitude p_1 is small, one can keep in the determinant only the quadratic terms in p_1.

To give an example, we consider a single degree of freedom in each unit cell. In this case we obtain the secular equation in the finite form:

$$1 = p_1^2 \lambda^2\, \mathrm{Re} \sum_{m=-N}^{N-1} [\, \cos(k_3 + 2\pi(m+1)/N) + \cos(k_3 + 2\pi m/N) - \cos(2\pi/N) - 1\,]^2$$
$$\times [\, \tilde{\omega}^2(k_3 + 2\pi(m+1)/N) - \omega^2\,]^{-1} [\, \tilde{\omega}^2(k_3 + 2\pi m/N) - \omega^2\,]^{-1} . \qquad (19)$$

The reader can easily understand why the cosinusoidal perturbation enters as a quadratic term. The smaller the perturbation, the closer are the solutions ω^2 to the folded "unperturbed" frequencies $\tilde{\omega}^2(k_3 + 2\pi m/N)$ (m labels the N-times folded branches). However, when N is large enough and $\mathfrak{B}^2(q_3)$ have their imaginary part, the spectral summation in (19) gives a continuous function. In this case, the folded branches form bands and eq. (19) gives the dispersion relations of few individual localized and resonant modes induced by the modulation of the impurity concentration. We believe that these are the most relevant features in the dynamics of real superlattices.

Acknowledgment

This work was performed in the framework of the Scientific Cooperation Agreement between the Consiglio Nazionale delle Ricerche (Italy) and the Consejo Superior de Investigaciones Cientificas (Spain).

References

1. Lord Rayleigh: Proc. London Math. Soc. 17, 4 (1887).
2. G.W. Farnell: in Physical Acoustics, Vol. VI, W.P. Mason ed., (Academic Press, New York 1970), p. 109. An updated extended review is presented by A.A. Maradudin: Festkörperprobleme 21, 25 (1981).
3. W. M. Ewing, W.S. Jardetzky and F. Press: Elastic Waves in Layered Media (Mac Graw-Hill, New York 1957).
4. E. Dieulesaint and D. Royer, in Handbook of Surfaces and Interfaces, Vol. 2, L. Dobrzynski ed. (Garland STPM Press, New York 1978), p. 158.
5. A. A. Maradudin, E.W. Montroll, G.H. Weiss and I.P. Ipatova: Solid State Physics Suppl. 3 (Academic Press, New York 1971).
6. A. A. Maradudin, R.F. Wallis and L. Dobrzynski: Handbook of Surfaces and Interfaces, Vol. 3 (Garland STPM Press, New York 1980).
7. J.P. Toennies and K. Winkelmann: J. Chem. Phys. 66, 3965 (1977).
8. G. Brusdeylins, R.B. Doak and J.P. Toennies: Phys. Rev. Lett. 44, 1417 (1980).
9. G. Brusdeylins, R.B. Doak and J.P. Toennies: Phys. Rev. Lett. 46, 437 (1981).
10. G. Brusdeylins, R.B. Doak and J.P. Toennies: Phys. Rev. B27, 3662 (1983).
11. R. B. Doak, U. Harten and J.P. Toennies: Phys. Rev. Lett. 51, 578 (1983).
12. J. P. Toennies: J. Vac. Sci. Technol. A2, 1055 (1984).
13. N. Cabrera, V. Celli and R. Manson: Phys. Rev. Lett. 22, 369 (1969).
14. R. Manson and V. Celli: Surf. Sci. 24, 495 (1971). For a critical review see F.O. Goodmann: Crit. Rev. Solid State and Materials Sci. 7, 33 (1977).
15. G. Benedek and G. Seriani: Jpn. J. Appl. Phys. Suppl. 2, Pt. 2, 545 (1974).
16. S. Lehwald, J. M. Szeftel, H. Ibach, T.S. Rahman and D. L. Mills: Phys. Rev. Lett. 50, 518 (1983); J. M. Szeftel, S. Lehwald, H. Ibach, T.S. Rahman, J. E. Black and D. L. Mills: Phys. Rev. Lett. 51, 268 (1983).
17. M. Rocca, S. Lehwald, H. Ibach and T. S. Rahman: Surf. Sci. 138, L123 (1984).
18. A. S. Barker, J. L. Merz and A. C. Gossard: Phys. Rev. B17, 3181 (1978).
19. C. Colvard, R. Merlin, M. V. Klein and A. C. Gossard: Phys. Rev. Lett. 45, 298 (1980).
20. J. Sapriel, J. C. Michel, J. C. Toledano, R. Vacher, J. Kevarec and A. Regreny: Phys. Rev. B28, 2007 (1983).
21. M. V. Klein, C. Colvard, R. Fischer and H. Morkoz: J. Phys. (Paris) Coll. Suppl. 45, C5-131 (1984).
22. D. C. Gazis, R. Hermann and R. F. Wallis: Phys. Rev. 119, 533 (1960).
23. T. E. Feuchtwang, Phys. Rev. 155, 731 (1967).
24. V. Bortolani, F. Nizzoli and G. Santoro, in Lattice Dynamics, M. Balkanski, ed. (Flammarion, Paris 1978) p. 302.
25. R. E. Allen, G. P. Alldredge and F. W. de Wette: Phys. Rev. B4, 1648 (1971); 1661 (1971); 1682 (1971).
26. T. S. Chen, F. W. de Wette and G.P. Alldredge: Phys, Rev. B15, 1167 (1977).

27. I. M. Lifshitz and L. Rozenzweig: Zh. Eksp. Teor. Fiz. 18, 1012 (1948); Nuovo Cim. Suppl. 3, 732 (1956).
28. G. Benedek: Surf. Sci. 61, 603 (1976); G. Benedek and L. Miglio: in Ab-initio Calculation of Phonon Spectra, J.T. Devreese, V.E. van Doren and P.E. van Camp eds. (Plenum, New York 1983) p. 215.
29. S. W. Musser and K. H. Rieder, Phys. Rev. B2, 3034 (1970).
30. T. S. Rahman, J. E. Black and D. L. Mills: Phys. Rev. B25, 883 (1982).
31. G. Benedek, G. P. Brivio, L. Miglio and V. R. Velasco: Phys. Rev. B26, 487 (1982)
32. G. Armand: Phys. Rev. B14, 2218 (1976); G. Armand and P. Masri: Surf. Sci. 130, 89 (1983).
33. W. Goldammer, W. Ludwig, W. Zierau and C. Falter: Surf. Sci. 141, 139 (1984); W. Goldammer and W. Ludwig: J. Phys. (Paris) Coll., Suppl. 45, C5-119 (1984).
34. F. Garcia-Moliner and J. Rubio: J. Phys. C2, 1789 (1969); Proc. Roy. Soc. London A324, 527 (1971).
35. F. Garcia-Moliner: Ann. Phys. (Paris) 2, 179 (1977).
36. F. Garcia-Moliner, G. Platero and V. R. Velasco: Surf. Sci. 136, 601 (1984).
37. G. Benedek, G. Brusdeylins, R. B. Doak and J. P. Toennies: J. Phys. (Paris) Coll. Suppl. 42, C6-793 (1981).
38. G. Benedek: Surf. Sci. 126, 624 (1983); J. Electr. Spectr. 30, 71 (1983).
39. G. Benedek: in Dynamics of Gas-Surface Interaction, G. Benedek and U. Valbusa, eds. (Springer, Heidelberg 1982) p. 227.
40. G. Benedek: in Collective Excitations in Solids, B. di Bartolo, ed. (Plenum, New York 1983) p. 523.
41. B. Djafari-Rouhani, P. Masri and L. Dobrzynski: Phys. Rev. B15, 5690 (1977).
42. B. Djafari-Rouhani, L. Dobrzynski and P. Masri: Ann. Phys. (Paris) 6, 529 (1981).
43. G. Benedek and V. R. Velasco: Phys. Rev. B23, 6691 (1981).
44. R. E. Camley, B. Djafari-Rouhani, L. Dobrzynski and A. A. Maradudin: Phys. Rev. B27, 7318 (1983).
45. B. Djafari-Rouhani, L. Dobrzynski, O. Hardouin-Duparc, R. E. Camley and A. A. Maradudin: Phys. Rev. B28, 1711 (1983).
46. L. Dobrzynski, B. Djafari-Rouhani and O. Hardouin-Duparc: Phys. Rev. B29, 3138 (1984).
47. R. Tsu and L. Esaki: IBM J.Res.Dev.14, 61 (1970); G. A. Sai-Halasz, R. Tsu and L. Esaki: Appl. Phys. Lett. 30, 651 (1977).
48. E. Grünbaum, in Epitaxial Growth, Part B, J. W. Mathews, ed. (Academic Press, New York 1975).
49. P. K. Hansma, in Low-Temperature Physics - LT14, Vol. 5, M. Krusius and M. Vuorio eds. (North-Holland, Amsterdam 1975) p. 264. For a wider review of the subject: T. Wolfram, ed.: Inelastic Electron Tunneling Spectroscopy (Springer, Berlin 1978).
50. F. Proix, R. Racek and M. Balkanski: J. Vac. Sci. Technol. 10, 663 (1974).
51. R. Stoneley: Proc. Roy. Soc. London A106, 416 (1924).
52. J. G. Scholte: Mon. Notes Roy. Astron. Soc. Geophys. Suppl. 5, 120 (1947).
53. B. Djafari-Rouhani and L. Dobrzynski: Surf. Sci. 61, 526 (1976).
54. V. R. Velasco and F. Garcia-Moliner: Surf. Sci. 83, 376 (1979); V. R. Velasco: Phys. Stat. Solidi A60, K61 (1980).
55. V. Bortolani, F. Nizzoli, G. Santoro, A.M. Marvin and J. R. Sandercock: Phys. Rev. Lett. 43, 224 (1979).
56. J. R. Sandercock, F. Nizzoli, V. Bortolani, G. Santoro and A. M. Marvin: Le Vide - Les Couches Minces Suppl. 201, 754 (1980).
57. G. W. Farnell and E. L. Adler: in Physical Acoustics, Vol. IX, W. P. Mason and R. N. Thurston, eds. (Academic Press, New York 1972) p. 33.
58. V. R. Velasco and F. Yndurain: Surf. Sci. 85, 107 (1979).
59. A. D. Novaco: Phys. Rev. B22, 1645 (1980).
60. P. Bak and H. Fukuyama: Phys. Rev. B21, 3287 (1980).
61. U. Schröder, Solid. St. Comm. 4, 347 (1967); V. Nüsslein and U. Schröder: Phys. Stat. Solidi 21, 309 (1967).

62. C. M. Falco: J. Phys. (Paris) Coll. Suppl. 45, C5-499 (1984); C. M. Falco, W. R. Bennett and A. Boufelfel: present volume.
63. A. Kueny, M. Grimsditch, K. Miyano, I. Banerjee, C. Falco and I. Schuller: Phys. Rev. Lett. 48, 166 (1982); A. Kueny and M. Grimsditch: Phys. Rev. B26, 4699 (1982)
64. B. Djafari-Rouhani, A. A. Maradudin and R. F. Wallis: Phys. Rev. B, to be published.
65. H. Böttger: Principles of the Theory of Lattice Dynamics (Physik-Verlag, Weinheim 1983).
66. K. Ploog and G. H. Döhler: Adv. Phys. 32, 285 (1983).
67. G. Benedek and G. F. Nardelli: Nat. Bur. Stand. (U.S.A.) Publ. 287, 161 (1967); Phys. Rev. 155, 1004 (1967).
68. G. Benedek and A. A. Maradudin: J. Phys. Chem. Sol. 29, 423 (1968).
69. H. Bilz, D. Strauch and R.K. Wehner, in Handbuch der Physik, Vol. XXV/2d, L. Genzel, ed. (Springer, Heidelberg 1984).
70. G. Benedek and G. F. Nardelli, J. Chem. Phys. 48, 5242 (1968).

Pseudopotentials and Dynamical Properties of Metallic Surfaces

C. Calandra and A. Catellani

Istituto di Fisica, Universita' di Modena, Via Campi 213/A
I-41100 Modena, Italy

1. Introduction

In the last few years it has become possible to measure the dispersion of the surface phonons in the low-index metal surfaces by experimental techniques such as the inelastic scattering of surface atoms [1-4] or the high-resolution electron energy-loss spectroscopy [5-8]. The experimental data have been interpreted using empirical force constant models obtained from a fit to the bulk phonon spectrum [9-12]. This theoretical work has clearly shown that in many cases the force constants must be modified near the surface, in order to reproduce the experimental data. The modifications have been attributed to relaxation or other geometrical changes in the near surface region [13] or to the different configuration of the valence electrons in the outermost planes [14].

Such empirical findings show the importance of looking for an approach to the determination of the atom motion, which is exempt from empirical parametrizations and permits to take into account the changes of the force constants caused by the valence electron behaviour at the surface or by the modifications of the inter-atomic distances present in the outermost layers. It must involve a first principle calculation of the total energy of the system for any given geometrical configuration,and the consequent determination of the dynamical matrix.

The possibility of carrying out this program is severely limited by two factors. First, the computational effort required for the determination of the total energy is considerable. This fact stems from the lacking of translational symmetry normal to the surface and from the necessity of a self-consistent determination of the screening potential, which includes exchange and correlation effects. Second, our ability of constructing realistic atomic models of the surface geometry is rather limited. This is partly a consequence of our ignorance of the dynamical processes occurring at the surface,and partly a limitation of the experimental techniques, which aim at the determination of the surface structures.

The primary purpose of this paper is to present a first-principle self-consistent approach to the determination of the surface phonons. As it is well known,there are two possible methods of calculation: i) the direct method, where the results of independent total energy calculations for different atomic positions are compared directly [15-18] and ii) the perturbation method, in which the self-consistent solutions for small displacements from the equilibrium configuration can be simplified by linearization of the equations [19-21]. We have chosen the second approach,which has the advantage of the simplicity and is less demanding from the computational point of view.

An essential feature of our method is the use of a pseudopotential to describe the interaction between the valence electrons and the ion cores. This is essentially a model hamiltonian description of the solid under consideration, where rather than treating the metal as made up of nuclei and the full complement of atomic electrons, we regard it as consisting of inert ion cores (that is the nuclei and the core electrons) and the valence electrons. These move under the influence of their mutual Coulomb repulsion and their interaction with the cores. Such interaction is represented by a non-local and energy- dependent pseudopotential,and includes the Coulomb field of the nuclei as well as the Coulomb and exchange coupling between the valence and core electrons. These two interactions nearly cancel, resulting in an effective electron-ion coupling, which is quite weak and for which a perturbative approach to the determination of the total energy is appropriate [22-23].

Strictly speaking, a pseudopotential approach of this form holds for simple metals only. For simple metals we mean those with sp bonding electrons and with a free electron-like behaviour, like the alkalis, Mg and Al. Although experiments have not yet been carried out for these systems, the study of their dynamical properties using a first principle approach can throw light on the behaviour of the nearly free sp electrons at the surface. Since even the more complicated metal surfaces contain a large component of sp electrons, which are known to be responsible of the spilling of the valence charge density out of the solid, our conclusions may be applicable to a large number of systems.

The extension of the theory to treat noble metals can be performed along the lines of the so-called generalized pseudopotential method of HARRISON and MORIARTY [24-27], which allow to treat systems with a full valence d-shell or at the beginning of the transition series. Such an extension is under progress and will not be discussed in these lectures.

2. Total Energy

An expression of the total energy can be obtained by visualizing a process, in which the pseudocharge density $\rho(\underline{r},\lambda)$ depends on a parameter λ varying from zero to 1 and attaining the final value 1, as the ion charges get switched on from zero to their final values.

We write the pseudohamiltonian as

$$\hat{H}(\lambda) = \hat{T} + \lambda^2 \sum_{\alpha<\beta} \frac{Z_\alpha Z_\beta}{|\underline{R}_\alpha - \underline{R}_\beta|} e^2 + \lambda \sum_\alpha \sum_i v(\underline{r}_i - \underline{R}_\alpha) + \sum_{i<j} \frac{e^2}{|\underline{r}_i - \underline{r}_j|} \tag{1}$$

here $\underline{R}_\alpha$ ($\underline{r}_i$) are the coordinates of the ions (electrons) respectively, $v(\underline{r}_i - \underline{R}_\alpha)$ is the bare pseudopotential acting in the i-th electron due to the α-th ion, which for simplicity we assume to be local and T is the kinetic energy of the system.

The total energy is given by

$$E = \int_0^1 \frac{\partial}{\partial\lambda} < \Psi \mid \hat{H}(\lambda) \mid \Psi > d\lambda + E_o =$$

$$= \int_0^1 < \Psi \mid \frac{\partial}{\partial\lambda} \hat{H} \mid \Psi > d\lambda + E_o \tag{2}$$

here Ψ is the ground state pseudowavefunction, E_o is the energy in absence of the ions. The last equality in (2) comes from the Hellman-Feyman theorem [28]. From (1) we have

$$\frac{\partial \hat{H}}{\partial\lambda} = 2\lambda \sum_{\alpha<\beta} \frac{Z_\alpha Z_\beta}{|\underline{R}_\alpha - \underline{R}_\beta|} e^2 + \sum_\alpha \sum_i v(\underline{r}_i - \underline{R}_\alpha) \tag{3}$$

Inserting (3) into (2), we obtain

$$E = \sum_{\alpha<\beta} \frac{Z_\alpha Z_\beta}{|\underline{R}_\alpha - \underline{R}_\beta|} e^2 +$$

$$+ \int_0^1 d\lambda \sum_\alpha \sum_i \int \Psi^*(x_1 \ldots x_N)\, v(\underline{r}_i - \underline{R}_\alpha)\, \Psi(x_1 \ldots x_N)\, dx_1 \ldots dx_N + E_o \tag{4}$$

where $x_i \equiv (\underline{r}_i, s_i)$ are the coordinates of the i-th electron and the integration over x_i implies a sum over the spin values. Introducing the pseudocharge density of the system for a given λ :

$$\rho(\underline{r},\lambda) = N \sum_s \int \Psi^*(\underline{r},s;x_2\ldots x_N;\lambda)\, \Psi(\underline{r},s;x_2\ldots x_N;\lambda)\, dx_2\ldots dx_N \tag{5}$$

we can write (4) as

$$E = \sum_{\alpha<\beta} \frac{Z_\alpha Z_\beta}{|\underline{R}_\alpha - \underline{R}_\beta|} e^2 + \int_0^1\!\int d\lambda\, \rho(\underline{r},\lambda)\, V(\underline{r})\, d\underline{r} + E_o \quad . \tag{6}$$

Here we have indicated with $V(\underline{r})$ the total bare pseudopotential acting on the valence electrons

$$V(\underline{r}) = \sum_\alpha v(\underline{r} - \underline{R}_\alpha). \tag{7}$$

The trouble with applying (4) to actual systems is that $\rho(\underline{r},\lambda)$ must be known throughout the range of integration. This is generally not possible, unless the external potential is weak, so that the charge density can be calculated to the first order. In this case

$$\rho(\underline{r},\lambda) = \rho_o(\underline{r}) + \lambda\, \rho^{(1)}(\underline{r}) \tag{8}$$

where $\rho_o(\underline{r})$ is the valence pseudocharge of the unperturbed system and $\rho^{(1)}(\underline{r})$ is the first order correction. This quantity is given by

$$\rho^{(1)}(\underline{r}) = \int d\underline{r}'\, \chi(\underline{r},\underline{r}')\, W(\underline{r}') \tag{9}$$

here $\chi(\underline{r},\underline{r}')$ is the bare longitudinal response function and $W(\underline{r})$ is the self-consistent screened pseudopotential. Inserting (8), (9) into (4) we get

$$E = E_o + \int \rho_o(\underline{r})\, V(\underline{r})\, d\underline{r} + \tfrac{1}{2} \int d\underline{r} \int d\underline{r}'\, \chi(\underline{r},\underline{r}')\, W(\underline{r}')\, V(\underline{r}) + \\ + \sum_{\alpha<\beta} \frac{Z_\alpha Z_\beta}{|\underline{R}_\alpha - \underline{R}_\beta|} e^2 = E^{(el)} + E^{(c)} \tag{10}$$

that is the expression of the total energy to second order in the electron-ion pseudopotential. It is formally equivalent to the result of standard perturbation theory carried out to second order. Here $E^{(el)}$ indicates the total electron energy and $E^{(c)}$ is the repulsive energy of the ions.

2.1 Bulk Case

In the bulk case, the unperturbed system is a free electron gas of uniform density ρ_o. Because of the translational symmetry of such system we have

$$\chi(\underline{r},\underline{r}') = \chi(\underline{r} - \underline{r}') \tag{11}$$

This allows us to write the electronic part of the total energy as

$$E^{(el)} = E_o + \rho_o \overline{V} + \tfrac{1}{2} \sum_{\underline{q}} \chi(\underline{q})\, V(-\underline{q})\, W(\underline{q}) \tag{12}$$

where $\overline{V}$ is the average value of the potential in the solid, and we have written the second order contribution in terms of the Fourier transform of the response function and of the pseudopotentials. From (7) we have

$$V(\underline{q}) = v(\underline{q})\, S(\underline{q}) \tag{13}$$

where $v(q)$ is the Fourier transform of the single ion pseudopotential and $S(\underline{q})$ is the static structure factor

$$S(\underline{q}) = \frac{1}{N} \sum_{\alpha} e^{-i\underline{q} \cdot \underline{R}\alpha} \tag{14}$$

The screened pseudopotential $W(\underline{q})$ can be written in a similar way $W(\underline{q}) = w(\underline{q})\, S(\underline{q})$ with $w(\underline{q})$ single ion screened form factor given by

$$w(\underline{q}) = v(\underline{q}) / \varepsilon(\underline{q}) \tag{15}$$

and $\varepsilon(\underline{q})$ is the dielectric function of the electron gas.

With the above definitions, we can write the total electronic energy as

$$E^{(el)} = E_o + \rho_o \bar{V} + \tfrac{1}{2} \sum_{\underline{q}} \chi(\underline{q})\, v(\underline{q})\, w(\underline{q})\, S(\underline{q})\, S^*(\underline{q}) \tag{16}$$

which shows that the first term does not depend upon the structure of the system and all of the structure dependence is in the second order contribution. Following HARRISON [22], we can write the second order term as

$$E^{(2)} = \sum_{\underline{q}} S(\underline{q})\ S^*(\underline{q})\ F(\underline{q}) \tag{17}$$

where $F(\underline{q})$ is called the energy wavenumber characteristic. Because it is the only part of the electronic energy, that depends upon the ion positions, $E^{(2)}$ has to be calculated in order to determine the stable structure and the dynamical properties of the system under consideration. Recalling the definition of $S(\underline{q})$ we can write (17) in the form

$$E^{(2)} = \frac{1}{N^2} \sum_{\alpha,\beta} \sum_{\underline{q}} e^{i\underline{q} \cdot (\underline{R}_\alpha - \underline{R}_\beta)}\ F(\underline{q}) \tag{18}$$

showing that the total electronic energy is a sum of central pairwise interactions between the ions. The consequence of this fact for the dynamical matrix is that we can describe the atomic motion as due to central two-body forces. To include three-body effects one should go to higher orders in the perturbation expansion of the total electronic energy.

2.2 Surface Case

The unperturbed system is now an electron gas in the presence of a surface. Taking the z direction as the surface normal, we have that the unperturbed charge density is not uniform, but depends upon the distance from the surface $\rho_o(\underline{r}) = \rho_o(\underline{z})$.

The response function of such a system has a translational symmetry only parallel to the surface i.e.

$$\chi(\underline{r},\underline{r}') = \chi(\underline{r}_{//} - \underline{r}'_{//},\ z,\ z') \quad . \tag{19}$$

Rather than a semi-infinite crystal, we can consider a metal slab of size 2ℓ along z. In such a case, the unperturbed system is given by a confined electron gas. If the slab is taken thick enough that the two surfaces can be considered as non-interacting, the system can reproduce the results of a semi-infinite crystal. As in the bulk case, it is convenient to work with the Fourier transforms of the quantities entering the expression of the electronic energy. For the screened potential we write the Fourier expansion as

$$W(\underline{r}) = \sum_{\underline{Q}} \sum_{q} W_{\underline{Q}}(\underline{q})\ \cos qz\ \ \eta_q\ \ e^{i\underline{Q} \cdot \underline{r}_{//}} \tag{20}$$

where $\underline{Q}$ is a two-dimensional wave-vector, $q=n\pi/2\ell$ with n an integer number and the factor $\eta_q = 1$ for $q \neq 0$ and $\eta_q = 1/2$ for $q = 0$.

Because of the lacking of translational symmetry normal to the surface the Fourier transform of the response function for a given wavevector $\underline{Q}$ is a matrix $\chi_{\underline{Q}}(q,q')$, whose size is equal to the number of $\underline{q}$ vectors needed in the Fourier expansion. Therefore we can write

$$E^{(2)} = \frac{1}{2} \sum_{\underline{Q}} \sum_{q} \sum_{q'} \chi_{\underline{Q}}(q,q') \; W_{\underline{Q}}(q') \; V_{\underline{Q}}(q) \; \eta_q \; \eta_{q'} \tag{21}$$

with the Fourier transform given by expressions of the form

$$V_{\underline{Q}}(q) = \frac{2}{N\,\Omega_o} \int d\underline{r}\; e^{-i\underline{Q}\cdot\underline{r}_{//}} \cos qz \;\; V(\underline{r}) \tag{22}$$

where N is the number of surface atoms, $\Omega_o = 2A_c\,\ell$ and A_o is the surface unit area.

In the following,we assume a two-dimensional periodicity parallel to the surface with one atom per unit cell. It is convenient to label the vectors giving the locations of the atoms in our slab with two indexes: an index indicating the site of the two-dimensional lattice and an index giving the atomic plane to which the atom under consideration belongs. Therefore we will write the atomic sites as

$$\underline{R}_{\ell m} = \underline{R}_{\ell} + \underline{\tau}_m \tag{23}$$

where $\underline{R}_{\ell}$ is a two-dimensional lattice vector and $\underline{\tau}_m = (\underline{\tau}_{m//}, z_m)$ is a vector giving the location of the origin of the two-dimensional lattice on the m-th plane.

With the above definitions the Fourier transform of the bare pseudopotential is

$$V_{\underline{Q}}(q) = \frac{1}{N} \sum_{\ell m} e^{-i\underline{Q}\cdot\underline{R}_{\ell m}} \; v_{\underline{Q}}(q,z_m) \tag{24}$$

where $v_{\underline{Q}}(q,z_m)$ is the single ion form factor given by

$$v_{\underline{Q}}(q,z_m) = \frac{2}{\Omega_o} \int d\underline{r}_{//}\; e^{-i\underline{Q}\cdot\underline{r}_{//}} \int_0^{2\ell} dz\, \cos qz \;\; v(\underline{r}_{//}, z-z_m) \; . \tag{25}$$

In the same way we can write the Fourier transform of the self-consistent potential as

$$W_{\underline{Q}}(q) = \frac{1}{N} \sum_{\ell} \sum_{m} e^{-i\underline{Q}\cdot\underline{R}_{\ell m}} \; w_{\underline{Q}}(q,z_m) \tag{26}$$

Inserting (25) and (26) into (21) we obtain

$$E^{(2)} = \frac{1}{2} \sum_{\underline{Q}} \sum_{q} \sum_{q'} \chi_{\underline{Q}}(q,q') \frac{1}{N^2} \sum_{\ell} \sum_{\ell'} \sum_{m} \sum_{m'} e^{-i\underline{Q}\cdot(\underline{R}_{\ell m} - \underline{R}_{\ell' m'})} \; .$$

$$. \; w_{\underline{Q}}(q,z_m) \;\; v_{-\underline{Q}}(q',z_{m'}) \tag{27}$$

This result is formally similar to (17), the second member being again the sum on the couples of ions. However, the interactions do not depend upon the difference of the ionic positions only: they depend also upon the distance between the ions and the surface.

To make a connection with the approximations adopted in the empirical models, it is convenient to divide $E^{(2)}$ into two parts. The first arise from the diagonal terms $q = q'$. In this case, the response function and the interactions between two atoms are the same as in the bulk. (This is only approximately true, since some surface effect is present in the screened pseudopotential). The force acting on a given atom depends upon z_m since a certain number of neighbours are missing at the surface and in the outermost planes. This is precisely the approximation that is commonly used in semi-empirical force constant calculations.

The second part is due to the terms with $q \neq q'$. This represents the contribution due to the change in the atomic interactions at the surface. Strictly speaking, this is the only "true" non-central part of the atomic interactions. It can be seen as a sum of three body terms, where the surface acts as a third body. As it is clear from what we have said on the symmetry of χ, all the effects of the electron charge rearrangement near the surface (surface screening effects) are included in the off-diagonal contributions. In terms of the force constants models, the second part of $E^{(2)}$ can be regarded as the responsible of the modifications of the surface force constants from their bulk values.

3. Determination of the Selfconsistent Potential

To proceed further, we need the response function of an electron gas in presence of the surface, and we have to evaluate the screened potential. Within the RPA approximation, the response function is given by

$$\chi(\underline{r},\underline{r}') = 2e^2 \sum_{\underline{k}} \sum_{\underline{k}'} \frac{\phi^*_{\underline{k}}(\underline{r})\ \phi_{\underline{k}'}(\underline{r})\ \phi^*_{\underline{k}'}(\underline{r}')\ \phi_{\underline{k}}(\underline{r}')}{\varepsilon_k - \varepsilon_{k'}} (f_{\underline{k}} - f_{\underline{k}'}) \tag{28}$$

here ϕ_k and ε_k are the single particle eigenfunctions and eigenvalues for the unperturbed system and f_k are the usual zero-temperature Fermi factors. The main difficulty with (28) is that, unlike the bulk case, the determination of ϕ_k and ε_k for a confined jellium is a complex numerical problem, which requires a self-consistent determination of the valence charge density in the proximity of the surface [29]. To avoid cumbersome numerical problems, it is convenient to resort to some simple quantum mechanical model, which reproduces the behaviour of the surface screening. One of the most common is the infinite barrier model, which has been used in the discussion of impurity screening [30], chemisorption [31] and structural stability [32]. This model is convenient, because it allows an analytical evaluation of the response function. The main shortcoming is that it does not reproduce accurately the spilling of the valence electron charge out of the solid. This is a consequence of the fact that the width of the surface region is proportional to k_F^{-1} rather than to the screening length. However, for bulk densities corresponding to $r_s = 4 \sim 5$ a.u. the two lenghts are roughly equal, and the infinite barrier models quite well the electron density obtained in more realistic calculations. Since this is the range of densities of the alkali metals, we will adopt the response function of the infinite barrier model in our calculations.

The selfconsistent potential is given by

$$W(\underline{r}) = V(\underline{r}) + V^H(\underline{r}) + V^{xc}(\underline{r}) \tag{29}$$

where $V^H(\underline{r})$ is the Hartree potential arising from the screening charge and $V^{xc}(\underline{r})$ is the exchange-correlation potential. In the following we assume $V^{xc}(\underline{r})$ to depend linearly on the screening charge through the relation

$$V^{xc}_{\underline{Q}}(q) = \sum_{q'} \eta_{q'}\ f^{xc}_{\underline{Q}}(q,q')\ \rho^{(1)}_{\underline{Q}}(q') \tag{30}$$

By Fourier transforming (29) and expressing the Hartree potential in terms of the screening charge we get [30]

$$W_{\underline{Q}}(q) = V_{\underline{Q}}(q) + \frac{4\pi e^2}{q^2+Q^2} \sum_{q'} \eta_{q'} \chi_{\underline{Q}}(q,q') W_{\underline{Q}}(q') +$$

$$+ \sum_{q'} \sum_{q''} f^{xc}_{\underline{Q}}(q,q') \eta_{q'} \eta_{q''} \chi_{\underline{Q}}(q',q'') W_{\underline{Q}}(q'') + \frac{2\Delta'_{\underline{Q}}(0)}{(q^2+Q^2)\ell} . \quad (31)$$

The last term in this equation comes from the difference between the normal derivative of the bare and of the screened pseudopotential at the boundaries of the slab.

We now make the assumption that $f^{xc}_{\underline{Q}}(q,q')$ be purely diagonal and given by

$$f^{xc}_{\underline{Q}}(q) = -\frac{4\pi e^2}{q^2+Q^2} G_{\underline{Q}}(q) . \quad (32)$$

Such an approximation is strictly valid only in the case of a uniform unperturbed charge density. Expressions for $G_{\underline{Q}}(q)$ in a uniform electron gas have been given by a number of authors [33,34]. With this assumption we obtain

$$W_{\underline{Q}}(q) = V_{\underline{Q}}(q) + \frac{4\pi e^2}{q^2+Q^2} \left[1 - G_{\underline{Q}}(q)\right] \sum_{q'} \eta_{q'} \chi_{\underline{Q}}(q,q') W_{\underline{Q}}(q') +$$

$$+ \frac{2\Delta'_{\underline{Q}}(0)}{(q^2+Q^2)\,\ell} \quad (33)$$

In terms of the dielectric screening matrix

$$\varepsilon_{\underline{Q}}(q,q') = \delta_{qq'}/\eta_{q'} - \frac{4\pi e^2}{q^2+Q^2} \left[1 - G_{\underline{Q}}(q)\right] \chi_{\underline{Q}}(q,q') \quad (34)$$

we can write the following expression for the screened pseudopotential

$$W_{\underline{Q}}(q) = \sum_{q'} \eta_{q'} \varepsilon^{-1}_{\underline{Q}}(q,q') \left[V_{\underline{Q}}(q') - \frac{2\Delta'_{\underline{Q}}(0)}{(q^2+Q^2)\ell}\right] \quad (35)$$

where we have introduced the inverse dielectric matrix defined by the relation

$$\sum_{q''} \varepsilon^{-1}_{\underline{Q}}(q,q'') \, \varepsilon(q'',q') \, \eta_{q''} = \delta_{qq'}/\eta_{q'} . \quad (36)$$

Equations (34-36) allow to determine the screened potential in a selfconsistent way, including exchange and correlation effects. The quantity $\Delta'_{\underline{Q}}(0)$ can be determined as indicated by NEWNS [30]. If we assume this quantity to be zero and we take a diagonal screening matrix, we recover the standard expression for the bulk case.

4. Dynamical Matrix

We are now in position to calculate the dynamical matrix of our slab. The slab dynamical matrix at a given point $\underline{Q}$ of the two-dimensional Brillouin zone is written as a sum of two separate contributions

$$\underset{\sim}{D}(\underline{Q}) = \underset{\sim}{D}^{(c)}(\underline{Q}) + \underset{\sim}{D}^{(el)}(\underline{Q}) \tag{37}$$

where the superscripts refer to the Coulomb and electronic energy respectively. In this way, we neglect the overlap interaction between core electrons of nearest neighbour ions, an approximation which is commonly done in this sort of calculations. A general matrix element of $\underset{\sim}{D}$ depends upon two plane indexes m and m' and upon the cartesian coordinates (α , β). Its expression is

$$D_{\alpha\beta}(\underline{Q};m,m') = \frac{1}{M} \sum_{\ell} e^{i\underline{Q}\cdot(\underline{\overset{0}{R}}_{\ell m} - \underline{\overset{0}{R}}_{0m'})} \phi_{\alpha\beta}\begin{pmatrix} \ell & 0 \\ m & m' \end{pmatrix} \tag{38}$$

where $\underline{\overset{0}{R}}_{\ell m}$ indicates the equilibrium position of the ℓ-th ion on the m-th plane of the slab and $\phi_{\alpha\beta}\begin{pmatrix} \ell & 0 \\ m & m' \end{pmatrix}$ is the component $\alpha\beta$ of the force-constant tensor of an atom at ℓ m with respect to the atom at 0m'.

Because of the invariance under arbitrary translations, the force constants must satisfy the condition

$$\sum_{\ell'} \sum_{m'} \phi_{\alpha\beta}\begin{pmatrix} \ell' & 0 \\ m' & m \end{pmatrix} = 0. \tag{39}$$

The sum includes $\phi_{\alpha\beta}\begin{pmatrix} \ell & 0 \\ m & m \end{pmatrix}$ which would represent the reaction of an atom to the force exerted upon it by the crystal as a whole. To avoid the mathematical difficulties present in the evaluation of $\phi_{\alpha\beta}\begin{pmatrix} \ell & 0 \\ m & m \end{pmatrix}$ it is convenient to subtract this expression from (38) in the calculation of the dynamical matrix elements when m = m'. We have therefore that

$$D_{\alpha\beta}(\underline{Q};m,m') = \frac{1}{M} \sum_{\ell}{}' \left(e^{i\underline{Q}\cdot\underline{\overset{0}{R}}_{\ell}} - 1\right) \phi_{\alpha\beta}\begin{pmatrix} \ell & 0 \\ m & m \end{pmatrix} + - \frac{1}{M} \sum_{\ell} \sum_{m'\neq m} \phi_{\alpha\beta}\begin{pmatrix} \ell & 0 \\ m' & m \end{pmatrix} \tag{40}$$

while we keep (38) for m $\neq$ m'. In this way we are sure that the crystal equilibrium exists, and that acoustic modes tend to zero in the long wavelength limit.

With the definitions given in the previous section we have that

$$\phi_{\alpha\beta}\begin{pmatrix} \ell & 0 \\ m & m' \end{pmatrix} = \frac{\partial^2 E^{(c)}}{\partial u^{\alpha}_{\ell m}\, \partial u^{\beta}_{0m'}} + \frac{\partial^2 E^{(2)}}{\partial u^{\alpha}_{\ell m}\, \partial u^{\beta}_{0m'}} \tag{41}$$

The Coulomb term has been evaluated with standard techniques using the Ewald method.

5. Applications

Using the expressions given in the previous sections, we have calculated the phonon dispersion curves of a slab of 15 (001) atomic planes for Na and K along the high-symmetry directions of the two-dimensional Brillouin zone of the (001) surface.

To perform such calculations three main ingredients are required. The first is the geometry of the system, specifying the two-dimensional periodicity, the location of the atoms in the unit cell and the interplanar distance in the outermost planes. The second is the bare ion pseudopotential, in our cases the Na and K pseudopotentials, to be consistently screened by our dielectric function. The third is some appropriate function G_Q (q) for the exchange and correlation corrections, to be introduced into the dielectric matrix.

The geometry of the problem was initially taken as the one corresponding to an ideal truncation of the crystal. Such unrelaxed configuration is rather unlikely so that we have performed calculations assuming some relaxed geometries too. To establish the differences between single and multilayer relaxations, we calculated the phonon dispersion curves for both cases. As to the pseudopotential used to represent the electron-ion interaction, we have used the Heine-Abarenkov model potential [35]. Unlike the semi-empirical potentials frequently used in bulk calculations, this is a non-local and energy-dependent pseudopotential. As shown by BERTONI et al. [36] one can dispose of the non-locality through some kind of averaging over the Fermi sphere, which allow to derive a reliable local form factor as in local theories. It should be noted that the model potential, being derived from the properties of the isolated ion, is exempt from any parametrization of bulk and surface properties, and it allows to perform a true first principle calculation. Among the several forms proposed for the exchange and correlation corrections, the most reliable are those by WASHISTA and SINGWI [33] and ICHIMARU and UTSUMI [37]. In our calculations we have adopted the form of reference [33]. Calculations performed using the form given in reference [37] lead to similar results [38].

In performing our calculations we have found the following important results: i) the inclusion of the off-diagonal elements of the response matrix is crucial for a realistic calculation of the surface phonon frequencies [39]; ii) the Hartree theory, neglecting exchange and correlation corrections, leads to imaginary values for the surface phonon frequencies [40].

Figures 1 and 2 show the dispersion of the calculated surface branches for the unrelaxed surfaces of Na and K respectively. These modes were discriminated from bulk features on the basis of their location in energy and of their localization i.e. the vibration amplitude at the surface layer or in the second plane. Surface modes have been plotted against the projected bulk phonon spectrum to emphasize the correspondence between modes localized near the surface and absolute or relative gaps of the bulk spectrum.

The distribution of the modes is essentially the same for the two metals, so that we can illustrate their nature by commenting the results for K. Along the $\bar{\Delta}$ direction we notice two main branches, which come off the corresponding edges of the bulk continua. The lowest one is primarily associated with vibrations along the z direction, and therefore has a shear vertical character (SV); the upper one is associated with vibrations in the direction transverse to $\underset{\sim}{Q}$ and parallel to the surface, and can be identified as a shear horizontal mode (SH).

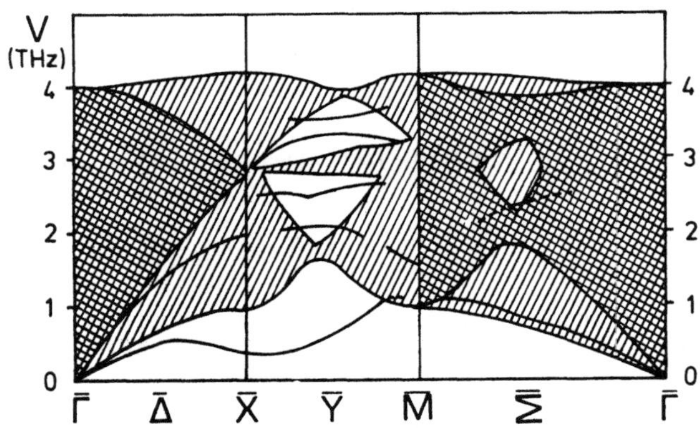

Fig. 1: Dispersion of the calculated surface modes for an ideal Na (001) surface along the high-symmetry directions of the two-dimensional Brillouin zone

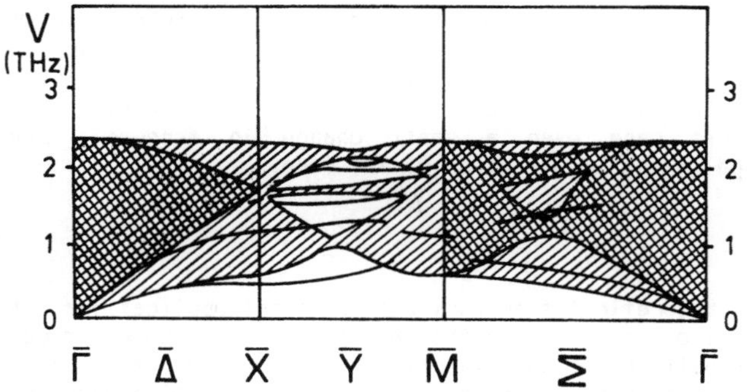

Fig. 2: Dispersion of the calculated surface modes for an ideal K (001) surface

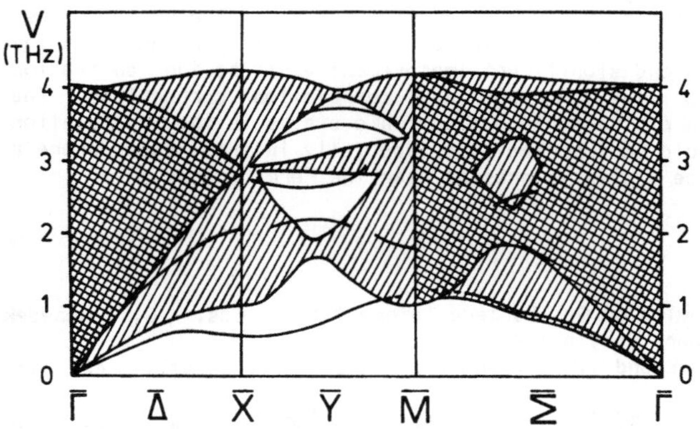

Fig. 3: Dispersion of the calculated surface modes for the Na (001) surface, assuming a multilayer relaxation with the interatomic spacings from [32]

Both modes mix with bulk states at low $\underline{Q}$, while, on moving toward $\bar{X}$, they show an increasing localization near the surface. The $\bar{Y}$ direction shows two large gaps of the projected bulk structure, which accomodate several surface modes. Some of them are well defined near the central portion of the gap, but do not persist over to $\bar{X}$ and $\bar{M}$. The lower band along the $\bar{\Delta}$ direction persists over most of the $\bar{Y}$ direction, but it acquires some SH character. Before $\bar{M}$ it enters into the bulk continuum, where it loses its surface nature. The higher band along $\bar{\Delta}$ crosses the absolute gap of triangular shape lying between 1.1 and 2.15 THz near the lower vertex. Then it becomes a resonance and gets mixed with the bulk continuum. Two main bands are found along the $\bar{\Sigma}$ directions, the lowest has mainly SV character, but the contribution of the horizontal vibrations is not negligible. A more detailed description of the spectrum has been given elsewhere [41].

For the purpose of the present paper it is interesting to notice that our results are qualitatively very similar to those obtained by semi-empirical force constant models [11], but significant differences arise in the frequency location of the SV modes and in the compositions of their eigenvectors [41]. To a large extent, such differences seem to be imputable to the neglecting of the surface screening effects in the semi-empirical models.

As an example of the sort of modifications which can be introduced by the changes in the interplanar distances caused by relaxation, we give in fig. 3 the calculated phonon spectrum for Na (001), assuming a multilayer relaxed configuration. The interatomic distances were obtained through a total energy minimization procedure to determine the equilibrium configuration of the surface and of the outermost planes. It is seen that the spectrum is not qualitatively changed by relaxation. Some branches, in particular the lowest one, are rised by relaxation; the effect is more significant for the modes having a pronounced SV character, and is less important for horizontal or longitudinal bands. However even small modifications in the eigenvalues may cause a significant change in the eigenvectors compositions. This is particularly true for the bands running near the edge of a gap, since in this case even a small change in frequency may determine a significant change in the localization of the mode.

6. Conclusions

We have presented the general theoretical framework and the first applications of the pseudopotential perturbation approach to the determination of the dynamical matrix of metal surfaces. We believe the same theory could be extended to treat a number of other problems of great physical interest, like the dynamical properties of overlayers, the determination of the stable surface structures and the effects of reconstruction on the phonon spectrum.

A number of improvements to the present formulation can be indicated. The most important would be to use the response function of a jellium slab, rather than the infinite barrier model. A second significant improvement would be the derivation of more accurate forms for the exchange and correlation corrections. The inclusion of the contribution of the surface electronic states in the calculation of the response function could be important too, specially in those cases where a significant density of surface states is present near the Fermi energy.

REFERENCES

1 J.P. Toennies: in "Dynamics of Gas-Surface Interaction", Eds. G. Benedek and U. Valbusa (Springer, Berlin 1982), 196
2 G. Brusdeylins, R.B. Doak and J.P. Toennies: Phys. Rev. Lett. 46, 437 (1981)
3 G. Brusdeylins, R.B. Doak and J.P. Toennies: J. Chem. Phys. 75, 1784 (1981)
4 R.B. Doak, U. Harten and J.P. Toennies: Phys. Rev. Lett. 51, 578 (1983)
5 H. Ibach: Phys. Rev. Lett. 24, 1416 (1970)
6 H. Ibach: Phys. Rev. Lett. 27, 253 (1971)
7 H. Ibach and D.L. Mills:"Electron Energy-Loss Spectroscopy and Surface Vibrations" (Academic Press, New York 1982)
8 S. Lewhald and H. Ibach: in "Vibrations at Surfaces" Eds. R. Caudano, J. M. Gills and A.A. Lucas (Plenum Press, New York 1982), 137.
9 D. Castiel, C. Dobrzymski and D. Spanjaard: Surface Sci. 59, 252 (1976)
10 V. Bortolani, F. Nizzoli and G. Santoro:"Proceedings of the International Conference on Lattice Dynamics" Ed. M. Balkansky (Flammarion, Paris 1977),302
11 J.E. Black, D.A. Campbell and R.F. Wallis: Surface Sci. 115, 161 (1982)
12 J.E. Black, F.C. Shanes and R.F. Wallis: Surface Sci. 133, 199 (1983)
13 S. Lewhald, J.M. Szeftel, H. Ibach, T.S. Rahman and D.L. Mills: Phys. Rev. Lett. 50, 518 (1983)
14 V. Bortolani, A. Franchini, F. Nizzoli and G. Santoro: Phys. Rev. Lett. 52, 429 (1984)
15 H. Wendel and R. Martin: Phys. Rev. B19, 5251 (1979)
16 J. Ihm and M.L. Cohen: Phys. Rev. B21, 1577 (1980)
17 A. Zunger: Phys. Rev. B21, 4785 (1980)
18 K. Kunc and R.M. Martin: Physica 117-118B, 511 (1983)
19 R.M. Pick, M.H. Cohen and R.M. Martin: Phys. Rev. B1, 910 (1970)
20 L.J. Sham: Phys. Rev. 188, 1451 (1969)
21 C.M. Bertoni, V. Bortolani, C. Calandra and E. Tosatti: Phys. Rev. Lett. 28, 1578 (1972)
22 W.A. Harrison: "Pseudopotential in the Theory of Metals" (Benjamin, New York 1966)
23 V. Heine: Solid State Physics 24, 1 (1970)

24 W.A. Harrison: Phys. Rev. 181, 1036 (1969)
25 J.A. Moriarty: Phys. Rev. B5, 2066 (1972)
26 J.A. Moriarty: Phys. Rev. B16, 2537 (1977)
27 J.A. Moriarty: Phys. Rev. B26, 1754 (1982)
28 A.S. Bamzai and B.M. Deb: Rev. Mod. Phys. 53, 95 (1981)
29 N.D. Lang and W. Kohn: Phys. Rev. B1, 4555 (1970)
30 D.M. Newns: Phys. Rev. B1, 3304 (1970)
31 J.P. Muscat and D.M. Newns: Surface Sci. 99, 609 (1980)
32 R.N. Barnett, U. Landman and C.L. Cleveland: Phys. Rev. B28, 1667 (1983)
33 P. Vashista and K.S. Singwi: Phys. Rev. B6, 875 (1972)
34 K.S. Singwi and M.P. Tosi: Solid State Physics 36, 177 (1981)
35 V. Heine and D. Weaire, Solid State Physics 24, 249 (1970)
36 C.M. Bertoni, V. Bortolani, C. Calandra and F. Nizzoli, J. Phys. F: Metal Phys. 5, 419 (1975)
37 S. Ichimaru and K. Utsumi, Phys. Rev. B24, 7385 (1981)
38 C. Calandra and A. Catellani, unpublished results.
39 C. Beatrice and C. Calandra: Phys. Rev. B28, 6130 (1983)
40 C. Calandra, A.Catellani and C. Beatrice: Proceedings of ECOSS-6, Surface Sci. (in press).
41 C. Calandra, A. Catellani and C. Beatrice: Surface Sci. in press.

One-Phonon Scattering of He Atoms from the Ag(111) Surface

V. Bortolani, A. Franchini, and G. Santoro

Dipartimento di Fisica, Universita di Modena and Gruppo Nazionale Struttura della Materia, I-41100 Modena, Italy

1. Introduction

The inelastic scattering of neutral atoms has proved to be a valuable tool for the study of surface phonon spectra and of the dynamics of the gas-surface interaction [1]. The time of flight spectra of scattered He atoms have yielded dispersion relations of surface phonons over the entire two-dimensional Brillouin zone (BZ) for several ionic crystal surfaces [2]. Recently this technique has been successfully applied to the study of noble metal surfaces [3].

To deal with this problem,we will first discuss the theory of the He-metal interaction [4]. For metals having a small corrugated surface one can use the distorted wave Born approximation (DWBA) [5]. The atom surface interaction is treated in Sec. 2 where explicit formulae for the cross-section are derived in terms of the phonon displacement field. The interaction potential which enters in the cross-section is treated in Sec. 3. We divide this interaction into an attractive potential of the Van der Waals [6] type V and a repulsive part V due to the interaction of the He atom with the corrugated charge distribution of the metal surface. The dynamical problem is discussed in Sec. 4 within the framework of a force-constant parametrization of bulk phonons [7]. With the force constants so derived, we determine the surface displacement field and the surface phonons by solving the dynamical problem of a slab. In principle,one should deal with a semiinfinite crystal in order to avoid the interference effects arising from the two surfaces. The splitting in energy caused by the second surface, however, is important only for long wavelength acoustic phonons. For short wavelength acoustic and optical modes, we found that a slab formed by 45 atomic planes is sufficient to reduce to zero the splitting in phonon energies. We will consider the force constants in the surface region different from those of the bulk, because of the different role played by the conduction electrons in the surface region. We will show that the change of the surface force constants introduces resonances in the continuous part of the spectrum,and allows us to explain the peaks observed in the time-of-flight experiments.

2. DWBA Cross-Section

Explicit formulae for the inelastic one-phonon scattering cross-section in the DWBA have been derived some years ago by CABRERA ET AL [8]. In this section we present the essential points of the derivation. For a two-body potential $V(\vec{r}-\vec{r}_1-\vec{u}_1)$ between the gas atom at $\vec{r}$ and the atoms in the solid at $\vec{r}_1+\vec{u}_1$ the interaction is:

$$\sum_{1} \vec{u}_1 \cdot \nabla_{\vec{r}_1} V(\vec{r}-\vec{r}_1) \qquad (1)$$

to first order in the phonon-induced displacements u . For the repulsive part of the interaction, only the atoms in the surface layer contribute appreciably to the sum in Eq. 1. The l-dependence of the u_1, for a phonon of lateral momentum $\vec{Q}$, is than given by $\exp(i\vec{Q}\,\vec{R}_1)$ (capital letters denote vectors in the plane of the surface). The matrix element of the interaction in Eq. 1 must be taken between eigenstates of the static potential:

$$\sum_l V(\vec{r}-\vec{r}_l) \qquad (2)$$

Only the zero Fourier component of the potential $V_0(z)$ is appreciable for smooth metallic surfaces such as the (111) surface of the noble and transition metals. The eigenfunctions of $V_0(z)$ are then of the form $\chi_{k_z}(z)\exp(i\vec{K}\cdot\vec{R})$ where, for an atom of energy E and mass m:

$$k_z^2 + K^2 = 2mE \qquad (3)$$

The matrix elements for transitions from $(\vec{K}_i,k_{iz})$ to $(\vec{K}_f,k_{fz})$ contain the integral:

$$\sum_l e^{i\vec{Q}\cdot\vec{R}_l} \nabla_{\vec{r}_l} \int d^2R\, e^{i(\vec{K}_i-\vec{K}_f)\cdot\vec{R}} V(\vec{r}-\vec{r}_l) \qquad (4)$$

which vanishes unless $\vec{Q}=\vec{K}_f-\vec{K}_i+\vec{G}$ where $\vec{G}$ is a reciprocal lattice vector. Ultimately then, we need to compute:

$$N \int dz\, \chi_{iz}^*(z)\, \chi_{fz}(z) \int d^2R\, e^{-i(\vec{Q}+\vec{G})\cdot\vec{R}}\, \nabla_{\vec{r}_l} V(\vec{r}-\vec{r}_l) \Big|_{z_l=0} \qquad (5)$$

where N is the number of atoms for unit surface area and χ_{iz}, χ_{fz} are eigenfunctions of the laterally averaged potential:

$$V_0(z) = N \int d^2R\, V(\vec{R},z) \qquad (6)$$

The Fourier components of the potential then determine both the dynamic interaction and the static potential. ∇_l can be replaced by $(i(\vec{Q}+\vec{G}),\partial/\partial z)$. By taking the square modulus of Eq. 6 and by summing all processes in the volume $d^2K_f dk_{fz}=(2mE_f)^{\frac{1}{2}}m\, dE_f$ and dividing by $dE_f d\Omega_f$ we obtain the differential reflection coefficient or cross-section per unit surface area. In this manner we obtain:

$$\frac{d^2R}{dE_f d\Omega_f} = \frac{1}{NM} \sum_{\vec{Q},j} \frac{(2mE_f)^{\frac{1}{2}} k_{iz}}{\omega(\vec{Q},j)}\, n\{\omega(\vec{Q},j)\}\, e^{-2W}\, \times$$

$$|\vec{e}(\vec{Q},j)\cdot\langle\chi_{iz}|(i\vec{Q},\partial/\partial z)V(\vec{Q},z)|\chi_{fz}\rangle|^2 \times$$

$$\delta(\vec{K}_f-\vec{K}_i-\vec{Q})\,\delta\{E_f-E_i-\omega(\vec{Q},j)\} \qquad (7)$$

Here we have introduced the phonon frequencies $\omega(\vec{Q},j)$ and the polarization vectors $\vec{e}(\vec{Q},j)$. $n(\omega)$ is the Bose factor, phonon creation (energy loss) corresponds to $\omega(\vec{Q},j)<0$; phonon annihilation (energy gain) corresponds to $\omega(\vec{Q},j)>0$. $V(\vec{Q},z)$ is the 2-dimensional Fourier transform of the potential and $\vec{Q}$ now is considered in the extended zone scheme. The Debye Waller factor 2W is given by:

$$2W = \sum_{\vec{Q},j} \{|k_{iz}|+k_{fz}\}^2 \langle|\vec{e}(\vec{Q},j)|^2\rangle \qquad (8)$$

3. Atom-Surface Potential

The atom surface potential is generally divided as follows:

$$V(\vec{r}) = V_R(\vec{r}) + V_A(z) \tag{9}$$

The short range repulsive part V_R arises from the overlap of the electronic cloud of the atom and the surface electrons, while the long range attractive polarization potential V_A has the asymptotic form:

$$V_A(z) = -\frac{C}{z^3} \tag{10}$$

To determine the repulsive part, which gives the major contribution to the matrix elements in Eq. 8 we use the ESBJERG-NORSKOV approximation [9]:

$$V_R(\vec{r}) = \overline{\alpha\rho(\vec{r})} \tag{11}$$

where α can be determined from electron-He atom scattering data. $\rho(\vec{r})$ is the surface charge density averaged with the atom electrostatic potential. We have found that the averaging effects of the He atom on the surface charge density at the typical incident energies used in atom-scattering experiments (E =8-30 meV) are a minor correction so that in Eq.12 we will take directly the charge density $\rho(\vec{r})$.

We than evaluate $\rho(\vec{r})$ with a superposition of atomic charges. The reason for using a simple superposition of atomic charges is that the impinging He atom experiences the charge density at large distances from the surface, where the surface states and the bonding effects are negligible.
Equation 12 is evaluated at the classical turning point for which:

$$V_R(\vec{R},z_t) = E_i \tag{12}$$

Around this region the potential can be written as:

$$V_R(\vec{R},z) = E_i\, e^{-\beta\{z-z_t-D(\vec{R})\}} \tag{13}$$

where β is the softness parameter and $D(\vec{R})$ is the corrugation function. The corrugation is measured with respect to z_t evaluated at $\vec{R}$=0 so that D(0)=0. To determine β and D we have evaluated $\rho(\vec{r})$ with a thirty atom cluster. Since in the evaluation of the cross-section we need also the gradient of the potential, we have calculated numerically the normal and in-plane derivatives. In Fig. 1 are plotted the results for Ag(111) with R along the $[11\bar{2}]$ surface direction.
As one can see, the normal derivative of the potential has a Gaussian type behaviour, while the lateral derivative is the derivative of the same Gaussian. We have found an analytic form of the potential which reproduces very accurately the results of the full calculation.
This approximate expression can be cast in the form:

$$V_R(R,z) = V\sum_l e^{-\beta|\vec{r}-\vec{r}_l|} = V\sum_l e^{-\beta\{(\vec{R}-\vec{R}_l)^2+(z-z_l)^2\}^{\frac{1}{2}}} \tag{14}$$

In the last expression we have considered only surface atoms at z_l=0. From our calculations for noble metals, it turns out that the classical turning point z_t is very large (z >4 A). For this reason, if the coordinate $\vec{R}$ happens to be in the l-th surface cell $|\vec{R}-\vec{R}_l| \ll z_t$. In this cell we can expand the potential as:

$$e^{-\beta\{(\vec{R}-\vec{R}_l)^2+z^2\}^{\frac{1}{2}}} \simeq e^{-\beta z}e^{-\frac{\beta}{2z_t}(\vec{R}-\vec{R}_l)^2} \tag{15}$$

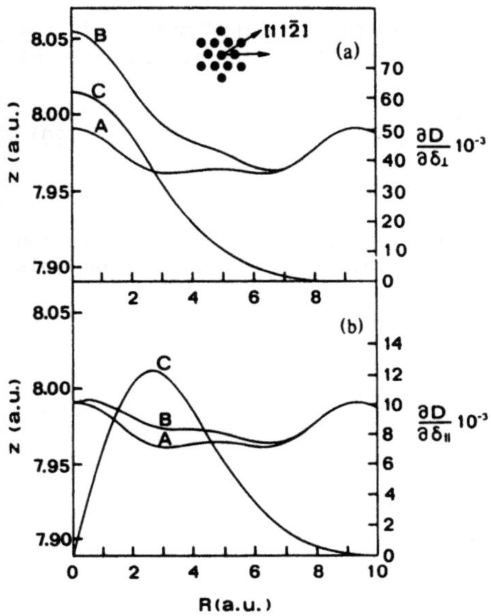

Figure 1 Corrugation profiles and their derivatives $\partial D(\vec{R})/\partial\delta$. Curve A: $D(\vec{R})$. Curve B: $D(\vec{R})$ obtained by moving an atom: in (a) along the z direction; in (b) along the $\vec{R}$ direction. Curve C represents the numerical derivative

This expansion has to be applied to all the nearest neighbors. For all the other neighbors, the contribution of the two-body potential to the total potential is negligible, so that this approximation can be performed without altering the accuracy of the total potential. V_R can thus be written as:

$$V_R(\vec{R},z) = V\, e^{-\beta z} \sum_l e^{-\frac{\beta}{2z_t}(\vec{R}-\vec{R}_l)^2} \qquad (16)$$

We emphasize that also the gradient of the potential reproduce to within 1% the results of the full calculation. The 2-dimensional Fourier transform of this potential has therefore the form:

$$e^{-Q^2/Q_c^2} \qquad (17)$$

This lateral cut-off gives a quantitative explanation of the Armand effect i.e. of the reduction in the cross-section produced by the cooperative movement of neigboring surface atoms. The cut-off factor reduces the scattering intensities for large lateral momentum transfer. We want to point out that this cut-off is just associated with the deformation of charge produced by a single moving atom and exists because of the softness of the repulsive potential.

We turn now to discuss the long-range attractive dipolar interaction. As shown by ZAREMBA AND KOHN [6] an approximate form of V_A is given by:

$$V_A(z) = -\frac{C_{VW}}{(z-z_{VW})^3} \qquad (18)$$

where for Ag c_{VW}=0.44 eV A^3 and z_{VW}=1.38 A. The divergence of V_A at $z=z_{VW}$ has been recently considered by NORLANDER AND HARRIS [10]. These authors found that the inclusion of exchange and correlation between the He atom and the surface regularize the divergence. In fact, according to these authors V_A has to be multiplied by the function:

$$f(x) = \{1 - 2x(1+x)+1\}\, e^{-2x} \qquad x = k_c(z-z_{VW}) \qquad (19)$$

that eliminates the divergence. k_c is a free parameter of the order of the inverse of the atomic radius. The laterally-averaged total potential can then be written as:

$$V_0(z) = V\, e^{-\beta z} - \frac{C_{VW}}{(z-z_{VW})^3}\, f\{k_c(z-z_{VW})\} \qquad (20)$$

The value of k_c can be obtained by fitting for a given surface the physisorption well or the bound states of the He-metal potential. In Fig. 2 is drawn this potential for He/Ag(111).

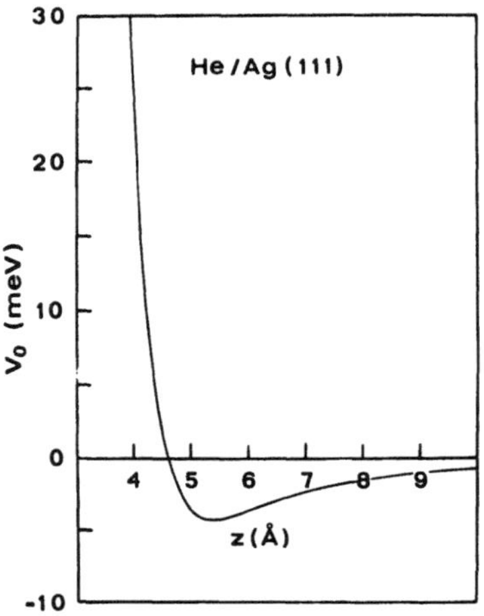

Figure 2 The laterally-averaged potential for He:Ag(111). The softness parameter β=2.2 A^{-1}, k_c=1 A^{-1} and V =296 eV

4. Calculation For The (111) Surface

The eigenvectors $\vec{e}(\vec{Q},j)$ and eigenvalues $\omega(\vec{Q},j)$ have been evaluated for a slab of 45 (111) ideal atomic planes, within a force-constant parametrization of the bulk dynamics. Central and angular forces have been included up to second neighbors. In this model, the surface interactions are determined by the nearest neighbor radial force constant β_1^S, the angular force constant δ_1^S and the tangential force constant α_1^S. By using surface force constants equal to those of the bulk, it is possible to explain the energy position and the fall off of the scattering intensity of the Rayleigh wave (RW) as a function of the momentum transfer $\vec{Q}$. In the bulk region, however, is absent any significant structure, in complete disagreement with the experiment [3] that shows the presence of a well-defined peak in this region. By considering the effect due to the termination of the electron gas at the surface,we can explain the origin of this new structure. The different screening produced by the electrons in the surface region can be introduced phenomenologically in our model by varying the surface force constants with respect to those of the bulk (hereafter denoted by the superscript B). By lowering the surface force constants,we are able to reproduce the observed peak [11]. The variation of the surface force constants could cause the softening of some mode,so that these variations cannot be performed arbitrarly. In Fig. 3 the surface stability diagram at fixed α_1^S is shown. The full line represents the values of the force constants for which the sagittal M mode becomes unstable. The dashed line corresponds to the instability of the M shear horizontal (SH) mode. As one can see,a small lowering of δ_1^S makes the surface unstable against the M SH mode,while the frequency of the M sagittal does not change appreciably. A significant lowering of the sagittal mode can be obtained by reducing the

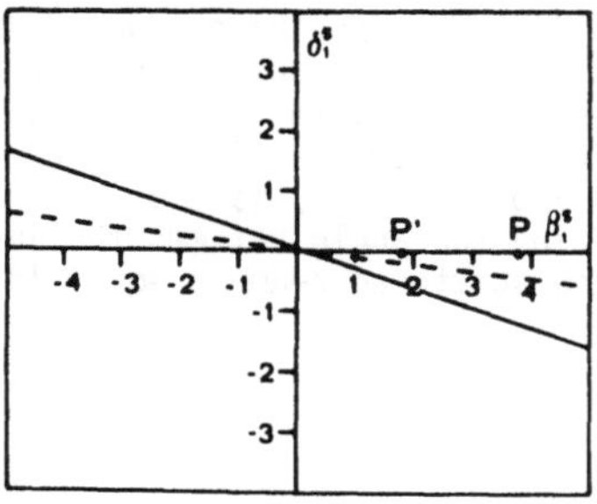

Figure 3 Stability diagram of the Ag(111) surface as a function of the nearest neighbor surface force constants (THz^2). β_1^S and δ_1^S at fixed $\alpha_1^S=\alpha_1^B$. P is for $\beta_1^S=\beta_1^B$ and P' is for $\beta_1^S=0.48\ \beta_1^B$

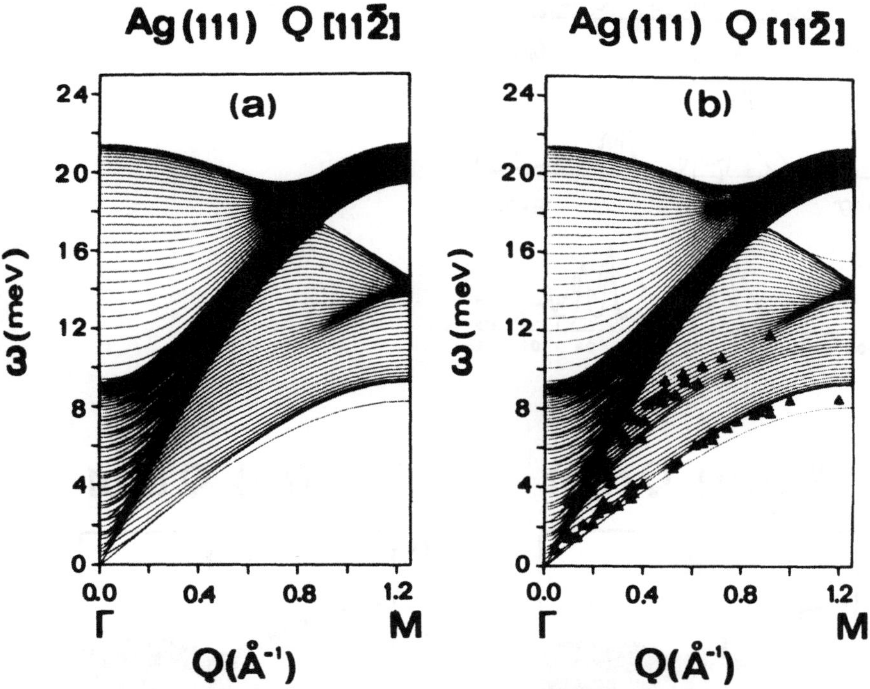

Figure 4 a) calcualated phonon dispersion curves for Ag(111) with $\vec{Q}$ [11$\bar{2}$]. The surface force constants are taken to be equal to their bulk values. b) same calculations performed with $\beta_1^S=0.48\beta_1^B$. Experimental points (triangles) from ref. [3]

radial force constant β_1^S and keeping $\alpha_1^S=\alpha_1^B$ and $\delta_1^S=\delta_1^B$. For $\beta_1^S=0.48\beta_1^B$ we found a well- defined resonance of longitudinal character embedded in the transverse bulk continuum which explains the experimental TOF data. In Fig. 4 are drawn the sagittal phonons of Ag(111) with $\vec{Q}$ along [11$\bar{2}$]. Fig. 4a refers to the ideal calculation. As one can see, there is no evidence of the resonances which becomes evident by lowering β_1^S as is shown in Fig. 4b. We want to stress that in both calculations the dispersion of the Rayleigh wave is practically the same, as is evident from Figs. 4a and 4b. The TOF spectra evaluated with Eq. 8 are presented in Fig. 5 for $\beta_1^S=0.48\ \beta_1^B$. The calculated TOF spectra accurately reproduce the experimental lineshape. Notice that also the behaviour of the scattering intensities versus $\vec{Q}$ is in agreement with the experimental data.

The calculated phonon spectrum for the Ag(111) surface with $\vec{Q}$ along [1$\bar{1}$0] is shown in Fig. 6. The direction [1$\bar{1}$0] is even more interesting than the [11$\bar{2}$] because in the former case there is no decoupling of sagittal and shear horizontal (SH)

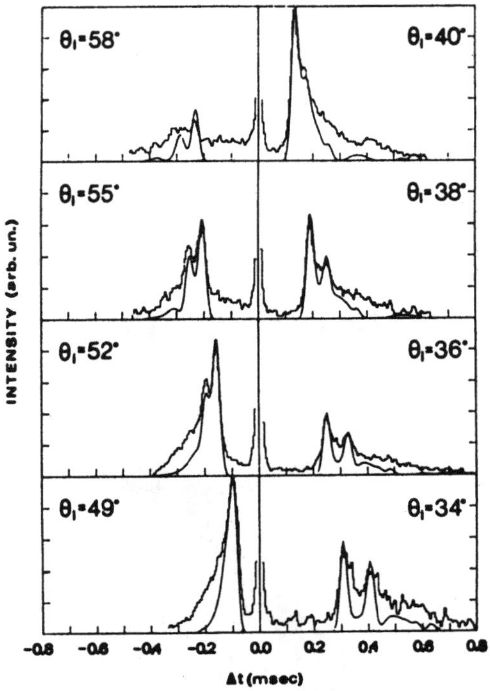

Figure 5 The TOF spectra evaluated in the distorted Born approximation for Ag(111) and $\vec{Q}$ [11$\bar{2}$]. The light lines represent the experimental data. The incident atom energy E_i is equal to 17.5 meV and the angle between the incident and scattered beam is equal to 90^0

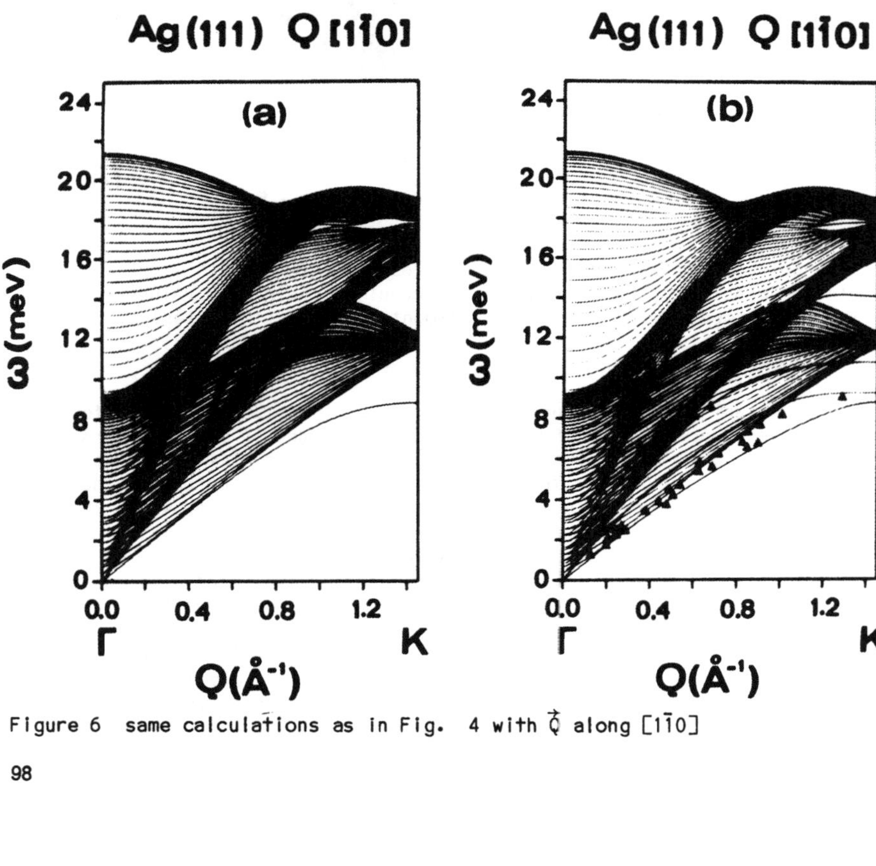

Figure 6 same calculations as in Fig. 4 with $\vec{Q}$ along [1$\bar{1}$0]

modes, which instead are mixed together. This situation allows the appearance of a long wavelength pseudo Rayleigh wave (PRW) in this direction [12]. We recall that for $\beta_1^S=\beta_1^B$ this mode is present in the long wavelength limit, and its velocity is very near to that of the upper transverse threshold [12]. Fig. 6b shows very clearly the existence of the PRW. The reduction of β_1^S lowers considerably this phonon branch for $Q<0.2$ Å^{-1}, so that it becomes very close to the true RW, which is only slightly affected by variations of β_1^S. Therefore the lowest set of experimental points is explained as due to the combined effect of the RW and of the PRW. Notice that although the PRW in the long wavelength limit is mainly polarized as SH (which is not detectable in atomic scattering), it becomes strongly polarized in the sagittal plane for large $\vec{Q}$ (>0.4 Å^{-1}). The upper set of experimental points coincides with the calculated dispersion curve of the longitudinal resonance (LR). Also this phonon branch is strongly affected by β_1^S and is lowered by reducing β_1^S. The LR gives the explanation for the upper set of experimental data, as in the [11$\bar{2}$] direction.

An explanation of the softening of the relevant force constant can be qualitatively understood by considering the pseudopotential expression for the total energy in the case of a semiinfinite crystal [13]. In the Fermi Thomas limit, the total energy for the bulk has the form $\sum \exp(-\lambda r_{ij})/r_{ij}$ where λ is the Fermi-Thomas screening length and r_{ij} is the distance between atom i and atom j. The inclusion of the surface produces new leading terms of the form:

$$
-\frac{e^{-\lambda r_{ij}^{+}}}{r_{ij}^{+}} - \frac{|z_i+z_j|}{R_{ij}^3}\,\frac{e^{-\lambda|z_i+z_j|}}{2\lambda}
$$

$$
r_{ij}^{+} = \{(R_i-R_j)^2+(z_i+z_j)^2\}^{\frac{1}{2}} \tag{21}
$$

where R_{ij} is the parallel component of $\vec{r}_{ij}$ and z_i and z_j are the normal components of the ionic position measured from the edge of the negative charge, located at $a_0/2$ in front of the outermost lattice plane. For the surface atoms $z_i=z_j=-a_0/2$ the bulk and surface terms in total energy strongly cancel, producing a reduction in the surface force constants. On the other hand, for atoms in the bulk the surface term becomes negligible. Moreover, Moriarty [14] has shown that for noble metals the effect of the s-d interaction is very important in the evaluation of the band-structure energy, and its effect increases in going from Cu to Au. According to these arguments, it is reasonable to expect that the lowering of surface force constants will be bigger in Au than in Ag, as we have found with our fitting procedure.

Acknowledgements

We like to thank V. Celli for usefull discussions. The calculations were performed with the support of Centro di Calcolo, Universita° di Modena.

References

1. R. Manson and V. Celli, Surf. Sci.26, 695 (1971)
2. M. G. Brusdeylins, R.B. Doak and J.P. Toennies, Phys.Rev.Lett. 16, 937 (1981); and to be published
3. R.B. Doak, U. Harten and J.P. Toennies, Phys. Rev. Lett. 51, 578 (1983).
4. V. Bortolani, A. Franchini, N. Garcia, F. Nizzoli and G. Santoro, Phys. Rev. B28, 7358 (1983).
5. V. Celli, in Dynamics of Gas-Surface Interaction, ed. G.Benedek and U.Valbusa (Springer, Berlin, 1982) p. 1
6. E. Zaremba and W. Kohn, Phys. Rev.B15, 1769 (1977)
7. V. Bortolani, A. Franchini, F. Nizzoli and G. Santoro , in Dynamics of Gas-Surface Interaction, ed. G.Benedek and U.Valbusa (Springer, Berlin, 1982) p. 196

8. N. Cabrera, V. Celli and R. Manson, Phys. Rev. Lett. 22, 346 (1969)
9. N. Esbjerg and J.K. Norskov, Phys. Rev. Lett. 45, 807 (1980)
10. P.Norlander and J.Harris, J. Phys: C: Solid St. Phys. 17, 1141 (1984)
11. V. Bortolani, A. Franchini, F. Nizzoli, G. Santoro, Phys. Rev. Lett. 52, 429 (1984).
12. G.W. Farnell, in "Physical Acoustics", vol. VI, ed. W.P.Mason and R.N. Thurston (Academic Press, New York, 1970) p. 109.
13. D.E. Beck, V. Celli,G. Lo Vecchio and A. Magnaterra, Nuovo Cim. B17, 230 (1970)
14. A. Moriarty, Phys. Rev. B6, 1239 (1972).
15. J.P.Toennies and coworkers, data presented at the Modena Meeting of the Surface Group of GNSM, December 1983, unpublished and to be published

Electron Energy-Loss Spectroscopy for Metal Surfaces

Fabrizio Nizzoli

Dipartimento di Fisica and G.N.S.M., Università di Modena
I-41100 Modena, Italy

1. Introduction

One of the most powerful experimental techniques for the study of the elementary excitations in solids has proved to be the electron energy-loss spectroscopy (EELS). Initially, during the 1960s, this technique was used to investigate excitation of plasma waves and of electronic transitions. The energy losses being of many eV, a resolution of typically 250 meV was considered to be more than enough in electron and plasmon spectroscopy. The order of magnitude of the phonon energies however is much smaller, so that the overall resolution that is needed in phonon spectroscopy must be better than 30 meV. Such performance has been obtained with a new generation of high-resolution spectrometers based on the use of cylindrical electrostatic deflectors [1]. One (or more) deflector works as monochromator and, in conjunction with an accelerating lens system, forms a source of highly monoenergetic primary electrons. The scattered electrons are analyzed in energy by another (or more) deflector. A resolution as good as 3.7 meV has been obtained with this kind of spectrometer [2].

The first evidence of energy losses due to vibrations of atoms (hydrogen) adsorbed on metal surface was reported by PROPST and PIPER [3]. Since then several EELS measurements of the spectral lines of chemical species adsorbed on metal substrates have been performed. The vibrational frequencies of an adsorbed species (internal modes of the adsorbed molecule and substrate-adsorbate modes) depend on the symmetry of the adsorption site, on the structure of the overlayer, on the type of adsorption (molecular of dissociative) and on the relative position of the adsorbate with respect to the surface. For these reasons EELS is currently used as a technique for the structural analysis of solid surfaces. More recently [4-6] EELS also proved to be useful in measuring surface phonons of the substrate itself.

Two different mechanisms contribute to the EELS spectra. At a crystal boundary, the vibrating surface or adsorbate atoms may give rise to an electric dipole moment. The long-range dipole field in the vacuum due to the oscillating dipole produces an intense scattering concentrated in a "dipole lobe" around the specular reflection angle [7]. The angular width of the dipole lobe is $\Delta\theta=\hbar\omega_o/4\pi E_i$, where ω_o is the phonon frequency and E_i is the energy of the impinging electrons. Because E_i is of the order of a few eV, $\Delta\theta < 0.1°$. Outside the dipole lobe (and the Bragg directions) the long-range scattering produces a very weak background.

The short-range repulsive part of the electron-surface potential is responsible for electron scattering at large angles off the specular reflection. This type of scattering is called impact scattering. In principle the dipole and impact scattering mechanisms coexist, however their different characters allow us to treat them separately. In the following we will therefore distinguish between the dipole regime and the impact scattering regime.

2. Dipole Scattering

The small angular width of the dipole lobe around the specular direction corresponds to a small exchange of parallel momentum $\underline{Q}$ between the electron beam and the surface phonons. If the electron energy is less than 10 eV, the surface phonon wavevector is less than 0.1 $Å^{-1}$ and the dipole EELS detects phonons of long wavelength, very close to the Γ point of the surface Brillouin zone. The basic theory of scattering of low-energy electrons from surface vibrations has been formulated by EVANS and MILLS [7] in a quantum mechanical framework and by LUCAS and SUNJIC [8] from a classical point of view.

The long-range part of the potential produced by a regular distribution of dipoles placed on the surface of a medium of dielectric constant ε is given by:

$$\phi(\underline{x})=-\sum_{\ell}\underline{P}_{\ell}\cdot\underline{E}_{\ell}(\underline{x}). \tag{1}$$

Here $\underline{P}_{\ell}$ is the dipole moment at the site $\underline{R}_{\ell}$ and $\underline{E}_{\ell}$ is the electric field produced by the dipole acting on the electron in the position $\underline{x}$, as shown in Fig.1. If we consider the electric dipole $\underline{P}(\underline{Q})$ produced by the excitations of a surface optical phonon of wavevector $\underline{Q}$ we can write:

$$\underline{P}_{\ell}=\underline{P}(\underline{Q})\exp(i\underline{Q}\cdot\underline{R}_{\ell}). \tag{2}$$

By using the image charge method [7,8] and performing the summation over ℓ in (1) the interaction potential takes the following form:

$$\phi(\underline{x})=\frac{4\pi n_o\varepsilon}{1+\varepsilon}\exp(i\underline{Q}\cdot\underline{x}-Qz)\left[\underline{P}(\underline{Q})\cdot\hat{n}-\frac{i\underline{Q}\cdot\underline{P}(\underline{Q})}{\varepsilon Q}\right] \tag{3}$$

where n_o is the surface density of dipoles and $\hat{n}$ is the unit vector normal to the surface. For metals $|\varepsilon|>>1$ so that only the first term of (3) is effective. By solving the time-dependent Schroedinger equation outside the crystal, one obtains the scattering cross-section. The x-axis is assumed to be normal to the surface.

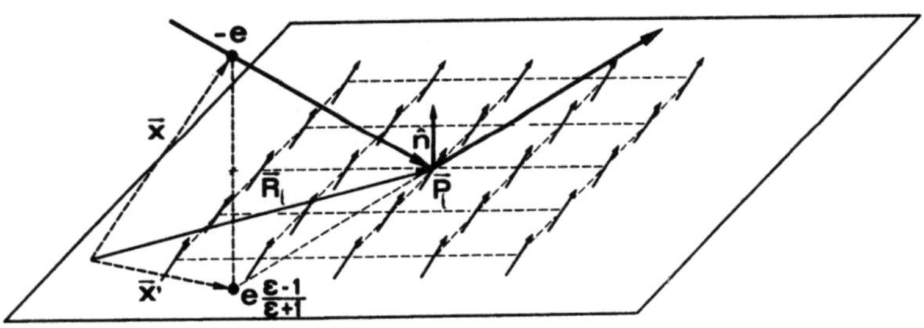

Fig.1. Geometry of the electron dipole scattering. The image charge is also shown

The cross-section is proportional to:

$$\frac{d\sigma}{d\Omega}\propto\langle|\underline{P}(\underline{Q})\cdot\hat{n}|^2\rangle. \tag{4}$$

The brackets mean thermal average. The dipole moment at the site $\underline{R}_{\ell}$ is defined by:

$$P_{\ell\alpha}=\sum_{\beta}e_{\alpha\beta}(\ell_z)u_{\beta}(\ell) \tag{5}$$

where $e_{\alpha\beta}(\ell_z)$ is the dipole moment effective-charge tensor and $u_{\alpha}(\ell)$ is the displacement operator given by:

$$u_{\alpha}(\ell)=\sum_{n}\frac{w_{\alpha}(\underline{Q},\ell_z,\omega_n)}{(2m_{\ell_z}\omega_n N_s)^{\frac{1}{2}}}\exp(i\underline{Q}\cdot\underline{R}_{\ell})(a_n+a_n^{\dagger}). \tag{6}$$

$\underline{w}(\underline{Q},\ell_z,\omega_n)$ is the phonon polarization vector on the ℓ_z-plane, N_s is the number of surface unit cells and m_{ℓ_z} is the mass of the atoms in the ℓ_z- plane. The annihilation and creation operators are indicated by a_n and $a_n^{\dagger}$. By inserting (6) and (5) in (2), the explicit expression of $\underline{P}(\underline{Q}).\hat{n}$ can be derived. Finally, by retaining only the one-phonon processes, we obtain the following expression:

$$\frac{d\sigma}{d\Omega} \propto \frac{N(\omega)}{\omega} \left| \sum_{\ell_z} \sum_{\substack{\text{unit}\\ \text{cell}}} \frac{e_{zz}(\ell_z) w_z(\underline{Q},\ell_z,\omega)}{(m_{\ell_z})^{\frac{1}{2}}} \right|^2 \tag{7}$$

where $N(\omega)$ is the Bose factor. For a metallic surface with an adsorbed layer, one can assume that the dipole is entirely due to the charges of the two outermost layers (ℓ_z=0,1) and that $e_{zz}(\ell_z=0)=-e_{zz}(\ell_z=1)$. Within this assumption and in the high-temperature limit we have the simplified expression:

$$\frac{d\sigma}{d\Omega} \propto \frac{k_B T}{\omega^2} \left| \sum_{\substack{\text{unit}\\ \text{cell}}} \left[\frac{w_z(\underline{Q},0,\omega)}{(m_0)^{\frac{1}{2}}} - \frac{w_z(\underline{Q},1,\omega)}{(m_1)^{\frac{1}{2}}} \right] e_{zz}(\ell_z=0) \right|^2 \tag{8}$$

This expression relates in a simple way the polarization vectors of the surface phonons to the EELS spectra. Three conditions must be satisfied in order to have a non-vanishing cross-section:

(1) The (zz) component of the charge tensor must be different from zero. Clearly, the clean surfaces of ionic crystal may be dipole active. The surfaces of covalent semiconductors are dipole active whenever the low symmetry of the surface, compared with the bulk, causes an appreciable charge-transfer between the two outermost layers. This is not the case for metal surfaces. However, as we will see later in a few examples, adsorbed molecules can provide a nonvanishing effective charge-transfer.

(2) Only modes with a nonvanishing displacement component normal to the surface can be detected. Actually this is an approximated result, valid in the limit of large dielectric constant [7]. This selection rule, which is similar to that of infrared spectroscopy, limits the number of surface modes which can be detected in small angle EELS.

(3) The previous conditions are satisfied but, for symmetry reasons, the sum over the surface unit cell in (7) or (8) vanishes. Although this selection rule does not allow the detection of some substrate-adsorbate modes, it can be useful for determining the symmetry of the adsorption sites and the structure of the submonolayers of adsorbed species [9]. Many examples exist in the literature of high resolution EELS spectra taken from gas coverd metal surfaces. Combinations of substrates such as W, Ru, Ni, Pt and adsorbates as O, H, N, CO, NO have been widely studied.

In these short notes, we focus our attention on a particular class of measurements, containing information not only on the vibrations of the adsorbed atoms against the substrate, but also on the surface phonons of the substrate itself. The basic idea is that an ordered submonolayer of adsorbed atoms gives the surface a new Brillouin zone (BZ), smaller than the BZ of the clean ideal surface. In this way, high symmetry points at the zone boundary of the clean surface become equivalent to the Γ point. It is therefore possible that acoustic surface phonons of the substrate couple with the adsorbate motion, giving rise to dipole-active long wavelength vibrations. Notice that the substrate-adsorbate interaction is never negligible, so that a certain frequency-shift of the substrate surface modes is expected. We consider now in some detail the (111) and (001) surfaces of nickel.

2.1. Dipole EELS in Nickel (111)

This surface has been studied by ANDERSON [4] and by IBACH and BRUCHMANN [5]. The surface adsorption geometry for two different overlayer of oxygen is shown in Fig.2. The two-dimensional BZ for the clean and covered surfaces are shown in Fig.3.

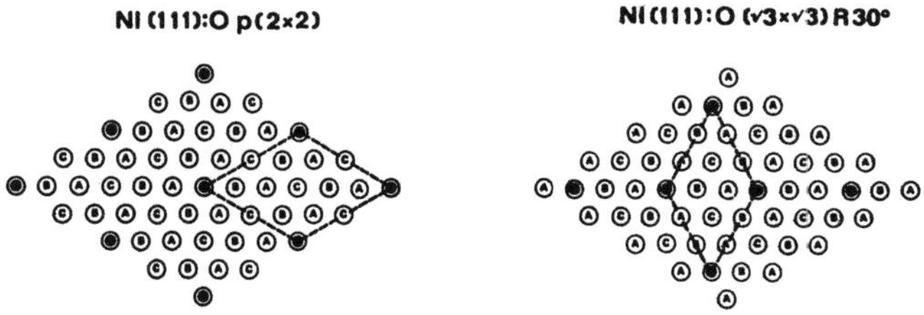

Fig.2. Geometry of Ni(111) with two different overlayers of oxygen. Black dots represent the oxygen atoms. The atoms A refer to the surface of the substrate, atoms B to the first layer underneath and atoms C to the second layer

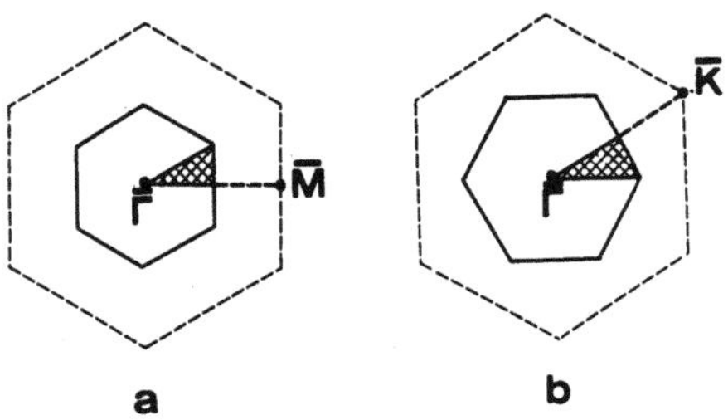

Fig.3. Two-dimensional BZ for: (a) p(2x2) and (b) ($\sqrt{3}$x$\sqrt{3}$)R30° overlayers. The two-dimensional BZ of the ideal (111) surface is also shown (dashed lines)

On the (111) surface a p(2x2) overlayer folds the M point of the BZ into the Γ point. The wavefront of the Rayleigh wave (RW) of nickel in M defines a plane normal to the surface which is not a symmetry plane. In this way, the sum in (8) over the atoms of the surface unit cell does not vanish. The RW of the substrate couples with the oxygen atoms and becomes dipole active. Experimental [5] and calculated [10] EELS spectra are shown in Fig.4. The experimental spectra have a resolution high enough to show a spectral line at 16.7 meV in the p(2x2) surface. This line is assigned to the RW of nickel. The spectrum of the ($\sqrt{3}$x$\sqrt{3}$)R30° surface does not show the analogous spectral line. In fact, in this case the wavefront of the RW (folded in Γ from the K point) defines a symmetry plane.

As a consequence, pairs of Ni atoms move in a direction perpendicular to the surface with a π phase-shift and the sum over the unit cell in (8) vanishes for symmetry reasons. The spectra of Fig.4 show other two structures: modes of symmetry A_1 and modes of symmetry A_2. These modes can be related to the normal

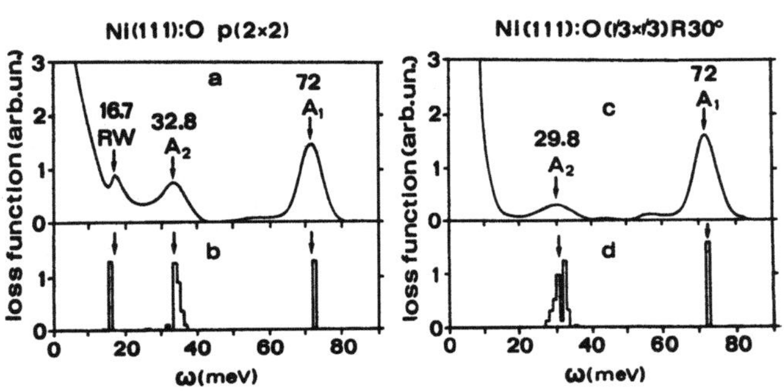

Fig.4. Experimental (a) and calculated (b) energy-loss spectra for Ni(111) with two different oxygen overlayers. After BORTOLANI et al. [10].

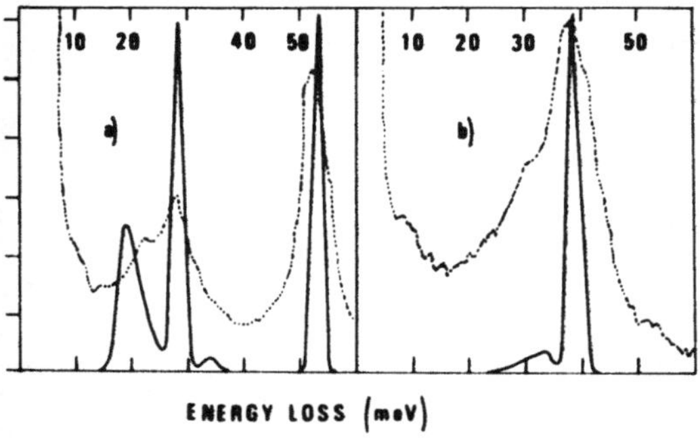

Fig.5 EELS spectra for: (a) the p(2x2) overlayer of oxygen and (b) for the c(2x2) overlayer. Theory [11]: continuous lines. Experiment [2]: dotted lines.

vibrations of the "molecule" Ni_3O, with a vertical motion of the oxygen atom with respect to the substrate. The low-frequency modes A_2 are degenerate with the Ni phonons, so that their position depends on the particular overlayer considered.

2.2. Dipole EELS in Nickel (001)

Oxygen overlayers on the Ni (001) surface have been experimentally investigated by LEHWALD and IBACH [2]. The experimental spectra are shown in Fig.5 and are compared with the calculations of BORTOLANI et al. [11]. It is interesting to examine these spectra in some detail in order to clarify the importance of the adsorbate-substrate interaction in this kind of measurements.

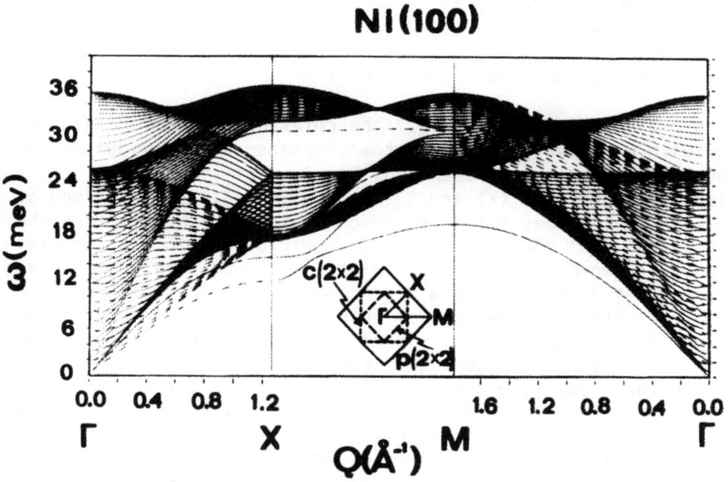

Fig.6 Calculated surface phonon frequencies for clean (001)Ni. The two dimensional BZ is also shown (clean surface: continuous line; overlayers: dashed lines)

The surface phonons of clean ideal Ni (001), calculated within a force constant parametrization of the bulk dynamics, are shown in Fig.6 for a slab of 65 atomic planes. The Ni-O unknown force constants have been obtained by fitting the position of the high-frequency (52 meV) experimental peak. In this way the lineshape of the spectra below the Ni cutoff (36meV) is well reproduced, as it is shown in Fig.5.

Let us describe the symmetry of the surface modes of (001) nickel, shown in Fig.6. S_1(M) and S_4(X) are vertically polarized modes with two Ni atoms moving normal to the surface (Rayleigh wave). S_1(X) represents a planar mode with a couple of atoms moving inwards and the other couple moving outwards. S_6(X) is

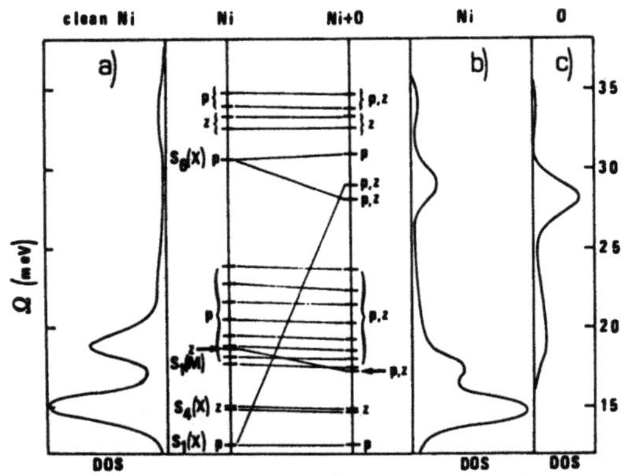

Fig.7 DOS for vibrations normal to the surface in the Γ point of the p(2x2) surface BZ for the free Ni surface (a), for the Ni surface when oxygen has been adsorbed (b) and for oxygen (c). The energy position of the phonons which give relevant contributions to the DOS is also shown, together with the shift and the splitting of the levels due to oxygen adsorption. The planar character of the modes is indicated by p and the surface-normal character by z. After BORTOLANI et al.[11]

still a planar mode but all the four Ni atoms move in-phase either inwards or outwards. Notice that the presence of the p(2x2) oxygen overlayer gives rise to a reduced surface BZ with the original X and M points equivalent to the Γ point, while the c(2x2) overlayer folds only the M point in Γ.

Consider first the p(2x2) overlayer. In Fig.7 the relevant modes of clean Ni and of the Ni surface with adsorbed oxygen are shown, together with the density of states of surface phonons (DOS). It can be seen in Fig.7 that the modes of z character on the clean surface, as $S_4(X)$ and $S_1(M)$, are only slightly shifted by the presence of the oxygen, which does not move along the z-axis. Furthermore, the Ni atoms move out of phase so that there is no dipole contribution to the computed cross-section. The double degenerate mode $S_1(X)$ splits in a mode with unperturbed polarization and in a mode at 29.1 meV with a z component (see Fig.7b). For both modes, the four atoms in the unit cell move out of phase and there is no dipole. We have verified that all the oxygen vertical vibrations, which are dipole active, originate from the coupling between oxygen and substrate modes of the same symmetry as $S_6(X)$. In the X point, apart from the localized state $S_6(X)$, there are other modes with this symmetry lying in the continuum of bulk phonons in the ranges 18-24 and 34-35 meV (see Fig.6). These modes are double degenerate and are split by the interaction with the oxygen. In the low-energy mode the four Ni atoms gain an in-phase z-motion and the oxygen strongly vibrates along z.

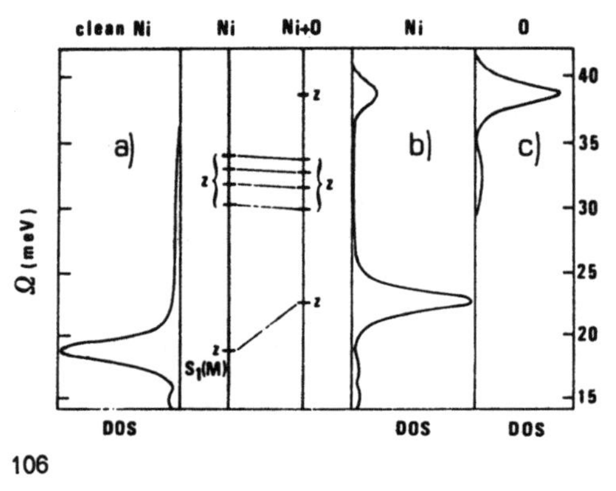

Fig.8 As in Fig.7 but for the c(2x2) overlayer. After BORTOLANI et al. [11].

The DOS results for the c(2x2) overlayer are shown in Fig.8 and the computed EELS spectrum is compared with the experimental one in Fig.5b. The $S_1(M)$ mode is not dipole active because the two Ni atoms in the unit cell move vertically out of phase. There is a dipole active strong mode at 38.6 meV and a weaker contribution in the range 30-34 meV, which accounts for the asymmetry of the experimental structure of Fig.5b. This continuous band is due to modes of the Γ point, which allow the oxygen to vibrate perpendicularly to the surface. In these modes the two Ni atoms of the unit cell move in phase.

3. Impact Scattering

The impact scattering of low-energy electrons from crystalline surface was first considerd from the theoretical point of view by ROUNDY and MILLS [12]. The generalization of the theory to the scattering of electrons from adsorbate vibrations has been developed more recently [13,14]. The general idea is that the scattering amplitude is proportional to the gradient of the electron-surface interaction potential. The reader will find some similarity to the scattering of atoms discussed in another contribution to this volume [15]. The main difference with respect to dipole EELS is that the longitudinal component of the displacements is also effective in producing scattering. In addition, it is also possible to detect surface acoustic phonons on clean substrates without the need of and adsorbate to generate a long-range dipole field in vacuum. Impact scattering is therefore a very powerful technique. The first experimental evidence of impact scattering in vibratioal EELS was observed by HO et al. [16] with an overlayer of H on W (001). This result is shown in Fig.9.

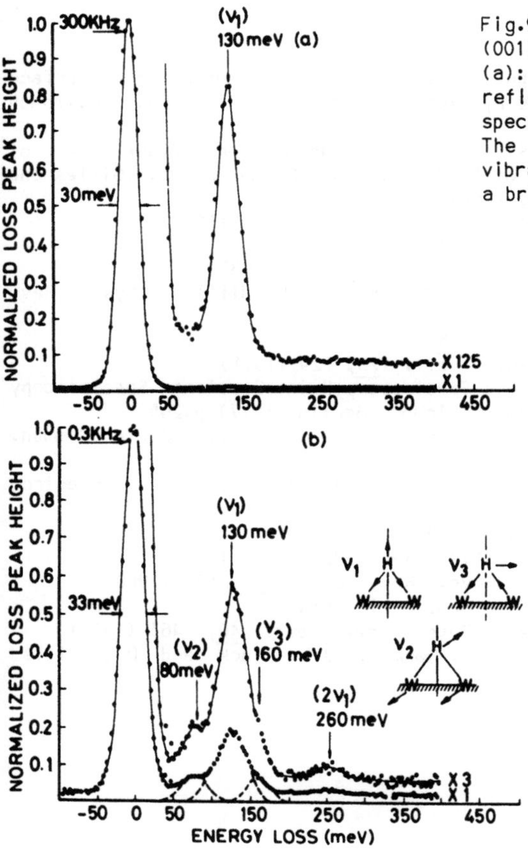

Fig.9 EELS measured spectra for H on W (001). Impact energy E_i=9.65 eV. Part (a): spectrum taken in specular reflection. Part (b): +17° off the specular direction toward the surface. The inset shows the fundamental vibrational modes of H corresponding to a bridge site. After HO et al. [16].

The only H-substrate stretching mode which gives rise to a dipole moment normal to the surface is the mode ν_1, as can be seen in the inset of Fig.9. Infact ν_1 is detected in specular reflection (Fig.9a). At angles away from the specular reflection direction, the modes ν_2 and ν_3 are resolved (Fig.9b). Because its motion is completely in the plane of the surface, ν_2 was forbidden in dipole coupling, whereas ν_3 was forbidden by the condition (3) mentioned in section 2. In fact, in the latter case the W atoms of the surface cell move with a π phase shift. Another interesting result of the paper by HO et al. [16] is the resonant-like behaviour of the amplitude of the impact scattering peaks versus the electron energy E_i. This has been justified by te calculations of TONG et al. [14], who found an oscillatory pattern of the cross-section in the energy-range 10-100 eV. For this kind of measurements it is therefore very important to choose the appropriate impact energy, which depends on the frequency and momentum of the surface phonon that has to be detected.

A great advance in the field has been recently achieved [6]. A proper choice of E_i (180-322 eV) allowed these authors to measure the dispersion relations of the RW along the symmetry direction X. The experimentalists of the Jülich group have also successfully performed impact measurements for the c(2x2) overlayer of oxygen on Ni (001) [17]. These experiments extend our knowledge of the surface vibrations of these systems to the entire BZ, without the limitations of the dipole scattering mechanism. In particular has been determined the frequency of the parallel vibration Ni-O, which is not excited in specular reflection. The new information is in favour of a distance between the Ni surface and the oxygen plane of about 0.9 Å, as suggested by previous calculations [11]. These example show the great potential of impact scattering EELS as a technique for surface analysis and surface strucutural determination.

References

1. D.Roy and J.D. Carette: "Design of Electron Spectrometers for Surface Analysis" in Electron Spectroscopy for Surface Analysis, Ed. H. Ibach (Springer, Berlin, 1977) p.13
2. S. Lehwald and H. Ibach: "EELS of Surface Phonons and Vibrations of Adsorbates Species" in Vibrations at Surfaces, Eds. R.Caudano, J.M.Gilles and A.A.Lucas (Plenum Press, New York, 1982) p.137
3. F.M. Propst and T.C. Piper: J.Vac. Sci. Technol. 4, 53 (1967)
4. S. Anderson: Surf. Sci. 79, 385 (1979)
5. H. Ibach and D. Bruchmann: Phys. Rev. Lett. 44, 36 (1980)
6. S. Lehwald, J.M. Szeftel, H. Ibach, T.S. Rahman and D.L. Mills: Phys. Rev. Lett. 50, 518 (1983)
7. E. Evans and D.L. Mills: Phys. Rev. B5, 4126 (1972)
8. A.A. Lucas and M. Sunjic, Phys. Rev. Lett. 26, 229 (1971)
9. H. Froitzheim: "Electron Energy Loss Spectroscopy" in Electron Spectroscopy for Surface Analysis, Ed. H. Ibach (Springer, Berlin, 1977) p.205
10. V. Bortolani, A. Franchini, F. Nizzoli and G. Santoro: Solid State Commun. 41, 369 (1982)
11. V. Bortolani, A. Franchini, F. Nizzoli and G. Santoro: J. Electron Spectrosc. Relat. Phenom. 29, 219 (1983)
12. V. Roundy and D.L. Mills: Phys. Rev. B5, 1347 (1972)
13. C.H. Li, S.Y. Tong and D.L. Mills: Phys. Rev. B21, 3057 (1980)
14. S.Y. Tong, C.H. Li and D.L. Mills: Phys. Rev. Lett. 44, 407 (1980)
15. V. Bortolani, A. Franchini and G. Santoro: this volume
16. W. Ho, R.F. Willis and E.W. Plummer: Phys. Rev. Lett. 40, 1463 (1978)
17. J.M. Szeftel, S. Lehwald, H. Ibach, T.S. Rahman, J.E. Black and D.L. Mills: Phys. Rev. Lett. 51, 268 (1983)

Electron Energy-Loss Spectroscopy of Surfaces and Absorbates

H. Ibach

Institut für Grenzflächenforschung und Vakuumphysik, Kernforschungsanlage Jülich, Postfach 1913, D-5170 Jülich, Fed. Rep. of Germany

1. Introduction

The physics of surfaces and interfaces relates to many areas of modern technology, such as heterogeneous catalysis, electrochemistry, tribology, semiconductor devices, solar devices etc. It is from those applications where many and important questions have been posed to the surface physicist and surface chemist. In addition, there is great interest in exploring properties of the condensed matter at the boundary of a solid, where one is dealing with phenomena between two and three dimensions. A summary of issues which have been addressed in surface science is given in table 1.

Table 1

Structure	Elementary excitations	Chemical Activity	Transport Phenomena
Clean Surfaces	Plasmons	Chemical bond of molecules to surfaces	Surface conductivity
Ordered Adlayers	Magnons	Decomposition and synthesis reactions	Surface diffusion
Ordered, stepped surfaces	Excitons	Metastable species	Energy transport
Disorder	Electron-hole pairs	Reaction path	Relaxation of excited states
Phase-transitions in two dimensions	2D-band structure		
	Surface phonons		

Many important questions also deal with the solid-liquid interface. Due to the general lack of spectroscopic and structural sensitive techniques, our present understanding of the solid-liquid interface is much behind the understanding of clean surfaces and the surface gas interface. There is, in fact, a fundamental reason for that: in order to be selectively sensitive to surface properties, one has to use techniques which in one or another way employ particles having a high scattering cross-section with matter. This, in most cases, excludes the presence of a liquid above the surface or studies of surfaces in high ambient gas pressure.

In the past 20 years, physicists have been rather ingenious in inventing new methods for the analysis of surfaces. Few of them never made their way out of the laboratory of the inventor. Many of them are still in use for one or another special purpose. Some techniques proved so useful that they more or less have become standard to any laboratory dealing with surfaces. In this particular lecture we will discuss **vibration spectroscopy** [1] and in particular electron energy-loss spectroscopy [2] for the purpose of studying surface vibrations. This techniques has been one of the more successful ones in surfaces, and has provided us with data that relate to many of the issues raised in table 1. The lecture will be organized in four more sections. In section 2 we will briefly describe the **principles of electron energy-loss spectroscopy** and the scattering mechanisms of electrons from surfaces. We then continue by looking into the application of the technique to **oxidation processes**, and we shall compare the oxidation of three materials which behave rather differently, namely **Pt**, **Ni**, and **Si**. In section 4 we present some data which relate to the **bonding of hydrocarbons** on transition metal surfaces.

Unlike photons, electrons also carry sufficient momentum to study the dispersion of the surface phonons, provided that the electron energy is not too small. In section 5 we describe the use of electron energy-loss spectroscopy for the study of the **dispersion of surface phonons** of clean metals and ordered adsorbed layers.

2. Electron Energy-Loss Spectroscopy

Of all the surface vibration spectroscopies which are known up to now, electron energy-loss spectroscopy has the greatest versatility. It can be used to study vibrations above approximately 100 cm^{-1}. Its sensitivity is so high that usually 10^{-2} of a monolayer of adsorbed species can be detected, in favourable cases even 10^{-3} of a monolayer. The main disadvantage of electron energy-loss spectroscopy compared to other vibration spectroscopies such as infrared reflection absorption spectroscopy or inelastic scattering of helium atoms is that the resolution is limited to about 30 cm^{-1} if the scattering cross-section of the investigated species is high. For weak scatterers a resolution of 60 cm^{-1} is more typical.

An advantage of electron energy-loss spectroscopy is that different scattering mechanisms can be employed. By scattering mechanisms we mean that in the theory of electron surface scattering limiting cases can be isolated which can be treated separately. These cases refer to different scattering parameters, such as energy and momentum parallel to the surface, and can therefore also be materialized experimentally. The traditionally most important scattering mechanism is named **dipole scattering** [2-4]. In dipole scattering, one considers the Coulomb interaction between the point charge of the electron and the dipole term in the expansion of the potential into the normal coordinates of the motions of surface atoms and adsorbed species on the surface. By virtue of the long-range nature of the dipole field, the inelastic scattering process has little effect on the trajectory of the scattered electron. Consequently, dipole interaction is observed in scattering angles near the specular reflected beam or other diffracted beams. There, the differential cross-section can be rather high. Since, for technical reasons, high-resolution electron spectrometers feature small acceptance angles, dipole scattering with its high differential cross-section is most easily observed. Dipole scattering is also important because the electrostatic screening provided by the substrate, in particular if it is a metal substrate, nearly eliminates the field of dipoles oriented parallel to the surface. This gives rise to a selection rule. The best formulation for this selection rule is within the language of group theory and it reads: in dipole scattering all those modes contribute to the vibration spectrum which belong to the **total symmetric representation** of the surface point group of the adsorbed species. As to whether and how much the symmetry of the surface has to be included in determining the appropriate point group is a matter of more subtle considerations which go beyond this summary. For a full discussion the reader is referred to chapter 4 of reference 1. A few examples of dipole active modes are shown in fig. 1. The advantage of the group theoretical formulation is obvious when the vibration of an adsorbed O_2 molecule with its axis aligned with the surface is considered. Neither is the O_2 molecule dipole active in the gas phase nor is the vibrational motion perpendicular to the surface. Yet the stretching vibration of adsorbed O_2 carries a dipole moment perpendicular to the surface. In physical terms, the dipole activity can be understood by a fluctuation of the π charges normal to the surface as a function of the position of the oxygen nuclei.

The local interaction of the electron with the valence and core electrons of the surface atoms (including adsorbed species) is usually referred to as **impact scattering**. The scattering potential is the same as used in low-energy electron

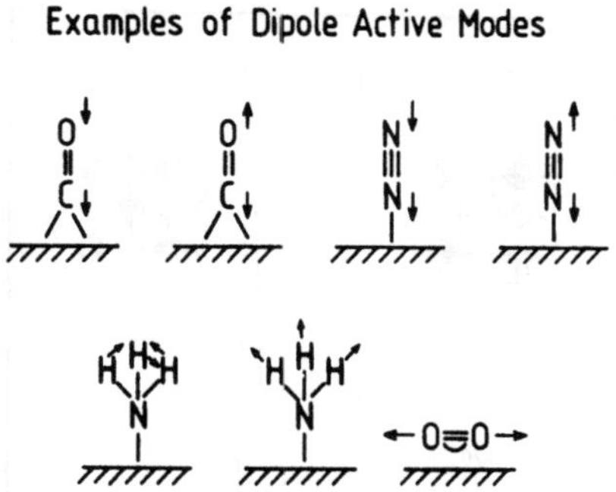

Fig. 1. Modes which bear a dipole moment perpendicular to the surface. The example of O_2 is particularly noteworthy since the dipole moment arises solely from the interaction with the surface

diffraction (LEED). For a formal description of the scattering mechanism, the dynamic scattering amplitude needs to be expanded into the displacements of the atom coordinates around their equilibrium positions. A formal theory has been developed [5]. Recently, theory and experimental intensity data have been compared for electron scattering from phonons, and excellent agreement has been found [6]. Impact scattering also features certain selection rules: When the scattering plane is aligned along a symmetry plane, modes which are odd with respect to the symmetry plane are not visible. The significance of impact scattering is that one is not restricted in energy and momentum exchange with the surface. Therefore, the full dispersion curve of surface phonons and phonons within adsorbed layers can be determined. For sufficiently dilute layers of adsorbed species, lateral interaction effects and therefore also dispersion, is negligible. Still impact scattering is useful in order to determine the full spectrum of normal modes of the species. In particular, one can exploit the effect that the normal modes have a different intensity for different scattering parameters. Therefore, by varying the scattering parameters in the spectra, one can let appear the full spectrum even when it is rather complex and comprises vibrational features which are too close in their frequency to be separated in a single scan.

Finally **resonance scattering** should be mentioned. In certain situations the electron and an adsorbed molecule can form a compound state with a life time large enough to significantly enhance the scattering cross-section for a particular energy. These resonances are well known in electron molecule scattering and have also been observed for surfaces [7,8]. For discussion of the selection rule applicable there see also [8,10].

3. Application to oxidation processes

It is well known that the chemical activity of materials with respect to oxygen can be very different. This is illustrated by comparing the materials **Pt, Ni**, and **Si**. Of the three, Pt is the least active material and does not form a stable oxide [11]. Nevertheless, oxygen adsorbs on platinum surfaces in a molecular state at low temperatures [12]. The vibration spectra of two molecular forms of oxygen on platinum, as well as of a very similar state on silver surfaces [14], indicate that molecular oxygen is π-bonded through a nearly complete charge-transfer of the metal electrons into the $1\pi_g^*$ orbitals of oxygen (Fig. 2). A similar π-bonded state of molecular oxygen has not been found on nickel, although it may exist as a metastable state. The lower stability of a π-bonded complex with respect to dissociation on nickel surfaces may be understood if one takes the lower ionization potentials of the d-electrons in nickel into account. This lowers the energy of the atomic state of adsorbed oxygen, and therefore also the activation barrier towards dissociation. Platinum also has the capability of dissociating oxygen near room temperature [13] which is the basic reason for its catalytic activity with respect to oxidation reactions.

Typical vibration spectra (dipole scattering) of ordered overlayers of atomic oxygen on nickel (100) are presented in Fig. 3. Nickel surfaces covered with oxy-

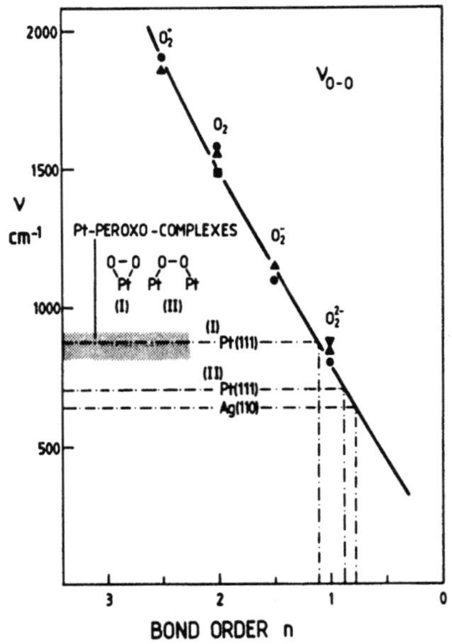

Fig. 2. Frequency of the O_2 stretching vibration vs. band order. For adsorbed O_2 the bond order is reduced because the antibonding π*-orbitals are filled

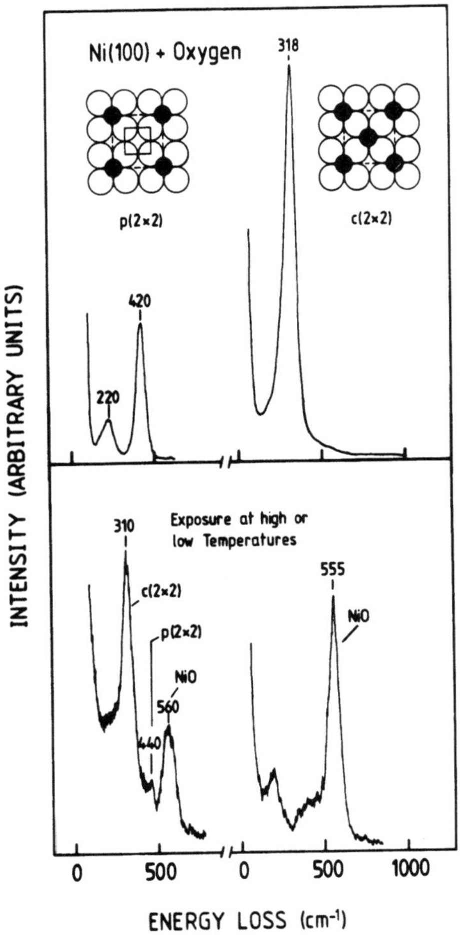

Fig. 3. Energy-loss spectra of various forms of oxygen on Ni(100) (dipole scattering). The upper two spectra correspond to the two ordered overlayers of oxygen with no trace of NiO. NiO clusters are formed either by prolonged exposure to oxygen at higher temperatures (~ 400-500 K) or by exposure with modest amounts of oxygen (~ 50 L) at 100 K followed by annealing to 400-500 K

gen overlayers are metastable with respect to oxidation. The oxidation process, which appears to nucleate at defects of surfaces, can however also be initiated on the Ni(100) surface by exposing the surface at low temperatures to oxygen, followed by subsequent annealing. The exposure at low temperatures leads to a high surface concentration of oxygen, which tends to loosen the strength of the nickel-nickel bonds as we shall see later. High surface concentration of oxygen can, of course, also be achieved at higher temperatures, provided the oxygen pressures are high enough.

Silicon has an even greater activity for oxidation. A metastable molecular state of oxygen can be isolated at low temperatures [15]. Once the oxygen is dissociated, it seems to prefer to move into the silicon-silicon bonds immediately [16].

4. Bonding of hydrocarbons on transition metals

One of the successes of vibration spectroscopy on surfaces was to reveal the nature of the bonding of unsaturated hydrocarbons on transition metal surfaces. Two limiting configurations may be considered for the bonding of an unsaturated hydrocarbons to a metal: **π-bonding** and **σ-bonding.** In π-bonding the electronic structure and the geometry of the adsorbed molecule is only weakly perturbed, and the interaction with the surface occurs via π-charge-transfer to empty sp or d-states of the metal. Stronger bonding to the surface involves formation of carbon-metal σ-bonds. This leads to a weakening of the CC-bonds within the molecule and to CCH bond angles which are characteristic of sp^3 hybrids. The corresponding changes in the electronic and geometric structure of the molecule are reflected in the vibration frequencies, in particular in the CH stretching frequencies and the CC stretching frequency. A complete vibrational analysis, including the study of deuterated molecules, has to be performed in order to determine the ν_{CC} mode. In achieving the mode assignment, the wide spectral range of EELS and the possibility of employing different scattering mechanisms has been crucial. Table 2 summarizes a few results which have been obtained with the adsorption of acetylene (C_2H_2) on various transition metal surfaces. One sees that, compared to the free molecule, both the ν_{CH} and the ν_{CC} stretching frequencies have been shifted downwards. Using the bond order vs. frequency correlations similar to the one used in Fig. 2 one finds that the bond order of acetylene on these transition metal surfaces is reduced to about 1.5 - 2. Very similar behaviour is found with other unsaturated hydrocarbons such as ethylene and propylene. No exception has been reported so far provided that the substrate is a transition metal. In the case of transition metals more to the left in the periodic table, immediate decomposition can also occur when an unsaturated hydrocarbon contacts the surface. Of course, these more reactive surfaces are reactive only as long as they are clean. Once a surface is covered with fragments of the molecules to which it had been exposed, or if the surface is covered with other adsorbates which act as poisons, weaker forms of bonding can materialize. For the noble sp-metals [23] π-bonded forms of unsaturated hydrocarbons have also been reported.

Table 2. CC and CH stretching frequencies of C_2H_2 adsorbed on various transition metals

Substrate	ν_{CC}	ν_{CH}	Ref.
Free molecule	1974	3374 3289	17
Pd(111)	1400	2990	18
Pt(111)	1310	3010	19
Rh(111)		2985 3085	20
Fe(110)	1240	2940	21
Ni(111)	1215	2915	22

5. Surface phonons

So far we have been discussing the **localized modes** of adsorbed atoms and molecules and have disregarded the **coupling to the phonon spectrum** of the substrate as well as **lateral interactions** between the molecules. For molecular vibrations this is a reasonably good approximation, since lateral interactions are relatively weak and the eigenfrequencies of chemically interesting small molecules are usually well above the phonon frequencies. For those vibrations which arise from the hindered translations and rotations of the molecule or the adsorbed atom ("molecule-substrate modes"), coupling to the phonons and lateral interactions cannot be neglected. When the layer of adsorbed molecules or atoms is disordered, the vibration spectrum resembles that of an amorphous solid; that is, one has a broad continuous band instead of sharp vibrational frequencies. The study of such systems is as difficult as studies of the amorphous solid. A quantitative approach to the problem can, however, be achieved when **ordered overlayers of adsorbates** are investigated. Fortunately, nature has provided us with ordered overlayers for many adsorbate systems. Then vibration amplitudes assume the general form

$$\mathbf{u_n}(\ell_z,\mathbf{q}_\parallel,\omega) = \mathbf{e}(\ell_z,\mathbf{q}_\parallel)\ e^{-i(\omega t-\mathbf{q}_\parallel\cdot\mathbf{r}_{\parallel\mathbf{n}})}$$

with $\mathbf{q}_\parallel$ the wave vector, $\mathbf{r}_{\parallel\mathbf{n}}$ the parallel component of the vector pointing to the equilibrium postion of an atom n, and ℓ_z the index of the layer. In inelastic scattering experiments, one then has wave vector conservation as well as energy conservation and complete dispersion curves $\omega(\mathbf{q}_\parallel)$ can be mapped out. The particles to be used in such experiments must have a large cross-section with surface atoms. So far, helium atoms and electrons have been used. The two techniques largely complement each other, in that helium scattering features a very high energy resolution, but is so far limited to the lower frequency range, while electron scattering has a significantly lower resolution, yet there is no upper limit in the frequencies which can be probed. Electron spectroscopy is therefore particularly suitable for the investigation of the dispersion curves of ordered overlayers of chemisorbed atoms and molecules. Substrate surface phonons can also be investigated, provided that their frequencies are not too small.

A schematic plot of the vibration spectrum of a surface with an ordered overlayer of atoms is shown in Fig. 4. On this plot of ω vs. $\mathbf{q}_\parallel$ all genuine surface modes contribute a discrete line, whereas bulk modes give rise to a band because of the arbitrary vertical component of **q**. The substrate usually has one or more acoustical surface modes (generalized Rayleigh waves) and other surface modes which can occur in gaps where no bulk modes exist. In general, surface modes do not exist within the bulk band. Occasionally, one has a situation where the surface mode belongs to a different representation than the bulk modes at the same frequency. Then, the two excitations do not interfere and the surface mode can retain its localized character even within the bulk band. Such a situation is depicted in Fig. 4 and applies to the [110] direction of (100) surfaces of fcc and bcc lattices. Ordered adsorbed overlayers contribute three optical branches per adsorbed atom in the surface unit cell, provided that the branches are above the bulk spectrum. When the interaction with the substrate becomes softer, or for a large mass of the adsorbed atom, part of, or the entire dispersion branch may submerge into the bulk continuum and then cease to exist as a localized mode.

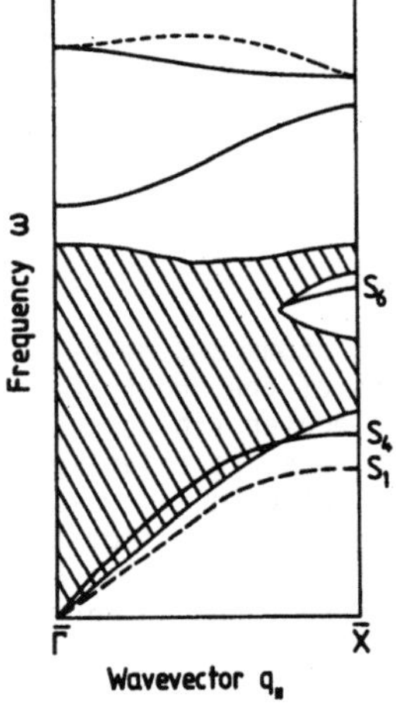

Fig. 4. Surface phonon dispersion curves for a c(2x2) overlayer on a (100) fcc or bcc surface along the [110]-direction (schematic). The upper three branches arise from the c(2x2) overlayer, the lower branches are substrate surface phonons. Modes polarized perpendicular to the sagital plane [(100)-plane] are shown as dashed lines. At $\bar{\Gamma}$ and $\bar{X}$ the overlayer modes are polarized parallel or perpendicular to the surface. The parallel modes are degenerate at $\bar{\Gamma}$ and $\bar{X}$

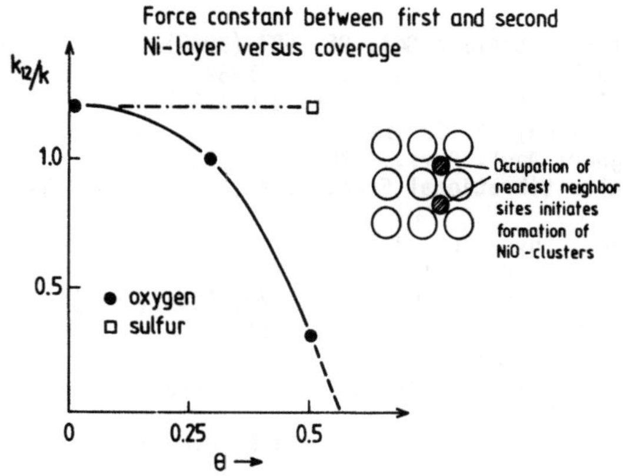

Fig. 5. Best fit coupling constants for the coupling between atoms of the first and second Ni-layer. the fit was performed to the S_4-dispersion for the p(2x2) and c(2x2) overlayers of oxygen and sulfur

Dispersion data of adsorbed atoms are available for the p(2x2) and c(2x2) oxygen layers on Ni(100) and the c(2x2) sulphur layer on the same surface [24,25, 26]. The dispersion has been calculated using the analytical Greens function method [27]. In order to match the data for the generalized Rayleigh wave S_4 for the surface covered with a c(2x2) oxygen overlayer, it was found necessary to substantially reduce the force constant which couples the Ni-surface atoms to the next layer of Ni-atoms. The best fit force constants for the first interlayer coupling vs. coverage are plotted in Fig. 5. One sees the strong tendency towards zero for coverage with oxygen, while for the c(2x2) overlayer of sulfur the coupling need not be changed [26]. It therefore seems that for oxygen coverages larger than $\Theta = 0.6$ the adsorbed phase becomes unstable with respect to oxidation. We see this as the basic reason for the production of NiO by exposing the surface at low temperatures (even with modest amounts of oxygen) followed by annealing. When the surface is exposed at low temperatures, the mobility of oxygen is so low that no ordered layer is formed. Then the local concentration of oxygen can exceed the limit indicated with Fig. 5 and nucleation of NiO clusters begins. The comparison of experimental data and theory on the adsorbate dispersion curves has also shown that the most significant contribution to the dispersion comes about through the **coupling via the substrate lattice**, while direct adsorbate interactions are less important, although they cannot be completely ignored.

Critical reading of the manuscript by T.S. Rahman is acknowledged.

References

1. A good overview over the current status of vibration spectroscopy, as well as a bibliography, is provided by C.R. Brundle and M. Morawitz **Vibrations at Surfaces** (Proceedings of the Third International Conference, Asilomar, Calif. USA, 1982) Elsevier, Amsterdam 1983
2. H. Ibach, D.L. Mills: **Electron Energy-Loss Spectroscopy and Surface Vibrations**, Academic Press, New York, 1982
3. A.A. Lucas, M. Sunjić: Progress in Surface Science 2, 75 (1972)
4. D.L. Mills: Surface Sci. **48**, 59 (1975)
5. S.Y. Tong, C.H. Li, D.L. Mills: Phys. Rev. Lett. **44**, 407 (1980); C.H. Li, S.Y. Tong, D.L. Mills: Phys. Rev. B **21**, 3057 (1980); S.Y. Tong, C.H. Li, D.L. Mills: Phys. Rev. B **24**, 806 (1981)
6. M. Rocca, H. Ibach, S. Lehwald, B.M. Hall, X.L. Xu, S.Y. Tong: to be published
7. G.J. Schulz: Rev. Mod. Phys. **45**, 453 (1973)
8. J.E. Demuth, D.S. Schmeisser, Ph. Avouris: Phys. Rev. Lett. **47**, 1166 (1981)
9. I.C. Walker, A. Stamatovic, S.F. Wong: J. Chem. Phys. **69**, 5532 (1978)
10. H. Ibach: J. Mol. Structure **79**, 129 (1982)
11. M. Peuckert, H.P. Bonzel: Surface Sci., in press

12. J.L. Gland, B.A. Sexton, G.B. Fisher: Surface Sci. **95**, 587 (1980)
13. H. Steininger, S. Lehwald, H. Ibach: Surface Sci. **123**, 1 (1982)
14. C. Backx, C.P.M. de Groot, P. Biloen: Surface Sci. **104**, 300 (1981)
15. J.E. Demuth, Ph. Avouris: Physics Today **36**, 62 (1983)
16. H. Ibach, H.D. Bruchmann, H. Wagner: Appl. Phys. A **29**, 113 (1982)
17. T. Shimanouchi: **Tables of Molecular Vibrational Frequencies**, Vol. I, NSRDS-NBS39 (1972); see also ref. [2]
18. L.L. Kesmodel, J.A. Gates: J. Chem. Phys. **76**, 4281 (1982)
19. H. Ibach, H. Hopster, B. Sexton: Appl. Phys. **14**, 21 (1977)
20. L.H. Dubois, D.G. Castner, G.A. Somorjai: J. Chem. Phys. **72**, 5234 (1980)
21. W. Erley, A.M. Barõ, H. Ibach: Surface Sci. **120**, 273 (1982)
22. S. Lehwald, H. Ibach: Surface Sci. **89**, 425 (1979)
23. C. Backx, C.P.M. de Groot, P. Biloen: Appl. Surf. Sci. **6**, 256 (1980)
24. H. Ibach: Proc. IX IVC-V ICSS, 17 (1983)
25. J.M. Szeftel, S. Lehwald, H. Ibach, T.S. Rahman, J.E. Black, D.L. Mills: Phys. Rev. Lett. **51**, 268 (1983)
26. M. Rocca, S. Lehwald, T.S. Rahman, H. Ibach: to be published
27. T.S. Rahman, J.E. Black, D.L. Mills: Phys. Rev. B **25**, 883 (1982); ibid. Phys. Rev. B **27**, 4059 (1983)

Contributions

High Resolution Inelastic He-Atom Scattering from Crystal Surfaces

U. Harten, J.P. Toennies, and Ch. Wöll

Max-Planck-Institut für Strömungsforschung, Böttingerstraße 4-8
D-3400 Göttingen, Fed. Rep. of Germany

High resolution inelastic time of flight He-atom scattering experiments have proved to be a powerful tool for measuring surface phonon dispersion relations[1] on insulators[2], semiconductors[3], and metals[4]. This is illustrated here by several examples from recent studies of noble metals.

Fig. 1 shows a schematic view of the apparatus used. By cooling the nozzle, the energy of the incident He-atoms can be varied from 70 meV (T = 300 K) down to 7 meV (T = 40 K). Incident and detection angles can be varied separately. Ten differential pumping stages are needed to keep the He-background in the detector low. The target chamber is equipped with standard devices for crystal preparation and analysis.

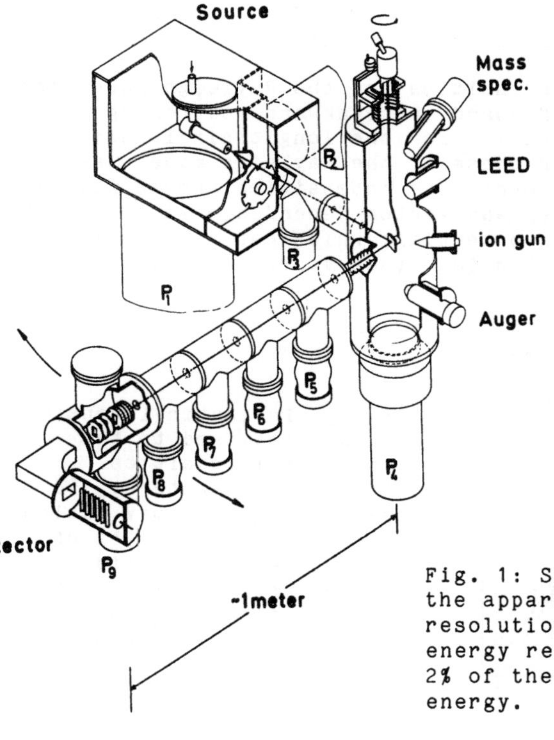

Fig. 1: Schematic view of the apparatus. The angular resolution is 0.1°. The energy resolution is about 2% of the incident beam energy.

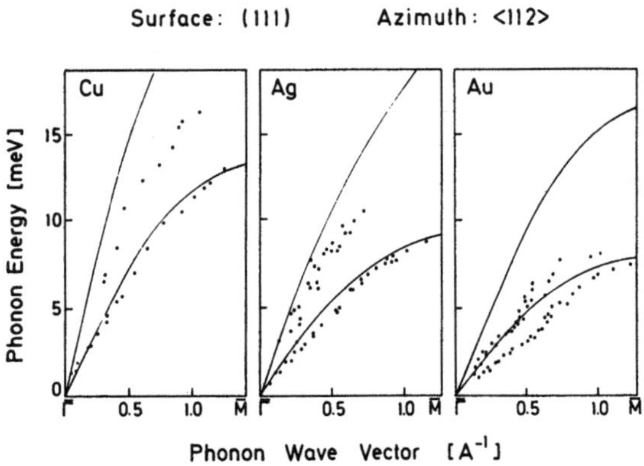

Fig.2: Measured phonon dispersion relations (points). Crystal temperatures: Cu:300 K, Ag:150 K, Au:100 K. The He beam energies ranged from 8 meV to 24 meV. Calculated curves from Armand[5].

A comparison of measured phonon dispersion relations for the (111) surface of Cu, Ag and Au is shown in Fig. 2. For rough orientation purposes we also show the calculated dispersion curves for the longitudinal edge of projected bulk bands (upper curves) and the transverse Rayleigh surface mode (lower curves). These are derived from a simple dynamical model calculation for Ag[5], which has been scaled according to the maximum bulk crystal frequencies. Above the Rayleigh mode, but still below the longitudinal edge, a second well-defined mode is observed in the measurements. In lattice dynamic calculations, this mode can be reproduced as a sharp surface resonance of longitudinal character by reducing the force constant between adjacent atoms in the outermost layer of the surface by about 50% of the bulk values (see V. Bortolani in this book). The frequencies of the longitudinal resonances are then in rough agreement with the measurements in Ag and Cu. For Au, the lowering in force constants is even greater[6].

Fig. 3 shows the intensities for the Rayleigh mode versus phonon wave vector for forward and backward scattering respectively. The intensities for the longitudinal resonance are always about one half of those of the Rayleigh mode. Taking only the Bose factor into account, the intensities should follow the dotted lines. The steeper wavevector dependence is due to the softness of the He-metal potential, which can be described by a cut-off factor in the potential[7] (see V. Bortolani).

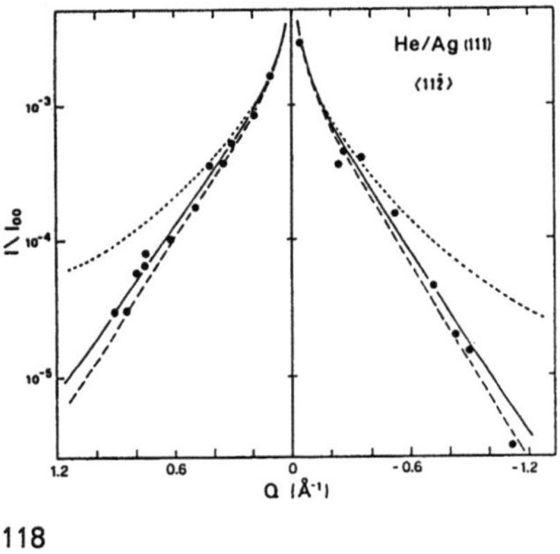

Fig.3: Inelastic intensities normalised to the specular intensities. Crystal temperature: 150 K. Beam energy: 17 meV. Calculated curves by Celli et al[6]: Broken and solid curves are for different cut-off factors. Dotted curve only includes the Bose factor.

1 J.P. Toennies, J. Vac. Sci. Technol. A 2, 1055 (1984)
2 G. Brusdeylins, R.B. Doak and J.P. Toennies, Phys. Rev. B 27, 3662 (1983)
3 G. Brusdeylins, J. Skofronick and J.P. Toennies, to be published
4 R.B. Doak, U. Harten and J.P. Toennies, Phys. Rev. Lett. 51, 578 (1983)
5 G. Armand, Solid St. Comm. 48, 81 (1983)
6 V. Bortolani, G. Santoro, U. Harten, J.P. Toennies,Ch. Wöll, to be published
7 V. Celli et al., Surf. Sci. 143, L 376 (1984)

Surface Lattice Dynamics of Ordered Overlayers on Metals

Talat S. Rahman

Cardwell Hall, Department of Physics, Kansas State University
Manhattan, KS 66506, USA

We have performed lattice dynamical calculations on several types of superstructures on the Ni(100), Ni(111) and Ru(001) surfaces, with the aim of studying the sensitivity of the electron energy-loss data on these surfaces to issues like the adsorption site symmetry [1], the adsorbate and substrate surface phonon and surface resonance dispersion [2], and the extended overlayer symmetry [3]. The model Hamiltonian consists of nearest neighbour pairwise interactions with central forces, although in cases where it is necessary, angle bending interactions have been included. The Fourier transformed Green's function for the system is constructed and the corresponding equations of motion obtained. The Green's function hierarchy thus obtained is solved by invoking an exponential ansatz, for the equation representing the motion of the atoms in the bulk of the metal, and by treating those for the motion of the atoms in the surface layers exactly. The phonon spectral density, which is directly related to the electron energy-loss spectrum, is calculated from these exact Green's functions.

With the above technique we have made contact with the electron energy-loss data for dipole scattering i.e. scattering in the near specular direction and for impact scattering i.e. large angle scattering. In the near specular direction where the selection rules allow the dominant contribution from dipole moment components polarized perpendicular to the surface, we find for $\sqrt{3} \times \sqrt{3}$ R 30° overlayers of oxygen on the Ni(111) surface that the calculated electron energy-loss spectrum [1] is sensitive to whether the oxygen is sitting on the threefold hollow site with a second neighbour nickel atom directly below it (hcp site), or the site with no such neighbour below it (fcc site). If the oxygen atom occupies the hcp site, then the resulting energy-loss spectrum contains two features at 177 cm^{-1} and 228 cm^{-1}, which arise from the coupling of the adsorbate motion to the substrate phonons and should appear as distinct features in the data. On the other hand, if the oxygen atom occupies the fcc site there is only one loss feature at 217 cm^{-1} which has origin in the S_3 surface phonon of the nickel substrate. Comparison with experimental data [4] leads us to believe that oxygen atoms occupy the fcc site. Similar investigations on the p(2x2) and p(1x2) overlayers of oxygen on the Ru(0001) surface [3] have shown that unlike LEED, electron energy-loss spectroscopy can distinguish between these two types of overlayers. In each case, the oxygen atoms occupies a threefold hollow site, but while the point group symmetry for the p(2x2) overlayer is C_{3v}, that for the p(1x2) overlayer is reduced to C_s. As a result, for the p(2x2) structure there is only one dipole active O-Ru mode polarized strictly perpendicular to the surface at 535 cm^{-1} while for the p(1x2) structure there are two dipole active modes with components polarized perpendicular to the surface at 430 and 585 cm^{-1} that appear in the electron energy-loss data taken in the near specular direction.

The dispersion of the surface phonons and resonances for the c(2x2) overlayers of oxygen on the Ni(100) surface along high symmetry directions has been recently studied by electron energy-loss spectroscopy, and comparison with theoretical results have led to certain predictions regarding the force constants between the atoms in the surface layers. Firstly, it is found that the coupling between the first and second layer nickel atoms in the presence of a c(2x2) overlayer of oxygen is reduced to 30 % of the strength of the coupling between the nickel atoms in the bulk. This is in contrast to the result on the clean nickel surface,where this particular coupling is enhanced by 20 % from its value in the bulk. This result is in qualitative agreement with ion scattering measurements [5]. Secondly, a small amount of oxygen-oxygen direct lateral interaction has to be included to give a good fit to the dispersion of the oxygen-nickel vibration parallel to the surface. In contrast to oxygen overlayers,which seem to affect the bonds between the nickel atoms in the top layers, sulfur overlayers [6] seem to leave these bonds unaltered.

References

1. T.S. Rahman, D.L. Mills and J.E. Black: Phys. Rev. B27, 4059 (1983)
2. J. Szeftel, S. Lehwald, H. Ibach, T.S. Rahman, D.L. Mills and J.E. Black: Phys. Rev. Lett. 51, 268 (1983)
3. T.S. Rahman, A.B. Anton, N.R. Avery and W.H. Weinberg: Phys. Rev. Lett. 51, 1979 (1983)
4. H. Ibach and D. Bruchmann: Phys. Rev. Lett. 44, 36 (1980)
5. J.W.M. Frenken, J.F. van der Veen and G. Allan: Phys. Rev. Lett. 51, 1876 (1983)
6. S. Lehwald, M. Rocca, H. Ibach and T.S. Rahman: to be published.

Bulk and Surface Phonons in Superlattices

B. Djafari-Rouhani

Laboratoire d'Etude des Surfaces et Interfaces, Associé au C.N.R.S.
Faculté Libre des Sciences, 13, rue de Toul., F-59046 Lille Cedex, France

The study of acoustic wave propagation in layered media [1] found its original motivation in the desire for an accurate description of the propagation of seismic shocks through Earth's crust, whose density and elastic properties vary with distance from the surface of Earth. With the development of techniques for fabricating artificial superlattices (SL), interest is beginning to turn to the study of vibrations in such periodically modulated structure. The SL has a larger periodicity in the direction perpendicular to the slabs, and therefore many phonon branches in the folded Brillouin zone. In the gaps between these branches may appear localized modes associated to defects, and in particular to the surface of a semi-infinite SL. Here we give a summary of the recent theoretical works on the vibrations in infinite and semi-infinite SL; the experimental aspects (Raman and Brillouin scattering) are reviewed by J. SAPRIEL and by C.M. FALCO et al [2]. The films in the SL are first treated as elastic continuum media, characterized by their elastic constants and their densities; this approach is complemented by use of simple atomic models to describe the dynamics of the SL.

1. Elastic waves in superlattices

The displacement field associated to a wave propagation in the SL can be obtained [3] by (i) solving the equations of motion of the elasticity theory in each film; (ii) writing at each interface the boundary conditions on the continuity of displacements and stresses (these relations link together the fields in two successive unit cells, and therefore introduce a transfer matrix); (iii) using the Bloch theorem. For a semi-infinite SL [3], a surface mode has to be sought, in the gaps, in the form of a wave decaying far from the surface and satisfying the conditions of vanishing stresses at the surface.

Assuming the films in the SL isotropic, explicit relations have been obtained for the dispersion of the bulk and surface modes [4-6] (the surface is parallel to the laminations); in this case, a decoupling occurs between shear horizontal [4,5] and sagittal waves [6]. The surface waves have also been studied numerically in other recent works [7,8]. In these calculations, the structure of the bulk bands is function of the elastic parameters of the films and their thicknesses; the surface modes are very dependent on the nature of the film at the surface, and on its thickness, which may be different from that of the corresponding film in the bulk.

The surface Rayleigh waves have also been investigated [9] in a new geometry where the surface is perpendicular to the laminations. The Rayleigh wave is folded at the Brillouin zone boundary leading to a second surface branch separated from the first by a gap. The magnitude of this gap can be modulated according to the elastic parameters and thicknesses of the films. The higher Rayleigh branch can disappear in the bulk bands when the films differ significantly in their elastic properties.

Finally, for very thin films, the SL becomes equivalent to an effective medium whose elastic constants are some combinations of those of the two layers. This approach has been introduced many years ago [10] for isotropic films, where the SL becomes an effective hexagonal medium.

2. Atomic models

Elasticity theory could fail to be valid if the films in the SL are composed of just a few atomic layers. Besides, the atomic models give the dispersion curves inside the whole Brillouin zone parallel to the layers. Also in semiconductor SL such models are needed to describe the phonon branches in the optical range.

Recently, the Green's function of a SL made from simple cubic crystals with interaction between nearest neighbors has been obtained in closed form [11], giving explicit relations for the dispersion of the bulk modes as well as of the surface modes. For a given $k_{//}$ (wavevector parallel to the layers), the SL bulk bands are wide (resp. narrow) in the frequency ranges belonging to the bulk bands of both (resp. one) constituents. The surface modes may exist everywhere in the Brillouin zone, and are dependent on the nature of the film at the surface and on the number of atomic layers it contains.

This calculation has been extended [4,12] to a linear chain SL, made from diatomic crystals, to describe qualitatively the phonons (at $k_{//} = 0$) in semiconductor SL like GaAs-AlAs. A numerical study of bulk phonons was also reported before [13] in this model. Let us consider a GaAs-AlAs SL with L and L' monolayers of GaAs and AlAs respectively in each unit cell; the first (L + L') folded phonon branches belong to the acoustical domain common to both GaAs and AlAs, the L next branches are in the optical domain of GaAs and the L' last ones in the optical domain of AlAs. These last (L + L') branches are very flat, as the optical domains of GaAs and AlAs do not overlap.

Finally, the Green's function method can be applied to other vibrational models and even to the study of electronic or magnetic properties of a SL. For example, the electronic structure of a metallic superlattice has recently been investigated in a tight binding formalism [14].

References

1. W.M. Ewing, W.S. Jardetzky and F. Press, Elastic waves in layered media (Mc Graw Hill, New-York, 1957); L.M. Brekhovskikh, Waves in layered media (Academic Press, New-York, 1960).
2. See also Proceedings of the First Int. Conf. on Dynamics of Interfaces, J. de Physique (Paris) C5, (1984).
3. B. Djafari-Rouhani, L. Dobrzynski, O. Hardouin-Duparc, R.E. Camley and A.A. Maradudin, Ref. 2, p. 151.
4. J. Sapriel, B. Djafari-Rouhani and L. Dobrzynski, Surface Sci., 126, 197, (1983)
5. R.E. Camley, B. Djafari-Rouhani, L. Dobrzynski and A.A. Maradudin, Phys. Rev. B27, 7318 (1983).
6. B. Djafari-Rouhani, L. Dobrzynski, O. Hardouin-Duparc, R.E. Camley and A.A. Maradudin, Phys. Rev. B28, 1711 (1983).
7. B.A. Auld, G.S. Beaupre and C. Herrmann, Elec. Lett., 13, 525 (1977).
8. A. Kueny and M. Grimsditch, Phys. Rev. B26, 4699 (1982).
9. B. Djafari-Rouhani, A.A. Maradudin and R.F. Wallis, Phys. Rev. B29, 6454 (1984).
10. S.M. Rytov, Sov. Phys. Acoust., 2, 68 (1956).
11. B. Djafari-Rouhani, L. Dobrzynski and O. Hardouin-Duparc, J. Elect. Spectr. and Rel. Phenomena, 30, 119 (1983) and Phys. Rev. B29, 3138 (1984).
12. B. Djafari-Rouhani, J. Sapriel and F. Bonouvier, to be published.
13. A.S. Barker Jr, J.L. Merz and A.C. Gossard, Phys. Rev. B17, 3181 (1978).
14. B. Djafari-Rouhani, L. Dobrzynski and P. Masri, Preprint.

Part III

Phase Transitions

Surface Reconstruction Phase Transformations

Roy F. Willis

Cavendish Laboratory, Department of Physics, University of Cambridge
Cambridge CB3 OHE, United Kingdom

1. INTRODUCTION

Clean single-crystal (metal) surfaces usually possess well-defined two-dimensional (2D) periodicities closely resembling the atomic ordering in the bulk. In some instances, however, the surfacelattice reconstructs into a phase with new symmetry. This can occur spontaneously with temperature, or it can be induced by small coverages of adsorbed atoms or molecules. The phase transformation may be reversible or irreversible. For example, Au (110) exhibits an irreversible reconstruction into a (2x1) superstructure once the clean surface has been prepared. This surface then undergoes a reversible order-disorder phase transition when the temperature is raised above a critical value, $T_c \simeq 700$ [K]. Similarly, a carefully cleaned W (100) surface reconstructs into a $(\sqrt{2} \times \sqrt{2})$ lattice[1] which also disorders reversibly at $T_c \simeq 300$ [K]. A small quantity of chemisorbed hydrogen switches the symmetry of the atomic displacements from domains with p2mg space group symmetry to domains with c2mm symmetry, while maintaining a $(\sqrt{2} \times \sqrt{2})$ translational periodicity [2]. Increasing the amount of hydrogen produces a commensurate to incommensurate phase transformation which is reversible with coverage [3]. Clean Mo(100) reconstructs into a surface lattice which is incommensurate with the bulk substrate structure over a wide temperature range below 300 [K] [4]. Si (111) forms a (2x1) reconstructed lattice which transforms upon heating irreversibly into a (7x7) superlattice [5]. Surprisingly, the (7x7) structure exhibits a reversible order-disorder phase transition.

These are the most studied surfaces to date. However, structural phase transformations have been observed on a number of metal surfaces, viz: Au, Ir, Pt (100), Pt (110) and Au (111) [6]. The stability of other metal surfaces, viz: Ag (110), Cu (110), Pd (110), Ni (110), Fe (110) and Cr (100) is also in doubt, notably in the presence of small quantities of certain adsorbates, but the evidence to date is much less definite [7]. In addition, the tendency of covalent semiconducting surfaces to reconstruct, viz. Si (100) (2x1), GaAs (2x1) etc. is well known [8].

Landau theory of 2nd order phase transformations of two-dimensional symmetry classes [9] is applicable to the surface reconstruction of clean surfaces. The above surface reconstruction phase transformations are rightly 2D in nature, since the atomic displacements only affect the outermost layers of the crystal. From this point of view, their behaviour should resemble quite closely the critical phenomena which has been observed in 2D epitaxial overlayer films of inert gases condensed on graphite [10]. However, clean surface reconstruction phase transformations have been the subject of very few critical exponent studies to date, and the number of phase transitions which have been reported in any detail is still very small [11]. This is because the observation of critical properties and the measurement of critical exponents in a continuous transition is very

1. Sometimes referrred to as c(2x2) or $(\sqrt{2} \times \sqrt{2})R45^o$ in Wood's notation [1]

difficult - requiring rather well-characterized samples, a highly surface sensitive probe and, (most important) high spatial coherence, in order to study the very long-range order fluctuations close to T_C.

Nevertheless, some progress has been made, particularly in the introduction of several new concepts in the statistical mechanical ideas which underpin the basic phenomenology of Landau theory of structural phase transformations in general, viz: the role of soft-phonon-mode periodic lattice distortion mechanisms, and non-linearly-coupled strongly anharmonic lattice vibrations [13]. It is my objective here to review this progress, making reference to particular systems which serve to illustrate these concepts rather well.

Section 2 introduces Landau theory of structural phase transformations and critical phenomena with specific reference to the reversible order-disorder phase transformations which occur on Au (110) (2x1) and W (100) ($\sqrt{2}$ x $\sqrt{2}$) surfaces. The theory is extended to incommensurate phases (section 3) in the context of the reconstruction of W (100) and Mo (100) surfaces. We conclude in section 4 with a brief discussion of other related systems of potential interest.

2. LANDAU THEORY OF STRUCTURAL PHASE TRANSFORMATIONS

Structural phase transformations occur when the symmetry of the crystal changes under the influence of a change in some external condition. A first-order phase transition is one in which there is a *sudden* rearrangement of the crystal lattice and the thermodynamic state of the solid changes discontinuously. Second (and higher) order phase transitions show a *continuous* change of state from one phase to the other at a critical temperature T_C. If we consider phases to be in thermal equilibirium (i.e. we neglect problems of nucleation and growth kinetics) then each phase can be defined in terms of its free energy: $F = H - TS$ with enthalpy $H = U + pV$. F is a continuous function of temperature T, pressure p, and variations in the volume V. It might also involve composition, but we only consider here perfect crystals with fixed composition[2]. Also, variations in V only introduce small and trivial effects. In most crystalline phase changes, the difference in the internal energy ΔU of the phases is much less than the latent heat of sublimation, which measures the total stability of the two phases. To predict the transition temperature T_C of a phase change, one must therefore know the difference in enthalpy H and entropy S between the two phases, and how they vary with temperature. This is the statistical mechanical basis of the thermodynamics of phase transformation phenomena [13]. That is, only the relative stability of phases at zero temperature can be discussed neglecting entropy.

An important concept in the theory of structural phase transformations (SPTs) is that of the soft phonon-mode [12]. Many SPTs are examples of 2nd order phase transitions in which the lattice distorts via a lattice instability associated with a *particular* normal mode of the crystal whose frequency "softens", such that the configuration of atomic displacements change continuously as the temperature approaches T_C i.e. $\Delta U \approx 0$ as one phase transforms into the other. The specific pattern of atomic displacements of wavevector $\vec{q}$ may be described by an "order parameter" $\sigma_i(\vec{q})$ which represents the amplitude of the distortion made up of mode components, i. Such transitions may be described phenomenologically by a minimization of the overall free energy of the system, expressed in terms of a power expansion (Landau 1937 [13]):

$$F = F_0 + \alpha u^2 + \beta u^4 + \gamma(\text{grad } u)^2 + \delta(\text{grad } \Phi)^2 + \ldots \qquad (1)$$

where u (amplitude) and Φ (phase) are the relevant order parameters of the soft

---

2. The enthalpy H of the surface will be increased by a chemical potential term μ for increasing adsorbate coverage in adsorbate-induced transitions.

modes. Landau made the important observation that, whereas first order phase transitions do not have any particular symmetry requirements, 2nd order SPTs may be classified according to the symmetry properties of the lattice. That is, the coefficients α, β, γ etc in (1) serve to distinguish between various symmetry classes of 2nd order phase transformations - often simply by their sign; positive, negative or zero values. (Specific computation of their values in terms of the enthalpy and entropy changes requires a microscopic model of interactive forces in order to predict T_c).

The concepts of soft-mode displacements and the symmetry requirements of the Landau expansion of the free energy form the basis of the statistical thermodynamics of most structural phase transformations. The surface is no exception, and we now go on to see how these ideas apply to the two-dimensional case.

2.1 ORDER-DISORDER TRANSITIONS AT SURFACES; CRITICAL EXPONENTS

We begin with the simplest case of a lattice distortion due to a single component (i = 1) normal mode. The free energy may be expanded in a power series of this single component order-parameter:

$$F = F_o + \tfrac{1}{2}r\sigma^2 + d\sigma^3 + u\sigma^4 \qquad (2)$$

Above T_c, the high symmetry phase is stable if the coefficients $d = 0$ and $r > 0$. Landau [13] was able to show that a continuous 2nd order phase transformation occurs at $T = T_c$ for $u > 0$ and u independent of temperature sufficiently close to T_c, with r changing sign through zero at T_c. The condition for a minimum in F is that $\partial F/\partial\sigma = 0$ and $\partial^2F/\partial\sigma^2 > 0$ such that the distorted phase stabilizes ($T < T_c$) when:

$$\langle\sigma\rangle^2 = (a/4u)\ (T_c - T) \qquad (3)$$

That is, above T_c the average of the order parameter $\langle\sigma\rangle = 0$ and the disordered (high symmetry) phase is stable, while below T_c, the lattice distortion increases continously according to a power law relationship:

$$\langle\sigma\rangle \sim (T_c - T)^\beta \qquad (4)$$

where β is known as the "critical exponent" of the order parameter. The transition is manifested in scattering (i.e. diffraction) experiments by the appearance of the Bragg reflections arising from the new superlattice structure, or in thermodynamic (i.e. heat capacity) measurements by singularities in such quantities as the specific heat or compressibility. These singularities are generally of power law form, and the powers are denoted by "critical indices" or exponents of the reduced temperature $t \equiv (T - T_c)/T_c$:

Specific heat, $C_p \sim |t|^{-\alpha}$ for $t \gtrless 0$ (5a)

Bragg peak intensity $I \sim |t|^{2\beta}$ for $t < 0$, and zero for $t > 0$ (5b)

(Note that first-order transitions at which Cp and I are discontinuous are described by $\alpha = \beta = 0$.) If the transition is continuous, however, then critical scattering is observed above T_c (Fig.1). If the superlattice Bragg peak occurs at a wavevector Q, the critical scattering intensity I(q) becomes increasingly concentrated near Q and the intensity at Q grows as $T \to T_c$. The normal mode fluctuations about T_c may be expressed in term of a Fourier transform of position (wavevector) and frequency-dependent susceptibility function $\chi(w'q')$, the primes representing fluctuations about the soft-mode oscillation. The imaginary part of this dynamic susceptibility gives the power dissipation in the system (dependent on one-phonon matrix elements coupling initial and final phonon states differing

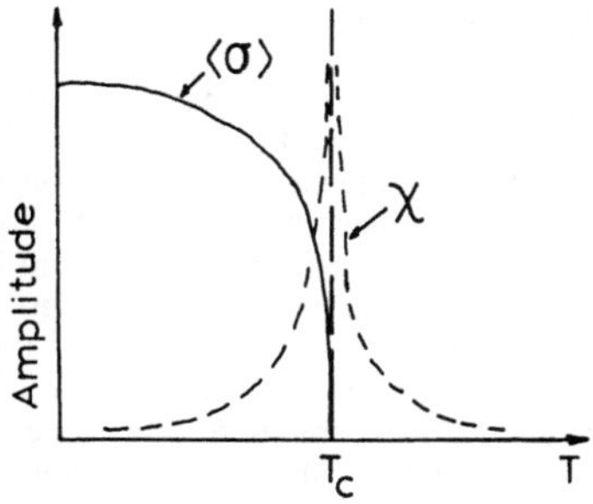

Fig.1 Variation of long-range-order (LRO) parameter σ, and susceptibility function χ near the phase transition temperature T_c.

in energy by $\hbar(w'-w)$ and is related to the static susceptibility $\chi(0)$ by the usual Kramers-Kronig relations [12]. The static susceptibility diverges at T_c reflecting a movement in weight of $\chi_{imaginary}$ to smaller w - a 'critical slowing down effect' representative of a particular soft-mode frequency going to zero at T_c. In a diffraction experiment, the intensity of $I(\bar{Q})$ diverges like:

$$I(\bar{Q}) \simeq |t|^{-\gamma} \text{ for } t > 0 \tag{5c}$$

this broadening of the diffraction spot profile in reciprocal space being measure of static susceptibility, which in Landau theory takes the form:

$$\chi(0) = (\partial^2F/\partial\sigma^2)^{-1} = 1/a(T - T_c)^{-\gamma} \text{ for } T > T_c \tag{5d}$$

$$= \tfrac{1}{2}a(T_c - T)^{\gamma} \text{ for } T < T_c \tag{5e}$$

If a measure of $\bar{q}$ in $I(\bar{q})$ be denoted by ζ^{-1} (inverse correlation length) then this width of the diffraction spot profile vanishes like

$$\zeta \sim |t|^{-\nu} \quad \text{for } t > 0 \tag{5f}$$

That is, ζ^{-1} is a measure of the range of correlated fluctuations in the system close to T_c.

The important points to emerge from this analysis are that, firstly, transitions which exhibit the same critical exponents belong to the same "universality class" of symmetry classification, and secondly, it is believed that the above critical exponents are not independent but satisfy "universal" scaling relations [9]:

$$\alpha + 2\beta + \gamma = 2 \tag{5g}$$

$$d\nu = 2 - \alpha \tag{5h}$$

where d is the dimensionality, d=2 for surfaces.

Thus, the problem of describing critical phenomena is one of determining the values of the critical exponents with sufficient accuracy to establish the universality class of the transition. Measurements of these exponents for various order-disorder transitions on surfaces has only recently begun - mainly for a few "lattice gas" overlayer systems [11]. However, we will now consider recent detailed diffraction studies [14] of the Au(110) (2x1) ↔ (1x1) order-disorder phase transition which appear to establish this particular clean surface phase transition as belonging to the two-dimensional Ising universality class with β=1/8, γ=7/4, ν=1 and α=0 (logarithmic singular behaviour at Tc) [15].

2.2 CLASSIFICATION OF THE Au(110)(2x1) ORDER-DISORDER TRANSITION

As well as cases in which a change of symmetry occurs by a continuous displacement of atoms (periodic lattice distortion), a 2nd order SPT may result from a change of ordering of the atoms amongst specific lattice sites. For example, when the number of lattice sites that can be occupied by atoms of a specific kind exceeds the number of available atoms, if the probability of filled and empty sites then becomes equal, all sites now become equivalent and new symmetry elements appear. In other words, the symmetry of the surface is increased and the lattice is disordered. The Landau order parameter now effectively describes the pattern of atomic sites which is responsible for this change in symmetry; the symmetry of the crystal is changed when the order parameter changes continuously to zero at T_c. In most cases, the phase of higher symmetry occurs at the higher temperatures i.e. the transition is from ordered to disordered phase with increasing temperature. At T_c itself there is really no clear distinction between displacive (soft mode) and order-disorder (diffusive hopping) atomic configurations, since it is always possible to express the appropriate Hamiltonian as a power series expansion in atomic displacements u_i (though the latter case will be more slowly convergent). Thus, a more precise (and formal) description of the order parameter characterizing any 2nd order (order-disorder) phase transition may be couched in the following group theoretical terms:-

Consider a function $\rho_o(\bar{r})$ representing the atomic density which is invariant under the operations of the space group of the crystal. As the structure changes, the atomic density changes to:

$$\rho(\bar{r}) = \rho_o(\bar{r}) + \Delta\rho(\bar{r}) \tag{6a}$$

and the symmetry of the new phase $\rho(\bar{r})$ must be a subgroup of that of $\rho_o(\bar{r})$. $\Delta\rho(\bar{r})$, which represents the displacements of atoms, can be expressed as a linear combination of normalised functions $\Phi_{ni}(\bar{r})$ which form the basis for the nth irreducible symmetry group of $\rho_o(\bar{r})$, of dimension equal to the number of i's for a given n. Landau theory tells us that continuous 2nd order phase transformations involve $\Delta_\rho(\bar{r})$ corresponding to a single irreducible representation of the high symmetry phase $\rho_o(\bar{r})$:

$$\Delta_\rho(\bar{r}) = \textstyle\sum_i C_i \; \Phi_i \; (\bar{r}) \tag{6b}$$

If we assume small values of the coefficients C_i, then the free energy can be expanded in a power series in C_i, to recover an equation identical in form to the Landau equation (2) earlier. The value of $\bar{q}=\bar{Q}$, which characterises the ordered phase, is that which minimizes ${}^{\delta F}/_{\delta\bar{q}}$ or, more specifically, $v(\bar{q})|\delta\bar{q}|_{\bar{Q}} = 0$ where v represents nearest-neighbour interactions summed over the first Brillouin Zone of the reciprocal lattice. If $\delta v(\bar{q})|\delta\bar{q}|_{\bar{Q}} = 0$ is satisfied for a particular $\bar{Q}$, then, by symmetry considerations, it will also be satisfied for those vectors which can be obtained from $\bar{Q}$ by the point-group operations, which leave the lattice invariant. This set of vectors (called the "star" of $\bar{Q}$) represent points of high symmetry within the first Brillouin Zone if the distorted structure is *commensurate* within the bulk lattice i.e.,

$$\Phi_i(\bar{r}) = U_{iq}(\bar{r}) \exp(i\bar{q}.\bar{r}) \tag{6c}$$

where $U_{iq}(\bar{r})$ has the periodicity of the lattice and the irreducible representations are characterised by vectors $\bar{q}=\bar{Q}$. With the exception of $\bar{Q}=0$, all such points are on the Brillouin Zone boundary (if the minimum occurs at $\bar{Q}=o$, then this implies that the ordered state is translationally invariant i.e., a liquid-gas transition). A more exact criterion for an ordered commensurate phase is the Lifshitz condition [13]: "that the anti-symmetric part of the representation, formed by the direct product of the representation R with itself, does not contain the vector representation".

Following the treatment of Schick [9], we can now apply these ideas to the (p2mm) space group of the rectangular array of sites encountered on the (110) surface of a fcc crystal viz. Au(110). The Brillouin Zone, together with the three stars $\bar{t}$, $\bar{p}$ and $\bar{s}$ which satisfy the Lifshitz condition, is shown in Fig. 2.

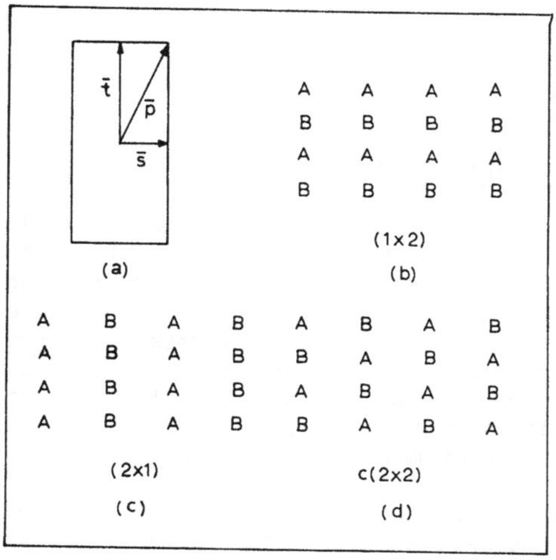

Fig.2 (a) Brillouin Zone of rectangular lattice (p2mm) of (110) face of fcc Au(110) with ordered configurations (b) (c) and (d) which are in the class of the anti-ferromagnetic Ising model (Schick, 1981)

Each star contains a single wavevector, so each corresponds to a one- dimensional order parameter. The three ordered configurations are shown in Fig.2 (b,c,d). Sites in which the densities ρ_i are equal are indicated by the same letter. The three possible states are denoted (1x2),(2x1) and c(2x2). If we assume nearest-neighbour interactions only, then the free energy expansion in each case is identical to that of the ferromagnetic Ising model. That is, writing $r=a(T-T_c)$ in (2), the Landau expression for a single-component system becomes:

$$F = F_o + 1/2a(T-T_c)\,\sigma^2 + u\sigma^4 \qquad (7)$$

with $u={}^{T_c}/_{12}$ from mean field theory of an equivalent spin lattice.

The critical properties of the 2D Ising model are, of course, well known [16]. In particular, $\beta = 1/8$, $\gamma = 7/4$ and $\nu = 1$, the specific heat diverging logarithmically ($\alpha=0$). That is, as T_c is approached from above, the intensity of the critical scattering diverges like $t^{-7/4}$, while the width in q-space of the critical scattering vanishes like $\xi^{-1} \sim t$. As T_c is approached from below, the intensity of the 1/2 order Au(110)(2x1) Bragg peaks will vanish like $I_{Bragg}(\bar{Q}=\bar{s}) = t^{1/4}$. Hence, the star of $\bar{Q}$ consists of a single vector $\bar{s}$, and there is only a single function cos $(s.r_i)$ in the basis of the ID representation of the space group of (p2mm) of the fcc (110) rectangular lattice (Fig.2c). Further, the order parameter has a single component (7) which can be written:

$$X(\bar{s}) = {}^{1}/_{A}\sum n_i \cos \bar{s}.\bar{r}_i \qquad (8)$$

The ensemble average of $X(\bar{s})$ is $\rho(\bar{s})$, which is real, and the form of the density distribution just below T_c is:

$$\rho_i = \rho + \rho(\bar{s}) \cos (\bar{s}.\bar{r}_i) \qquad (9)$$

which is the density distribution shown in Fig.2c. In the case of the Au(110)(2x1) reconstruction, the positions marked B represent missing columns of atoms alternating between occupied columns A. The structure deduced from grazing incidence X-ray diffraction studies [17] is shown in profile in Fig.3.

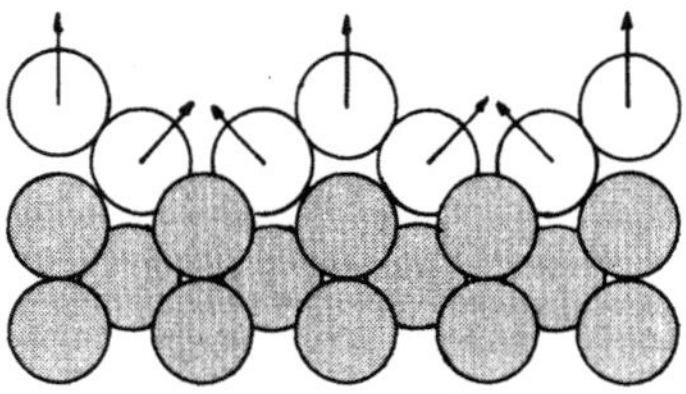

Fig.3 Diagram of Au(110)(2x1) in profile to illustrate missing rows; the bulk atoms are shaded and the atomic displacements in the surface are exaggerated (after Robinson, 1983).

2.3 LEED SPOT-PROFILE OF Au(110)(2x1) PHASE TRANSITION

The Au(110) surface is reconstructed at ambient temperature to form a (2x1) superlattice in which the surface unit cell doubles its periodicity along the [100] crystallographic direction (fig.2c). This produces 1/2 order diffraction peaks which gradually disappear indicative of a 2nd order PT at $T_c \sim 400^oC$. The wavevector of the reconstructed lattice $\bar{k} = {}^{2\pi}/_a\ ({}^1/_2,0)$ suggests an order parameter of the form: $u(\bar{r}) = A\hat{e}\cos(k.r)$ with $\hat{e}$ a unit vector parallel to $\bar{k}$. As we have seen above, when the symmetry operations of the (110) lattice are applied to $u(\bar{r})$, the amplitude of the displacements $A \to \pm A$, which implies a phase transformation isomorphous with the 2D Ising model universality class.

The variation in the intensity and angular profiles of the ${}^1/_2$-order diffraction spots with temperature can be used to derive experimentally the critical exponents of the order-disorder phase transition [14,18]. Assuming the missing row model (fig.2c), the surface is alternate rows of atoms and empty lattice sites. The scattering intensity (I) of the electron beam off the (2x1) lattice system is:

$$I = \sum_{l}\sum_{r} e^{-2W} \langle f_l f_{l+r} \rangle\, e^{i\bar{K}.\bar{r}} \qquad (10a)$$

where f_l is the scattering factor of the lth site, $\bar{K}$ is the change in wavevector $(k-k')$ of the electrons and e^{-2W} is the Debye Waller factor which accounts for any thermal fluctuations about the equilibrium position of any rth atom at the $\bar{r}$ position. If we now chose to define variable S_l which is +1 if an atom is present in a site, and $S_l = -1$ if the site is empty, then we may write: $f_l = {}^a/_2 + {}^a/_2 S_l$, where a is the scattering factor for a Au atom, such that:

$$I = \sum_{l}\sum_{r} e^{-2W} \langle {}^{a^2}/_4\ (1 + S_{l+r} + S_l + S_l S_{l+r}) \rangle\, e^{i\bar{K}.\bar{r}} \qquad (10b)$$

which, if there are 50% of all sites effectively occupied[3] reduces to

$$I = \sum_{l}\sum_{r} e^{-2W}\, e^{i\bar{k}.\bar{r}} \langle {}^{a^2}/_4 (1+S_l S_{l+r}) \rangle \qquad (10c)$$

In order to examine the ${}^1/_2$ order spots, we introduce a reciprocal lattice vector $\bar{g}$ with a fractional sum of indices, e.g. $({}^1/_2,0,0)$ $(1{}^1/_2,1,0)$, and define a probability $P_L = S_L\, e^{i\bar{g}.\bar{l}}$, which takes values of +1 for a correct occupation of site l for a (2x1) structure and −1 for an incorrect occupation. If $\bar{K}$ are close to $\bar{g}$ (i.e. examine ${}^1/_2$-order spots only) then the first term in I (10c) vanishes giving:

$$I = \sum_{l}\sum_{r} e^{-2W} ({}^{a^2}/_4)\, e^{-i\bar{g}.2\bar{l}} \langle P_l P_{l+r} \rangle\, e^{i(\bar{k}-\bar{g}).\bar{r}} \qquad (10d)$$

where we define the correlations in fluctuations in density from some average value at positions l=1 and l+r=2 as: $\langle P_1 P_2 \rangle = \underline{P}(r_{12}) + \langle P \rangle^2$, where $\langle P \rangle$ is the

3. A model of a rough surface of steps and terraces (which is nearly always the case on metal surfaces) provides effectively for 50% filled and 50 empty sites above T_c [19].

average probability related to the long range order and $\underline{P}(r_{12})$ is density-density correlation function-linked to short range order fluctuations:

$$I = \sum_r N e^{-2W} (a^2/4) [\underline{P}(\bar{r}) + \langle P\rangle^2] e^{i(\bar{K}-\bar{g}).\bar{r}} \qquad (10e)$$

where N is the total number of atoms in the layer. Thus, (10e) relates the Landau long-range order parameter σ with $\langle P\rangle$ and the susceptibility X with P(r) to give two terms which determine the intensity distribution around the $^1/_2$-order superlattice spots:

$$I = N(a^2/4)\, e^{-2W} \;(N\sigma^2\, \delta(\bar{K}-\bar{g}) + \sum \underline{P} e^{i(\bar{K}-\bar{g}).\bar{r}} \qquad (10f)$$

The first term gives a delta function at $\bar{K}=\bar{g}$ with intensity dependent on the long-range order parameter $\sigma^2 \sim t^{2\beta}$; this term is dominant for $T<T_c$ so that β can be determined after correcting for the Debye-Waller factor temperature-dependence. The second term, while present at $T<T_c$, is the only term present at $T>T_c$ so that the critical exponent of the susceptibility $X \sim t^{-\gamma}$ can be determined. The variation in full width at half maximum gives the inverse correlation function $\xi^{-1} = t^{-\nu}$.

In essence, (10f) tells us that the LEED diffraction pattern is basically a structure function of the sample which is of the form:

$$I(K)/I^o(K) = N^{-1}\int d\bar{r}\; G(\bar{r})\exp(-i\bar{K}.\bar{r}) = N^{-1}S(\bar{K}) \qquad (10g)$$

where the structure factor $S(\bar{K})$ is the Fourier transform of the density-density correlation function $G(\bar{r})$ which relates fluctuations between two sites. At T_c, $S(\bar{K})$ expresses phonon correlations in the system:

$$S_c(\bar{q},t) = I(t)\; \delta(\bar{q}-\bar{q}_o) + X(\bar{q}-\bar{q}_o,T) \qquad (10h)$$

and the effect of the sum of these two terms on the actual LEED spot profile is illustrated in fig.4.

Thus, in order to determine the critical exponents, the entire diffraction beam profile must be measured for a succession of temperatures about T_c. After deconvolution of an instrument response function (IRF), the diffraction spot profile can be resolved into a contribution, due to the long-range order δ-function (for $T<T_c$ only) and a contribution due to fluctuations. It is

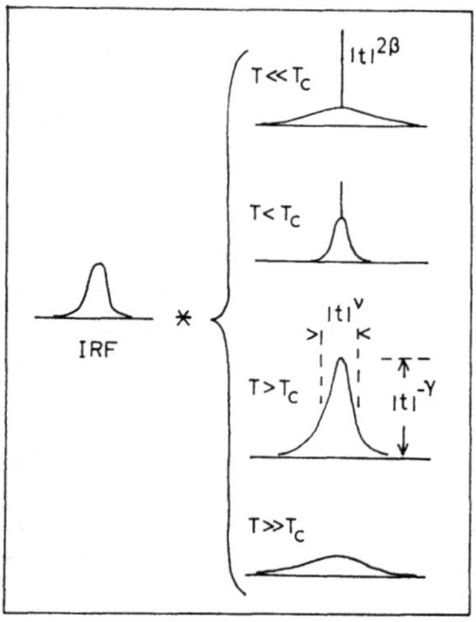

Fig.4 Diffraction spot profile showing long-range order below T_c and short-range order diffuse scattering above T_c in terms of the Fourier Transform of sample pair correlation function critical exponents, convoluted with a Gaussian instrument response function IRF

diffucult to establish the IRF in many circumstances, principally due to intrinsic spot broadening from a rough surface. The conventional LEED camera has a coherence length of only 80Å or so, which is a severe handicap. However, recently, measurements have begun to be made with "high resolution" LEED systems [20]. Campuzano, Foster and Willis [21] have employed a system with a spatial resolution believed to be in excess of 2000Å; their measurements for the order-disorder transition on Au(110)(2x1) ↔ "(1x1)" are shown in fig.5.

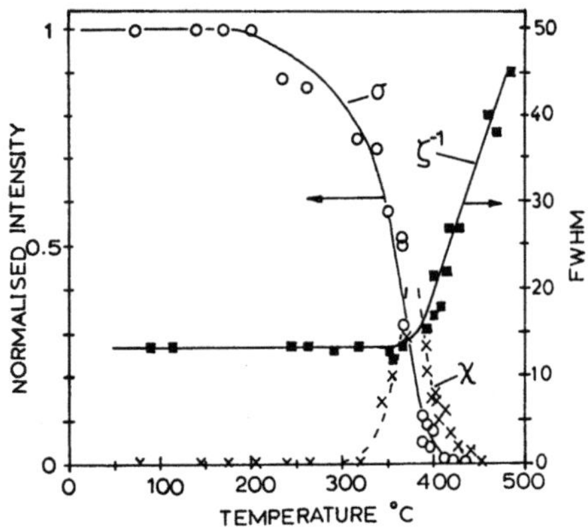

Fig.5 LEED measurements of variation in 1/2-order spot intensity and profile for Au(110)(2x1) ↔ (1x1) phase transition; σ is LRO parameter, χ is susceptibility and ξ^{-1} inverse pair correlation length near T_c (Campuzano, Foster & Willis, 1984)

The δ-function is assumed to be broadened into a Gaussian function of finite width due to finite surface roughness effects and, in this analysis the IRF is assumed to be sharper than this intrinsic width and is ignored. The short-range order fluctuations contribution is fitted to the Fourier transform of pair correlation functions; ORNSTEIN & ZERNIKE [22] developed a classical theory based on the expansion of the Fourier transform of correlation functions such that, near a critical point and for large r, the critical fluctuations are represented by a correlation function of the form:

$$h(r) \sim \exp(-r/\xi)/r^{(d-1)/2}$$

where d is the dimensionality and ξ, the correlation length, is a measure of the range of h(r) i.e., its Fourier transform is a Lorentzian distribution to a first approximation. As the critical point is approached, ξ becomes very large i.e. the correlations become infinitely long-ranged[4].

By carefully fitting the variation in the $^1/_2$-order Au(110)(2x1) LEED diffraction profile with these three parameters: Gaussian height, Lorentzian height and width; Campuzano et al [14] were able to derive values for the critical exponents $\beta = 0.13 \pm 0.02$, $\gamma = 1.75 \pm 0.10$, $\nu = 0.93 \pm 0.09$ in agreement with the 2D Ising model within experimental accuracy.

2.4 X-Y MODEL WITH CUBIC ANISOTROPY W(100)(√2x√2)

The reconstruction observed on W(100) (and H/W(100)) is a considerably more complicated case [24]. The point group of the (100) face of a cubic crystal is (p4mm) so that the appropriate star of $\bar{Q}$ is generated by applying to $\bar{Q} = \bar{p} = {}^{\pi}/_a(\hat{i}+\hat{j})$ which lies at the corner $\bar{M}$ of the first Brillouin Zone, the point group operations of the square (fig.6).

4. At the critical point, where $\xi\to\infty$, the OZ theory yields $h(r)\sim 1/r^{d-2}$ i.e. in 3D $h(r)\sim 1/r$ and ξ while long does not decay to zero. However, in 2D the theory gives the nonsensical result: $h(r)\sim\log r$ increasing with distance. This has lead FISHER [23] to suggest that the result should be modified: $h(r)\sim 1/r^{d-2+\eta}$ with $\eta = {}^1/_4$ for the 2D Ising model.

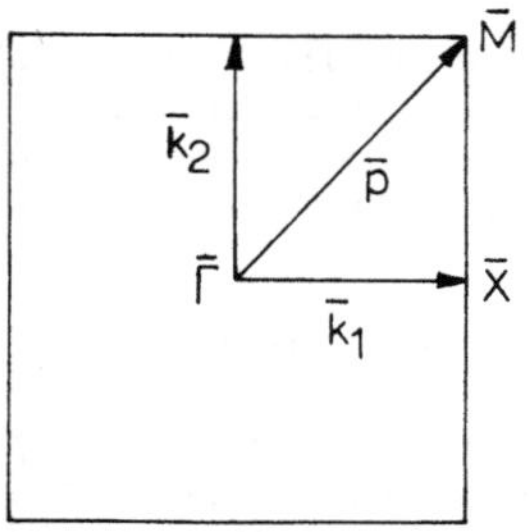

Fig.6 Brillouin Zone of the square lattice showing that the star of wavevectors $\bar{Q}$ consists of a single vector $\bar{p}$

The W(100) system is the realization of an XY model where displacements may be in either of two independent directions. If we let u_1, u_2 be degenerate displacement vectors in, say, the <11> and <1$\bar{1}$> directions, then, adding these displacements to the b.c.c. (100) lattice sites with periodicity p (Fig.6), produces a reconstructed ($\sqrt{2}$x$\sqrt{2}$) surface which now transforms as the 2-D irreducible representation of p4mm symmetry. Because the representation is two-dimensional, a two-component order parameter results, χ_1 and χ_2, with:

$$u1 = \chi 1 \; p1 \cos \bar{p}.\bar{r}$$
$$u2 = \chi 2 \; p2 \cos \bar{p}.\bar{r} \qquad (11a)$$

Note that there is only one wavevector $\bar{p}$ which determines the periodicity but there are two independent displacement directions $\bar{p}_1$, $\bar{p}_2$. The corresponding Landau expansion of the free energy, maintaining p4mm symmetry, takes the form:

$$F[T\chi 1\chi 2] = F(T,0,0) + {}^1/{}_2 r(\chi_1{}^2 + \chi_2{}^2) + u(\chi_1{}^2 + \chi_2{}^2)^2 + v(\chi_1{}^2\chi_2{}^2) + {}^1/{}_2(\nabla\chi_1)^2 + {}^1/{}_2(\nabla\chi_2)^2 \qquad (11b)$$

This free energy is precisely the form of another well-known magnetic spin model - the "2D X-Y model with cubic anisotropy". The first two terms (in χ) are those of the free energy of the isotropic planar rotor or x-y model, which is described by a two-component magnetization spin parameter $\sigma = \sigma x\underline{i} + \sigma y\underline{j}$; this is seen by identifying $\chi 1$ with σ_x and $\chi 2$ with σ_y i.e., the first two terms are proportional to $\sigma.\sigma$ and $(\sigma.\sigma)^2$ which are isotropic. The third term, analogous to $\sigma_x{}^4 + \sigma_y{}^4$, singles out the x and y axes and destroys the isotropy - called the "cubic anisotropy" term [9]. The cubic anisotropy depends on the magnitude of the third term "coupling" coefficient v(T), i.e., if v(T) is positive, then the periodic lattice distortion (fig.7) can occur in one of two directions, $\bar{p}1$, or $\bar{p}2$. The experimentally proposed distortion (fig.7), in which $\chi 1 \neq 0$, $\chi 2 = 0$ (or $\chi 2 \neq 0$ for $\chi 1 \neq 0$) requires that v(T)>0 i.e., orthogonal domains [25].

The observed ($\sqrt{2}$x$\sqrt{2}$) structure of clean W(100) has two orthogonal domains: one with atomic displacements along <11>, as shown in Fig.7; the other domain has atomic displacements along <1$\bar{1}$>. The symmetry is p2mg which introduces a glide

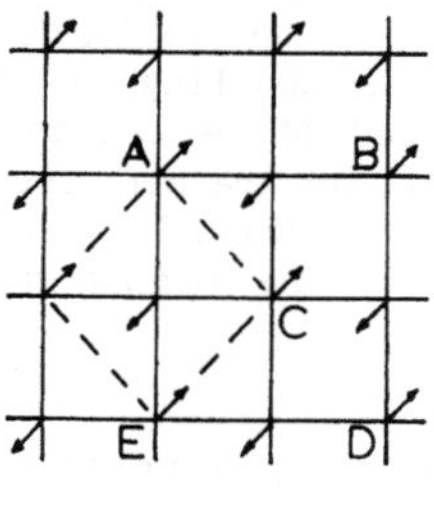

Commensurate $C(2x2)_{\langle 11\rangle}$
W(100)

Fig.7 ($\sqrt{2}$x$\sqrt{2}$)R45° (dashed) or $c(2x2)_{<11>}$(ABCDE) reconstructed lattice of W(100) showing p2mg (glide plane) symmetry of a single domain with atomic displacements along <11> directions of magnitude $u_1 \sim 0.15$Å; T_c 300°K [25]

plane asymmetry in the intensity distribution of the 1/2 order diffraction beams in the LEED pattern [25]. In the Landau expansion (11b), $v \neq 0$ defines an XY system with cubic anisotropy. The reason for v taking a positive value is to ensure that when F is minimized, X_1 (or alternatively X_2) is zero. Should v be negative then F is minimized when $|X_1| = |X_2|$[5].

The 2D X-Y model with cubic anisotropy W(100)($\sqrt{2}$x$\sqrt{2}$) has other unusual critical properties for which specific predictions have been made [24]. In particular, the adsorption of a small coverage of hydrogen causes symmetry switching from atomic displacements along <11> to <10> directions associated with vf(T) in eqn. (11b) changing sign. Evidently, the anisotropy is controlled by coverage and at the point of vanishing anisotropy (v(T)=0) the surface should exhibit a Kosterlitz-Thouless "melting type" transition [30].

The temperature-dependence of the ($\sqrt{2}$x$\sqrt{2}$) to (1x1) reconstruction on W(100) may be usefully dealt with by model, effective lattice-dynamical Hamiltonian methods [11,13,24]. Roelofs et al [27] have carried out Monte Carlo simulations and renormalization group studies on the symmetry class of the X-Y model with cubic anisotropy to investigate the nature of this phase transition. They find that just above Tc, the surface structure consists of ($\sqrt{2}$x$\sqrt{2}$) domains randomly distributed such that no long-range order exists i.e., the direction of the atomic displacements in the different domains are uncorrelated, varying randomly from <11> to <1$\bar{1}$>. The correlation length $\xi(T)$ within each domain is very large for T just above Tc, decreasing rapidly with increasing T. Thus, on both sides of Tc, the reconstruction is a displacive transition but, in the immediate neighbourhood of Tc, the fluctuations have strong order-disorder character. This dual character of the transition is now a well-established feature of many structural phase transition, and is particularly pronounced in lower dimension systems [28]. More important, it implies that one should be wary of comparing theoretical expectations of the W(100)(1x1) ground state structure with experimental data obtained for the high temperature "(1x1)" phase indicated by LEED, which is dependent on the spatial coherence of these fluctuating domains [29].

Although there have been extensive studies aimed at providing experimental evidence for the reconstructed $\sqrt{2}$x$\sqrt{2}$ phases of W(100) and H/W(100) [31], little has been reported to date concerning values of the critical exponents, which are "non-universal" due to the (weak) cubic anisotropy term.

3. LANDAU THEORY OF INCOMMENSURATE SURFACE PHASES

With increasing hydrogen coverage, the H(100)($\sqrt{2}$x$\sqrt{2}$) phase becomes incommensurate with the undistorted lattice. The clean surface of Mo(100) also shows a commensurate-incommensurate phase-transition. Superlattice satellite spots appear about the 1/2-order ($\sqrt{2}$x$\sqrt{2}$) positions which are non-integer multiples of the Bragg reciprocal lattice wavevectors (fig.8).

Various mechanisms have been proposed for the C-IC transitions [24], but to date the experimental evidence is not sufficient to unequivocally establish the actual mechanism. However, a most appealing proposal is that due to FASOLINO, SANTORO and TOSATTI [32] which is based on phonon soft-modes coupling effects. Basically, FST propose that the incommensurate reconstructed phase arises not because of long-range interactions between the atoms [9], but rather from a coupling between two normal modes; one with displacements perpendicular to the surface ($\bar{M}_1$), and one in the plane of the surface ($\bar{M}_5$). The mechanism is illustrated in Fig.9.

---

5. If we had chosen to define our order parameter in terms of displacements along $\bar{k}_1$, $\bar{k}_2$ rather than $\bar{p}$ (Fig.6), then v in the Landau expansion (11b) would take the opposite sign to that used here for the appropriate structures [11,18,24].

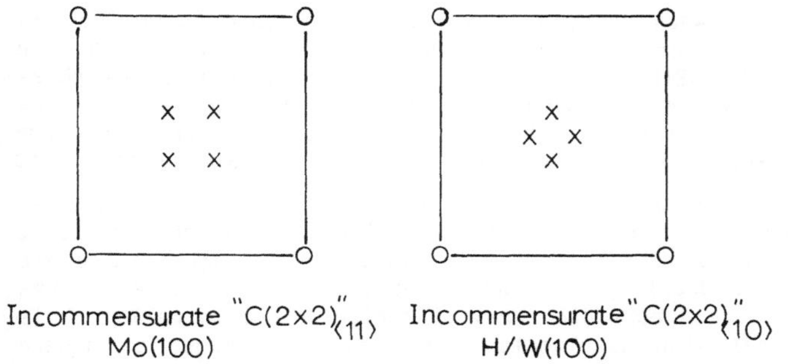

Fig.8 Showing positions of superlattice spots characterizing the incommensurate phases of H/W(100) and Mo(100).

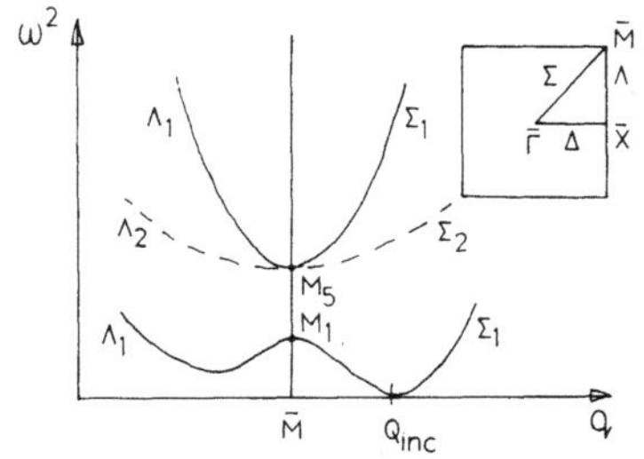

Fig.9 Phonon soft mode coupling between M5 and M1 symmetry modes to produce incommensurate wave-vector Qinc. The symmetry directions in the 2D Brillouin Zone are shown inset (Fasolino, Santoro & Tosatti, 1980)

Off the symmetry point $\bar{M}$ we have the mode $\bar{M}_1$ coupling to one of the doubly degenerate modes $\bar{M}_5$ active along either the Σ or Λ symmetry directions with atomic displacements in directions <11> or <10> in the surface plane. Off the symmetry point $\bar{M}$, these orthogonal modes can couple and soften to low frequency at some wavevector Q_{inc} which is incommensurate with the undistorted (100) lattice. Thus, the (1x1) undistorted surface has two possible modes of deformation; the $\bar{M}_5$ modes alone can go soft at $\bar{M}$ to give a phase-transition into the commensurate $(\sqrt{2}\times\sqrt{2})_{<11>}$ state (with displacements along <11> as is observed experimentally [25]). Alternatively, the $\bar{M}_5$ and $\bar{M}_1$ modes can soften and couple to give an incommensurate phase with displacements along <11>, as is observed for clear Mo(100) or along <10>, as is observed for H/W(100) [2]. Figure 9 shows the $\bar{M}_1$ mode softening, but the principle applies to the $\bar{M}_5$ softening preferentially, as is the case for W(100) and Mo(100) surfaces.

The phase-transition can be described in terms of a two component order parameter displacive phase-transformation in which the order parameters ($\rho_j(\vec{q})$, j=1,2) are complex rather than real quantities [33]. These complex order parameters are coupled and complex conjugates of one another:

$$\rho 1 = {}^1/\sqrt{2}\, A \exp(i\Phi),\ \rho 2 = {}^1/\sqrt{2}\, A \exp(-i\Phi) \qquad (12a)$$

where A is the amplitude and Φ the phase (real quantities) of the periodic lattice distortion wave. The Landau free energy expansion now takes the form:

$$F = Fo + a(T-Tc)\ \rho 1\ \rho 2 + 4u\ \rho 1^2\ \rho 2^2 \qquad (12b)$$

or, substituting (12a) into (12b):

$$F = Fo + {}^1/2a\ (T-Tc)\ A^2 + uA^4 \qquad (12c)$$

i.e., we recover expression (7) for a single component system in terms of an order parameter A which is the amplitude of the incommensurate periodic lattice-disortion wave in the surface.

This coupling exists only when one moves away from symmetric (commensurate) wavevectors for the reconstruction. The mechanism is a special case of a more general mechanism proposed by HEINE and McDONNEL [34] for incommensurate phase formation in bulk ferroelectrics. However, if the incommensurate state is represented by a sinusoidal distortion with a single incommensurate wavevector Q_{inc}, this is known to overestimate the energy of the incommensurate state and leads to a first order C-IC transition [24]. Basically, the sinusoidal distortion tends to "square-off" due to the addition of higher harmonic periodic displacement Fourier components which effectively introduce strongly anharmonic displacements. That is, the coupling interaction between the two modes occurs over a localised region to form a periodic array of antiphase domain walls. This coupling of orthogonal model with displacements ρ1 and ρ2 produces a "square wave" system of antiphase domains separated by localized domain walls, as illustrated in fig.10a and b [35].

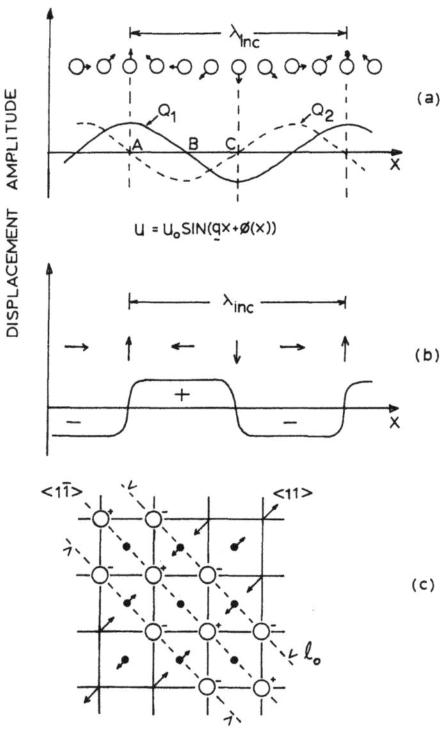

Fig.10 Atomic displacement amplitudes of (a) coupled modes Q_1 and Q_2 in quadrature, (b) "squared-off" antiphase (±) domain structure, and (c) atomic displacements at domain wall of width l_o in Mo(100) showing horizontal (→) and vertical displacements (±) (Heine, 1984)

The actual atomic displacements at a domain wall for the Mo(100)<11> incommensurate phase is also shown (fig.10c); that for H/W(100) has been discussed in detail by WILLIS [29,36].

HEINE and McDONNEL [34] have shown that phase gradient interaction terms in q-space, equivalent to $H.[(\bar{\nabla}\rho_1)\rho 2 - \rho 1(\bar{\nabla}\rho_2)]$ in real space, lower the overall free energy of the system due to a relief of stress caused by the non-uniform displacements of (say) distortion $\rho 1(\bar{q})$. In the case of the $\bar{M}_5$ displacements (fig.9), we have a non-uniform (oscillating) ρ1 wave with atoms moving to the left in one domain and to the right in the neighbouring antiphase domain (fig.10b). In the domain wall region, we have a non-zero value of the phase gradient $\nabla\rho_1$ resulting in stress which is relieved locally by coupling (parameter H) to a mode in which the surface atoms move perpendicular to the surface, the latter constituting a second ρ2 distortion wave. We see that the displacement of one mode is maximum when the other is minimum or zero; i.e. the amplitude of the second mode $\rho 2(\bar{q})$ automatically adjusts itself so as to minimize the overall free energy of the system. In other words, the system takes advantage of two possible modes of lattice instability.

Finite slab calculations [32] indicate that these reconstructions are strongly localized in the surface region, and the amplitude of the distortion at the second layer (small dot, fig.10c) is typically only a few percent of that in the first layer (the latter ~ 0.15A in W(100) [25]). Nevertheless, the substrate atoms must move in concert with the toplayer atoms, as shown in fig.10c.

By analogy with other C-IC transitions occuring in physisorption systems [37] as well as transition metal dichalcogenides [33], WILLIS [29,36] has suggested that a periodic array of extremely localized domain wall distortions may best describe the C—IC transition on H/W(100) in terms of a "soliton superlattice" arising from the strongly anharmonic nature of the coupling.

3.1 'SOLITON SUPERLATTICE' STRUCTURE OF INCOMMENSURATE PHASE

At the incommensurate phase transformation (fig.10), there are strong non-linear lattice dynamical effects which, in the Landau expansion of the free energy, are described by extra "Umklapp terms" involving pth-order anharmonicity i.e. eqn. (12c) takes the form [38]:

$$F = Fo + {}^1/_2 a(T-T_c)\, A^2 + uA^4 + VA^p \cos p\Phi \qquad (13a)$$

in amplitude A and phase Φ coordinates. The phase of the distorted wave now is one of p different choices of the phase Φ which serves to minimize (13a). In the simplest case of a sinusoidal incommensurate distortion: $u = uo \sin(\bar{q}x + \Phi(x))$, where $u = uo \sin \bar{q}x$ is the wave commensurate with the undistorted lattice, the phase $\Phi(x)$ is the shift in the equilibrium position of the atoms with respect to some potential minimum - the "lock-in" potential energy V (13a). (In the case of epitaxial overlayers [38], V = substrate periodic potential). The "unperturbed" incommensurate wave distortion (V=0) is simply a phase relation linear in coordinate x; $\Phi(x) = \delta x$, with δ the degree of misfit between the distorted and undistorted phases. However, if the lock-in potential V strongly couples to the incommensurate distortion, we can have local variations in the phase which cause $\Phi(x)$ to vary in a stepwise fashion, fig.11.

Physically, we have regions or domains of site l which are commensurate with the undistorted lattice, separated by regularly spaced domain walls of site lo which represent "discommensurate" regions, over which the phase varies rapidly by $2\pi/_p$ (with p=4 [38]). The stepwise variation in the phase of the distortion is favoured by a gain in lock-in energy throughout the commensurate domain regions, which effectively lowers the free energy over that of the "unpinned" incommensurate wave distortion (Mcmillan [33]). The ground state configuration, which minimizes the free energy (eqn. 13a), is found among the solutions of the ID sine-Gordon equation of (non-linearly) coupled normal modes:

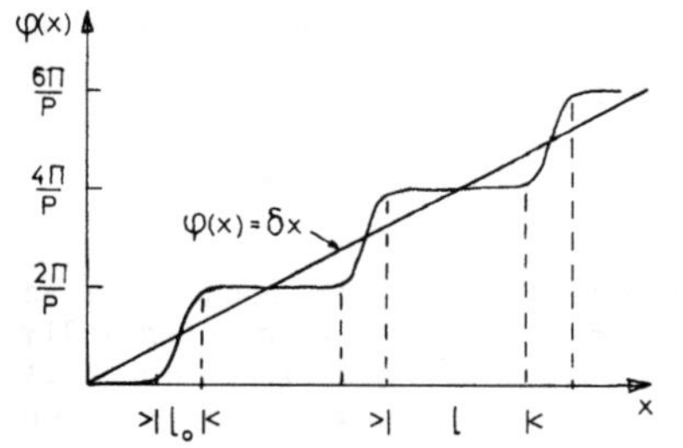

Fig.11 Antiphase commensurate domains of size l separated by narrow regions l_o of rapidly varying phase $\Phi(x)$ (McMillan, 1976)

$$d^2\Phi/dx^2 = pV \sin p\Phi \qquad (13b)$$

one solution of which is the solitary wavepacket distortion or "soliton", which describes the rapid phase variation over the narrow region of the domain wall discommensuration l_o separating two commensurate domains of size l:

$$\Phi(x) = {}^4/_p \tan^{-1} [\exp(p\surd Vx)] \qquad (13c)$$

Thus, solutions to eqn. (13) are regularly spaced solitons ("soliton superlattice"), the distance l being equal to one-half the incommensurate wavevector Qinc. The soliton superlattice is a compromise between an elastic energy term which favours $\Phi = \delta x$ (the term quadratic in A, eqn.13a) and the interaction energy with the substrate lattice ($VA^p \cos p\Phi$ "Umklapp term). The width of the soliton wall is $l_o = 1/p\surd V$ and the average misfit between domain walls: $\langle d\rangle = 2\pi/pl$, where $\langle d\rangle$ represents the degree of overall misfit i.e. the soliton density. The theory extends to 2-dimensions where the solitons become linear boundaries separating individual domains in a plane [39,40], as shown in fig.12.

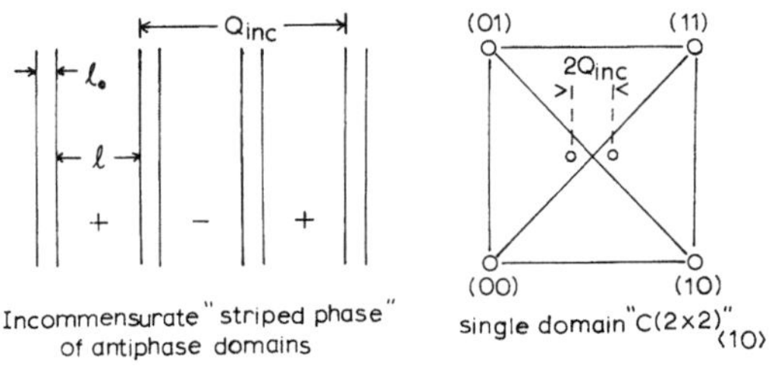

Fig.12 'Soliton superlattice' of antiphase domains of width l separated by domain wall discommensurate regions of width l_o. The incommensurate wavevector Q_{inc} is indicated by diffraction spot splitting; the single domain case is illustrated

In the H/W(100) system, the H-induced C—IC phase boundary occurs at coverages in the range, H/W ratio = 0.2 to 0.4 [41], the actual value depending on the density of steps and other defects in the surface which affect soliton ordering, l. As the coverage increases to H/W $\gtrsim$ 0.5, spot streaking occurs (but only over a limited range of q-space) indicative of a 1-dimensional disordering of the domain walls. WILLIS [29] has argued that this corresponds to an effective 'roughening transition' due to an increase in the configurational entropy of the domain walls. That is, the domain walls develop "kinks" at a critical spacing, $l_{cr} \simeq \pi^2 l_o^2$ [42]. From the coverage dependence of this transition, $l_{cr} \sim 4$ or 5 lattice spacings (i.e. 12 to 15A) and $l_o \sim$ a single lattice spacing, which is in accord with recent LEED spot profile measurements of the minimum domain size [43]. This minimum domain size comes about due to wall-wall energy repulsion effects, which increase exponentially with decreasing l. A model of a single domain "striped phase" with wall kinks, which would account for the observed I-D streaking of the incommensurate satellite spots, is shown in fig.13 (cf fig.12).

The H-induced H/W(100) C-IC phase-transition is therefore characterised by domain wall formation. Unlike the single component soft-mode lattice distortion (eqn.12), the number density of these domain walls $\langle d\rangle$ becomes the effective order parameter, together with that of the single Fourier component of the atomic displacements in the domain walls. That is, near the C-IC phase boundary, where the soliton density is initially low, the free energy of a regular array of solitons (single domain, fig.11) takes the form [38,39]:

$$E_{walls} = ({}^{4\surd V}/_{\pi} - \delta)\langle d\rangle + {}^{16\surd V}/_{\pi\langle d\rangle} \exp\left(\frac{-2\pi\surd V}{\langle d\rangle}\right) \qquad (13d)$$

The first term is proportional to the soliton density, and may be considered as the energy of a single domain wall. The second term, which decays exponentially with distance l between solitons, is an effective wall-wall repulsion term. It can be shown that when V becomes small enough (or δ large enough), the soliton

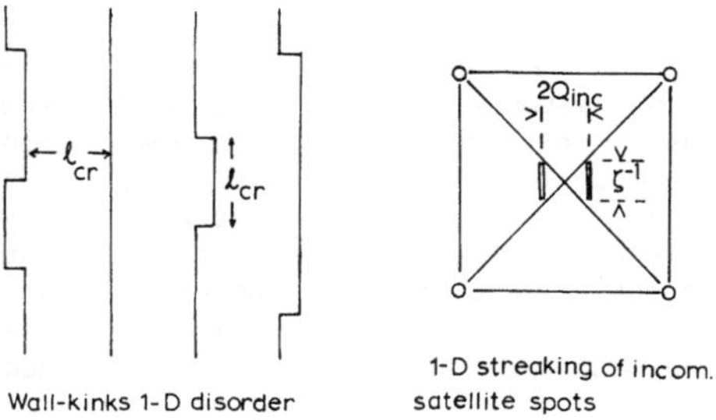

Fig.13 Wall-kinks down to a minimum distance $l_{cr} \sim l$ which would account for the I-D streaking of incommensurate satellite spots (ξ^{-1}) over a limited region of k-space observed in LEED; ξ^{-1} is the wall-wall correlation of length

SOLITON DENSITY: $\langle d \rangle = 2\pi / p\ell$

Critical domain size: $\ell_{cr} = \pi^2 \ell_o^2 \qquad \ell_o = 1/p\sqrt{V}$

Energy of soliton array: $E_{walls} = \alpha\langle d\rangle + \beta\langle d\rangle \exp -(\kappa\langle d\rangle^{-1})$

energy becomes negative and the commensurate phase becomes unstable with respect to the spontaneous formation of domain walls. That is, we observe a continuous 2nd-order transition into the incommensurate phase[6].

3.2 INCOMMENSURATE PHASE FLUCTUATIONS

What can we say about the lattice dynamical modes of the incommensurate phases on H/W(100) and Mo(100), and the nature of the fluctuations at the incommensurate-disorder phase-transition boundary? To date - very little, except to relate the behaviour to low-frequency modes which been observed in bulk incommensurate ferroelectrics [46].

From the point of view of lattice dynamics, the incommensurate phase is characterized by new low-frequency vibrations (10 to 10^{-1}cm^{-1}), namely by fluctuations in the amplitude and phase of the incommensurate modulation wave - the so-called 'amplitudons' and 'phasons'. These fluctuations in the incommensurate phase give rise to two separate susceptibility response functions below the critical incommensurate phase-transition temperature T_{IC}: viz. $X_L(q_s,w)$ and $X_T(q_s,w)$ are the frequency and wavevector-dependent susceptibilities of the corresponding longitudinal and transverse fluctuations. X_L corresponds to amplitude fluctuations parallel to the direction of the ordering vector ξ and X_T corresponds to phase fluctuations transverse to ξ. The wavevector dependence of these fluctuating modes close to T_{IC} is shown schematically in fig.14.

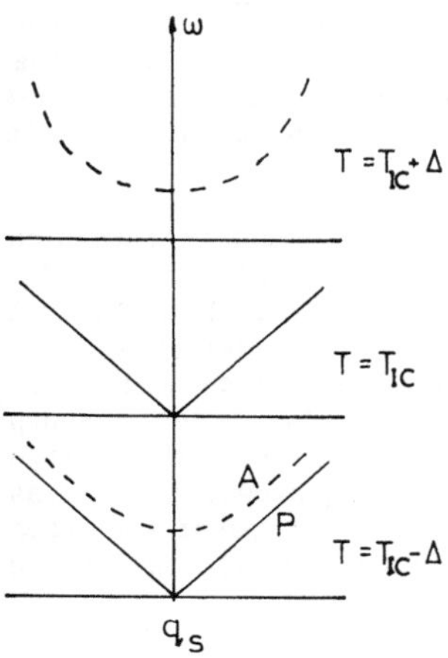

Fig.14 Wavevector dependence of soft-mode fluctuations in incommensurate phase: T_{IC} = critical temperature for IC-phase; A and P signify 'amplitudon' and 'phason' branch modes (Bruce & Cowley, 1978)

6. Equations (13) are based on a continuum rather than a discrete lattice model. AUBRY [44] BAK & POKROVSKY [42] have shown a discrete Hamiltonian introduces a narrow 'fluid phase' between the commensurate and incommensurate phase, due to the introduction of dislocations with the overlayer phase. Such a 'fluid phase' has recently been advocated from experimental LEED measurements for N chemisorption on Ni(110) [45].

Above T_{IC} we have the usual single-component soft mode behaviour, but below T_{IC} we have two separate branches - amplitudon A and phason P - both of which originate from a single soft-phonon branch characteristic of the normal lattice (soft) modes of the high temperature phase. The behaviour of the amplitudon branch remains similar to that of this soft mode below T_c, as in the case of a single-component order parameter displacive phase-transition. However, the new feature in a domain-wall incommensurate phase is the emergence of a phason branch at much lower frequency, which is expected to be heavily damped for such wavevectors q_s - and consequently, it is much more difficult to observe[7]. In the case of incommensurate overlayers [38,39], the 'phason' and 'amplitudon' branches take the form shown in Fig.15.

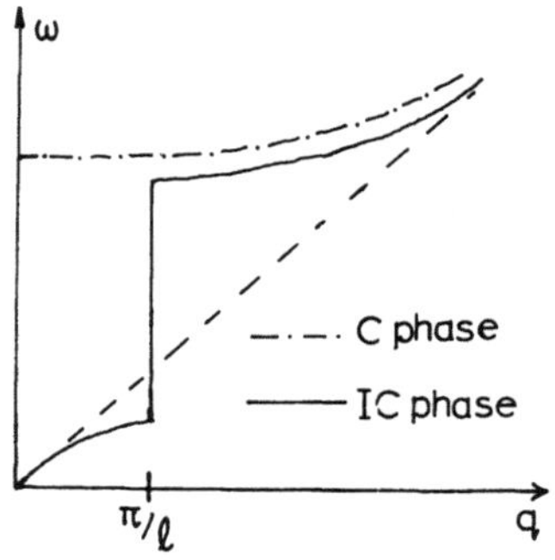

Fig.15 Dispersion relations of fluctuating modes in the commensurate (c-phase) and incommensurate (IC-phase) (McMILLAN, 1977; POKROVSKY & TALAPOV, 1978; NOVACO, 1980)

The IC phase has two branches in its dispersion curve separated by a 'band gap' at $q=\pi/l$ arising from the periodicity of the soliton superlattice. This is the spectrum of elementary excitations in 2D incommensurate systems near the commensurate-incommensurate transition [33,47,48] The low-lying modes are the 'phasons' (OVERHAUSER 1971 [49]) since they correspond to modulations of the phase of the periodic structure; the long-wavelength phason modes take the form of phonons in the soliton lattice. This branch is very soft because of the exponentially weak interactions between the walls (eqn. 13d). At the commensurate-incommensurate phase-transition, l goes to infinity and the soft phason branch disappears, leaving a gap at q=0 which represents the energy of harmonic oscillations of the atoms in the periodic commensurate potential. The upper branch (fig.15) modes are, therefore, more like ordinary phonons within the commensurate domains[8]. An important point is to realize that these phason modes give a contribution to the free energy just like phonons in an harmonic atomic lattice.

Thus, at such an incommensurate phase-transition, we must distinguish carefully between essentially long-range and short-range order fluctuations. For CDW-driven incommensurate phase-transitions, McMILLAN [33] arrived at a physical picture in which each domain can be treated as having its own normal modes with intercell interactions between modes. These modes have amplitude and phase which, below T_{IC}, can exhibit very long-range coherence to give rise to sharp superlattice diffraction spots. However, above T_{IC}, there still remains a large amplitude of local modes within domains, the latter decreasing slowly in size as the temperature increases. In effect, the long-range order parameter goes discontinuously to zero while short-range order persists above the phase transition temperature, as illustrated in fig.16.

7. Recent Russian work using backward-wave oscillator spectroscopy in the microwave spectral region claims to provide evidence for a phason mode in ferroelectric K_2SeO_4 [46].

8. It is not clear in this picture to what extent the upper branch modes are amplitude or phase modulated; OVERHAUSER [49] only considered phase variations within a charge-density-wave (CDW) driven model.

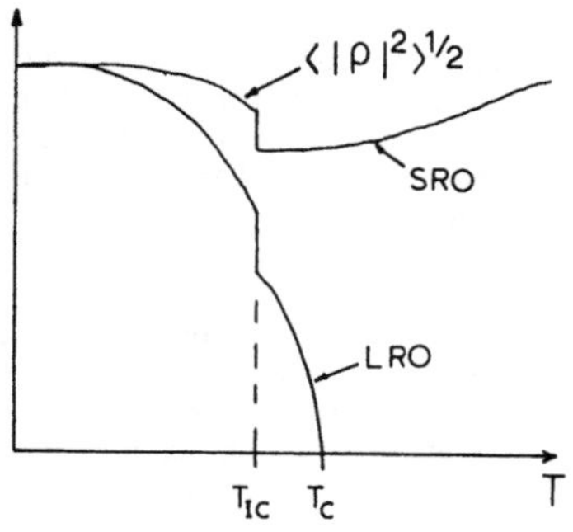

Fig.16 Variation of long-range order (LRO) parameter $\langle\rho\rangle$ and the square root of the short-range order (SRO) parameter $\langle|\rho|^2\rangle$ with temperature for incommensurate phase-transitions (McMILLAN, 1977)

This concept of a 'range of disorder', over which the correlation length changes slowly to zero, is one of the most important concepts to emerge from the most recent studies of structural phase-transition phenomena [28,40]. Clearly what you perceive to be the 'degree of disorder' near T_c will depend on the spatial coherence length of your experimental probe.

4. RELATED SYSTEMS

We have concentrated on the Au(110)(2x1) and W(100)($\surd$2x$\surd$2) surface reconstruction phase transformations, since these are the most studied systems to date. However, in this section, we review briefly other systems which may be related in their behaviour, viz. the other noble metal surfaces (Section 4.1) and the semiconductors (section 4.2).

4.1 Au,Ir,Pt(100) and Au(111) RECONSTRUCTED SURFACES

The reconstructed phases of the low-index surfaces of the noble metals produce much larger superlattice periodicities than Au,Pt(110)(2x1) and ressemble incommensurate 'epitaxial' overlayers. Various structural models have been discussed in the context of LEED analysis (VAN HOVE et al [6]). The most likely reconstructions on the (100) surfaces involve the formation of hexagonal close-packed domains on a cubic symmetry substrate. Clean metastable unreconstructed (100)(1x1) surfaces have been prepared, and the interesting point is that, as the temperature is increased, these surfaces show an irreversible transition into the distorted state. For example, the Ir(100) surface continuously reconstructs into a (1x5) lattice over a temperature range 700-1200[K], the reconstructed phase then remaining stable from 55-2100 [K]. The Au(100) surface converts into an incommensurate "(20x5)" structure at 373[K]. The metastable Pt(100)(1x1) transforms into a similar "(20x5)" IC-phase at 400[K] but, on further heating to 1100-1150[K], converts into a Pt(100) "hexagonal-rotated-7°" superlattice, which is then stable from 1450[K] down to liquid N_2 temperatures. Recent vacuum tunnelling microscopy observations of these surfaces [50] also suggests incommensurate domains of uniformly contracted and slightly rotated hexagonal top layers. The domain walls themselves contain atomic displacements compressed out of the surface plane.

The irreversible nature of these transformations is very reminiscent of that of the elemental semiconductors eg. Si(111)(1x1) → (2x1) → (7x7). One possible explanation is the role of dislocations, which serve to "pin" the final structure. The presence of transverse strains, in addition to longitudinal elastic (compressive) strains, favour a rotation of the domains which can lower the "lock-in" energy further with respect to the cubic symmetry of the substrate lattice[9]. The crucial ingredient here is the behaviour of the domain walls.

9. This is known as a NOVACO-McTAGUE [51] orientational instability from work on epitaxial inert gas overlayers on graphite.

Increasing temperature produces kinks in the atomic rows of the reconstructed overlayer, at places where the atoms intersect a wall; a succession of such kinks provides a means of effectively rotating the domains relative to the substrate in the incommensurate phase. This, in turn, provudes a mechanism for maximum lock-in energy with the substrate's interaction potential V, eqn. (13).

A second factor in stabilizing these high-order reconstructed phases is the presence of defects such as steps in the surface; the incommensurate "(20x5)" reconstruction on Pt(100) has been shown to vary its incommensurate periodicity with step density and orientation [6]. Also, steps can act as either a source or sink of atoms for the compressed IC overlayer structures which order on the terraces; the buckling at the domain walls, referred to above, appears to be produced by an excess of atoms.

Thus, these noble metal structures are similar to IC compressed epitaxial overlayers, which produce a 'soliton superlattice' of commensurate domains separated by narrow walls (FRANK and VAN DER MERWE, 1949; POKROVSKOY and TALAPOV, 1978 [38]). This is observed as a splitting in the LEED spots for the Au(100) and Pt(100) (20x5) structures. However, the incommensurate 'misfit' is not the same for the two cases; the Pt(100) IC-phase appears to have domain walls at regular intervals in one direction only, whereas Au(100) has regularly spaced domains with walls intersecting along orthogonal directions. Ir(100) shows a simpler commensurate (1x5) structure with no diffraction spot splitting. Au(111) can be interpreted as the superposition of three 120° rotated domains, each domain consisting of rectangular ($\surd$3x22) unit cells [6].

Step-terrace defects and NOVACO-McTAGUE rotation [51] serve to 'pin' these incommensurate distortions, and to stabilize the structure at high temperatures, preventing the IC phase reverting back into the commensurate structure. The soft-mode lattice dynamics and associated electronic mechanisms driving these distortions are not yet understood.

4.2 SEMICONDUCTOR RECONSTRUCTED SURFACES

As the recent review by KAHN [8] shows, this is already an enormous field. However, to date there has been little work as regards phase-transition behaviour. The Si(111)(7x7) to "1x1" transition at 1140[K] is the only case where such an attempt has been made [5].

BENNETT and WEBB (1981) used LEED and observed that excess diffuse scattering appears during the phase-transition whose temperature-dependence and magnitude was consistent with the loss of the 7th order superlattice intensities, indicating a transition into a disordered surface phase rather than a commensurate (1x1). The superlattice intensity showed the transition to be continuous and reversible. The properties of interest here are those which can be measured in diffraction intensities, which are a measured in diffraction experiments (section 2.3) viz: the superlattice diffraction intensities, which are a measure of the long-range order parameter; the diffuse scattering, which is related to the short-range order and wavevector-dependent susceptibility; and the relaxation time-dependence of the diffracted intensity in response to changes in temperature, which is related to 'critical slowing down' of the fluctuations near T_c. In these experiments, the long-range order critical exponent $\beta = 0.11 \pm 0.15$ which is of the order of the 2D Ising model value, 0.125. However, the behaviour of the diffuse scattering was not that expected for the 2D Ising system; the kinetics were asymmetric with sign of the temperature pulse and the relaxation time diverged near T_c but only for negative temperature increments.

If we consider the Landau symmetry classification of this transition [52], one observes 7 x 7 - 1 = 48 new superlattice spots in LEED, each of which in principle represents an independent component of the order parameter (the

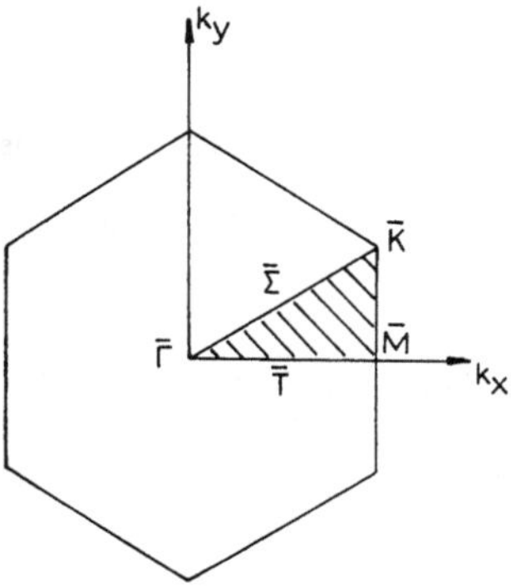

Fig.17 Surface Brillouin Zone of Si(111) surface

symmetry of the ideal Si(111) surface is C_{3V} with only 6 classes of operators). The (7x7) diffraction patterns suggest that there is a 48 component order parameter which contains more than one irreducible representation of the surface space group. According to Landau's rules, the 7 x 7 transition should therefore be first order, like ordinary crystal melting, which would provide support for structural models involving ordered arrays of defects, vacancies or other entities on the surface [53]. The other possibility is one of the models which arise due to a soft-mode driven periodic lattice distortion [54]. The (7x7) reconstruction can be explained in terms of a softening of six surface phonons around the K symmetry point along the Σ directions of the surface Brillouin Zone, fig.17.

If this is the case, then the corresponding 6-component order parameter is now an irreducible representation of the space group,and the phase-transition can, at least in principle, be continuous and 2nd order. However, as pointed out by TOSATTI [52], the role of topological defects, such as dislocations, stacking faults, etc. will affect the diffuse LEED intensity distribution close to the critical point. Also, a recent surface phonon soft-modes analysis by GOLDAMMER et al [54] suggests that the (7x7) reconstruction may in fact be incommensurate with the substrate lattice, although phonon softening occurs at Q_{inc} only a few percent of the $\bar{\Gamma}\bar{K}$ distance off the $\bar{K}$ point, which would be difficult to detect experimentally. However, this would again predict a structural phase-transition of 2nd order.

These calculations also indicate that the atomic displacements belonging to the soft-mode distortions are almost entirely restricted to the outermost two layers i.e. the surface vibrations are effectively decoupled from those characteristic of the bulk,to a large degree. If this proves to be true, then the elemental semiconductors do indeed fit into the overall scheme which we have depicted for the metal surface reconstructions, despite their strong covalent bonding nature. However, the "pinning" effects of topological defects in stabilizing the irreversible phases needs to be futher understood - not only in semiconductor, but in other low-dimensional phase-transition systems.

ACKNOWLEDGEMENTS:

The author wishes to acknowledge fruitful discussions with Profs. V. Heine and S.C. Ying, and the prompting by his students: principally E.F.J. Didham, M.S. Foster and B.J. Hinch; which greatly helped to clarify many of the ideas expressed here.

REFERENCES:

1 E.A. Wood: J. Appl. Phys. 35, 1306 (1964)

2 D.A. King, G.Thomas: Surface Sci. 92, 201 (1980); R.A. Barker, P.J. Estrup: J. Chem. Phys. 74, 201 (1980)

3 A.H. Smith, R.A. Barker, P.J. Estrup: Surface Sci. 136, 327 (1984)
4 T.E. Felter: J. Vac. Sci. Technol. A2, 1006 (1984)
5 P.A. Bennet, M.B. Webb: Surface Sci. 104, 74 (1981)
6 M.A. Van Hove, R.J. Koestner, P.C. Stair, J.P. Biberian, L.L. Kesmodel, I. Bartos, G.A. Somarjai: Surface Sci. 103, 189 (1981)
7 B.E. Hayden, K.C. Prince, P.J. Davie, G. Paolucci, A.M. Bradshaw: Solid State Comm. 48, 325 (1983)
8 For a review see: A. Kahn; Surface Sci. Reports 3, 193 (1983)
9 For a review see: M. Schick; Progr. Surf. Sci. 11, 245 (1981)
10 Recent reviews have appeared in: A. Thomy, X. Duval, J. Regnier; Surface Sci. Reports 1, 1 (1981). Phase Transitions in Surface Films, eds. J.G. Dash, J. Ruvalds (NATO-ASI B51 - Plenum, New York 1980)
11 See review by L.D. Roelofs, P.J. Estrup: Surface Sci. 125, 51 (1983)
12 A.D. Bruce, R.A. Cowley: Structural Phase-Transitions (Taylor & Francis, London 1981)
13 L.D. Landau, E.M. Lifshitz: Statistical Physics (Pergammon, Oxford, U.K. 1980)
14 J.C. Campuzano, M.S. Foster, R.F. Willis, W.N. Unertl: to be published.
15 H.E. Stanley: Introduction to Phase-Transitions and Critical Phenomena (Oxford, U.K. 1971)
16 S.K. Ma: Modern Theory of Critical Phenomena (W.A. Benjamin, Mass., U.S.A. 1976)
17 I.K. Robinson: Phys. Rev. Letters 50, 1145 (1983)
18 L.D. Roelofs; A.R. Korton, T.L. Einstein, R.L. Park: Phys. Rev. Letters 46, 1465 (1981)
19 J.C. Campuzano, G. Jennings, A.M. Lahee in "Proc. ECOSS VI" (Surface Sci. in press 1984)
20 For example, see W.N. Unertl, S.R. McKay in "Proc. Conf. Determination of Surface Structure by LEED" eds. P.M. Marcus, F. Jona (Plenum, New York 1984). Also, M. Henzler in this proceedings
21 M.S. Foster, J.C. Campuzano, R.F. Willis: Proc. Roy. Microscop. Society (London) in Press 1984
22 L.S. Ornstein, F. Zernike: Proc. Acad. Sci. Amst. 17, 793 (1914)
23 M.E. Fisher: J. Math. Phys. 5, 944 (1964)
24 See S.C. Ying's lecture in this proceedings
25 M.K. Debe, D.A. King: Phys. Rev. Letters 39, 708 (1977); T.F. Felter, R.A. Barker, P.J. Estrup: Phys. Rev. Letters 38, 1138 (1977)
26 P. Bak: Solid State Comm. 32, 581 (1979)
27 L.D. Roelofs, G.Y. Hu, S.C. Ying: Phys. Rev. B28, 6369 (1983)
28 A.D. Bruce: Adv. in Physics 29, 111 (1980)
29 R.F. Willis: Proc. ICSS-1, Berkeley, California (Springer, Berlin, in press 1984)
30 J.M. Kosterlitz, D.J. Thouless: J. Phys. C.; Solid State Phys. 6, 1181 (1973)
31 For a recent experimental review see, D.A. King: Physica Scripta T4, 34 (1983)
32 A. Fasolino, G. Santoro, E. Tosatti: Phys. Rev. Letters 44, 1684 (1980); Surface Sci. 125, 317 (1983)
33 W.L. McMillan: Phys. Rev. B12, 1187 and 1197 (1975); ibid B14, 1496 (1976); ibid B16, 643 (1977)
34 V. Heine, J.D.C. McConnel: Phys. Rev. Letters 46, 1092 (1981); J. Phys. C; Solid State Phys. 17, 1199 (1984)
35 V. Heine: Statics and Dynamics of Non-Linear Systems eds. H. Bilz, G. Benedek, R. Zeyher (Springer, Berlin 1984)
36 R.F. Willis: Many Body Phenomena at Surfaces eds. D. Langreth, H. Suhl (Academic, New York 1984)
37 J. Villain, M.B. Gordon: Surface Sci. 125, 1 (1983)
38 P. Bak: Rep. Prog. Phys. 45, 587 (1982)
39 J. Villain: Ordering in Strongly Fluctuating Condensed Matter Systems ed. T. Riste (NATO-ASI 50B - Plenum, New York 1980)
40 A.D. Bruce, R.A. Cowley, A.F. Murray: J. Phys. C; Solid State Phys. 11, 3577; 3591; 3609 (1978)
41 A.H. Smith, R.A. Barker, P.J. Estrup: Surface Sci. 136, 327 (1984)
42 P. Bak, V.L. Pokrovsky: Phys. Rev. Lett. 47, 958 (1981)
43 J.F. Wendelken, G.C. Wang: J. Vac. Sci. Technol. A2, 886 (1984)
44 S. Aubry: Solitons and Condensed Matter Physics eds. A.R. Bishop, T. Schneider (Springer, Berlin 1979)

45 M. Grunze, P.H. Kleben, W.N. Unertl, F.S. Rys: Phys. Rev. Lett. 51, 582 (1983)
46 See recent article by G.V. Kozlov, A.A. Volkov: Europhysics News 15, No7 (1984)
47 V.L. Pokrovsky, A.L. Talapov: Zh. Eksp. Teor. Fiz. 75, 1151 (1978); ibid 78, 269. Also with W.L. McMillan: Sov. Phys. JETP 51, (1) 134 (1980)
48 A.D. Novaco: Phys. Rev. B22, 1645 (1980)
49 A.W. Overhauser: Phys. Rev. B3, 3173 (1971)
50 G.K. Binnig, H. Rohrer, Ch. Gerber, E. Stoll: Surface Sci. 144, 321 (1984)
51 A.D. Novaco, J.P. McTague: Phys. Rev. Lett. 38, 1286 (1977)
52 Following the argument of E. Tosatti: Modern Trends in the Theory of Condensed Matter eds A. Pekalski, J. Przystawa (Springer, Berlin 1980)
53 For example, E. McRae: Proc ICSS-1 Berkeley, California (Springer, Berlin, in press 1984)
54 For example, W. Goldammer, W. Ludwig, W. Zierau, C. Falter: Surface Sci. 141, 139 (1984)

Structural Phase Transitions on W and Mo Surfaces

S.C. Ying*
Department of Physics, Brown University, Providence, RI 02912, USA

1. Introduction

In chemisorption systems, a rich variety of surface structures has been found as a function of temperature and adsorbate coverage. The detailed investigation of these surfaces phases and the transitions between them has been the subject of many recent studies both experimentally and theoretically. In most chemisorption systems, the occurrence of different phases can be understood, at least qualitatively, with a lattice gas model, in which the adsorption is site specific and the ordered phases are due to the lateral adsorbate-adsorbate interaction [1]. However, recent LEED studies [2] first indicated that in the case of W(001) and Mo(001) even the clean surfaces undergo a reversible structural phase transition as the temperature is lowered. The transition is from the (1x1) structure to c(2x2) structure on W(001) and an incommensurate structure in the case of Mo(001). The role of the adsorbate at low coverage is to enhance or suppress the inherent instability of the surface towards reconstruction as well as to alter the relative stability of different structures which are close in energy [3]. These LEED findings were corroborated subsequently by many other types of experiment, such as high energy ion scattering [4], electron energy-loss spectroscopy [5] and electron stimulated desorption [6]. By now, the existence of structural phase transitions for clean W(001) and Mo(001) surfaces is clearly established. These phase transitions are of a different nature from those driven by adsorbate-adsorbate interactions. In the subsequent sections we shall discuss the nature of these new phases and current theoretical models under study for understanding these surface structural phase transitions.

2. Driving Mechanisms for the Structural Transition

The driving mechanism for the structural phase transition on W(001) and Mo(001) is electronic in origin. Recent detailed electronic structure calculations [7,8,9] have shown that the reconstructed surface indeed has a lower total energy. We want to distinguish here between the short and long coherence length limits. In the case of near perfect Fermi surface nesting where small gaps open near the Fermi surface upon reconstruction, only a small number of states in $\underline{k}$ space are affected, and the coherence length is large. On the other hand, for W(001) and Mo(001), photoemission experiment [10] show that the Fermi surface defined by the surface state is very

* Supported in part by the M.R.L. program at Brown University funded through National Science Foundation and by an O.N.R. contract N00014-84-K-0411.

curved and the nesting is far from perfect. In this situation, the electronic structure over a wide region of the surface Brillouin zone is changed upon reconstruction,as opposed to the perfect nesting situation. This results in a small coherence length. Mcmillan [11] and Tosatti [12] have referred to this situation as "strong coupling charge density wave" (C.D.W.) An important consequence of the short coherence length is that over the range of temperatures covering the phase transition from the (1x1) to the reconstructed phase, the electronic contributions to the entropy is negligible compared with the phonon counterpart. Because of the negligible contribution of electronic entropy and the short-ranged nature of the interaction, an effective temperature-independent lattice Hamiltonian is useful for the study of the strong coupling transition. The only effect of surface electronic structure is to renormalize the interaction between the ions. FASOLINO, SANTORO and TOSATTI [13] have used this approach with considerable success. By analyzing the eigenmodes of a finite slab, they showed that most observed phases on W(001) and Mo(001) can be accounted for by an appropriate choice of surface force constants.

3. Monte-Carlo Simulations of Model Two-Dimensional Hamiltonian

For finite temperature studies of the phase transition, we need to include anharmonic terms in the effective lattice Hamiltonian. The finite slab calculations [13] demonstrated that these reconstructions are strongly localized in the surface region; the amplitude of the distortion at the second layer is typically only a few percent of the corresponding value at the first layer. It is therefore a good approximation to limit the anharmonic effects entirely to the first layer. Under this assumption, we have shown [14] that given a microscopic lattice dynamic Hamiltonian, it is possible to generate an effective two-dimensional lattice Hamiltonian for the study of equilibrium phase transition properties. This is accomplished through eliminating the subsurface layer degrees of freedom in the partition function.

At present, our knowledge about the surface force constants of W(001) and Mo(001 especially the anharmonic part, is not yet detailed enough to warrant carrying out this step at a quantitative level. Instead, we have carried out [15] [16] Monte-Carlo simulations and renormalization group studies of simpler two-dimensional Hamiltonians with the same symmetry. For the study of the (1x1) to c(2x2) reconstruction on W(001), we need to keep only the parallel components of displacement $\tilde{u}_x(\underline{\ell})$, $\tilde{u}_y(\underline{\ell})$ in the Hamiltonian. It is also convenient to define a reduced displacement $\underline{u}(\underline{\ell})$

$$u_\alpha(\underline{\ell}) = \tilde{u}_\alpha(\underline{\ell}) \cos \underline{q}_0 \cdot \underline{\ell}, \quad \alpha = x \text{ or } y \quad \text{with} \tag{1}$$

$$\underline{q}_0 = (\pi/a, \pi/a)$$

the phase factor $\cos(\underline{q}_0 \cdot \underline{\ell})$ alternates between -1 and +1 for neighbouring sites. This transformation is analogous to converting an antiferromagnetic system into a ferromagnetic one. The simplest Hamiltonian, keeping only the on-site and nearest neighbour interactions, has the following form

$$H = \frac{J}{2} \sum_{<\underline{\ell},\underline{\ell}'>} \left| \underline{u}(\underline{\ell}) - \underline{u}(\underline{\ell}') \right|^2 - \frac{a}{2} \sum_{\underline{\ell}} u^2(\underline{\ell}) + \frac{b_1}{4} \sum_{\underline{\ell}} u^4(\underline{\ell}) + b_2 \sum_{\underline{\ell}} u_x^2(\underline{\ell})\, u_y^2(\underline{\ell}) \qquad (2)$$

The model Hamiltonian in (2) belongs to the symmetry class of x-y model with cubic anisotropy. For the clean W(001) surface, it is known that the displacement vector of the c(2x2) phase is along <11> direction. This implies that $b_2 < 0$. Also, the transition temperature T_c is ~ 210°K [17]. Comparison with Monte-Carlo results indicates that the coupling J > 0.45 eV/Å^2.

One important result from the Monte-Carlo simulations concerns the development of short-range order above the transition temperature. Just above T_c, the atoms do not return to the equilibrium position immediately. Instead, there are strong short-range order with large domains of ordered region separated by walls. The direction of displacement in different domains are uncorrelated; varying randomly from <1,1> to <1, -1>, so that no net long-range order exists. The size of each domain is of the order of correlation length ξ (T), which is very large at T just above T_c. In this region, each atom is displaced from the (1x1) equilibrium position, even though long-range order does not exist. This is similar to an order-disorder transition. As the temperature increases further, the correlation length ξ (T) decreases rapidly, and eventually the atoms return to vibrational motion (highly anharmonic) about their (1x1) equilibrium positions. Thus, if one just focuses on the high and low temperature limits of the phases, it is a displacive transition. On the other hand, the immediate neighbourhood of T_c has a strong order-disorder character. This dual character of the transition is by now well established in the theory of structural phase transitions [18]. It is particularly pronounced in lower dimensions. When comparing different experimental information on W(001), it is crucial to ask whether a particular experiment is measuring short-range or long-range order. For example, LEED measures a relatively long-range order on the scale of a hundred angstroms. On the other hand, electronic structure depends on relatively short-ranged order, and should not show any immediate change above T_c as measured for example in photoemission experiments. This explains why theoretical calculations [19] of electronic structure of (1x1) W(001) should not be compared with experimental data in the high-temperature phase of W(001), even when LEED is indicating a (1x1) structure.

Another interesting feature that emerges from the Monte-Carlo simulation concerns the size effect. The anisotropy field for W(001) is known to be very weak, since it changes sign for small amount of chemisorbed hydrogen. For this weakly anisotropic case, the correlation length decreases slowly when T is increased above T_c. It is of the order of 20 lattice spacings even at 20 or 30% above T_c. This implies a strong size-effect on the reconstruction of W(001). Experimentally, WENDELKEN and

WANG [17] found a sharp transition on a sample with small step densities at a $T_C \sim 210°K$. On the other hand, BARKER and ESTRUP [3] found a smeared out transition at a $T_C \sim 250°K$. Our interpretation is that in the latter experiment, the step densities are higher and the typical good region on the surface has a linear dimension of only 20 - 30 Å. Thus, the transition is broadened and shifted to an apparently higher temperature. This is confirmed by our numerical simulations [20]. The same numerical work also indicates that the boundary imposes additional anisotropy fields, which may compete against or enhance the intrinsic anisotropy. This effect leads to the theoretical result that for a step along the <1,1> direction, the displacement is perpendicular to the step, whereas for steps along <1,0> direction, atomic displacements would be parallel to the step. These results are in agreement with experimental findings [17] [21].

4. Effect of Adsorbate on the Transition

We now come to the question of the role of adsorbates in the surface structural transitions. This depends crucially on the mobility of adatoms on the surface. When the lateral barriers which the adatoms experience are too high for thermal activation or tunneling, they are quenched in random positions, and generally this leads to the reduction of long-range order and even complete unobservability of the phase transition. In the opposite case of mobile adatoms, the adsorbates can reach thermal equilibrium themselves, and this corresponds to "annealed" impurities. For the most studied system H/W(001), both of these situations can occur, depending on the temperature of the system [3].

We discuss here briefly the situation with high-temperature "annealed" impurities. In this case, LAU and YING [22] have introduced the following Hamiltonian for the chemisorption system,

$$H = H_S [\underline{u}(\ell), \underline{p}(\ell)] + \frac{1}{2} \sum_{\underline{i},\underline{j}} J_{\underline{i},\underline{j}} \, n_{\underline{i}} n_{\underline{j}} + \sum_{\underline{i},\ell} n_{\underline{i}} W(\underline{R}(\underline{i}) - \underline{R}(\ell) - \underline{u}(\ell))$$

$$= H_S + H_{ad} [n_{\underline{i}}, \underline{u}(\ell)] \qquad (3)$$

Here, $\underline{R}(\ell)$, $\underline{u}(\ell)$ and $\underline{p}(\ell)$ stand for the equilibrium position, displacement and momentum vector of the substrate atoms, $n_{\underline{i}}$ the occupation number for the adsorption sites $\underline{R}(\underline{i})$ and W the pairwise interaction between adsorbate and substrate atoms. The lattice Hamiltonian H_S contains the inherent instability of the substrate towards reconstruction. The partition function is given by

$$Z = \int \pi_{\ell} \, d\underline{p}(\ell) \, d\underline{u}(\ell) \exp \left\{ -\beta H_S [\underline{u}(\ell), \underline{p}(\ell)] \right\} . \operatorname*{Tr}_{\{n_i\}} \exp \left\{ -\beta H_{ad} [n_i, \underline{u}(\ell)] \right\} \qquad (4)$$

In (4) the trace over the set $\{n_i\}$ is carried out with the restriction that the coverage θ is maintained at a constant value. From eq. (4), it follows that the whole system can be represented by an effective lattice Hamiltonian

$$H_{eff}\,[\underline{u}(\underline{\ell}),\,\underline{p}(\underline{\ell})] \;=\; H_s + F_{ad}\,[\underline{u}(\underline{\ell});\,T,\theta]$$

where

$$F_{ad}[\underline{u}(\underline{\ell});\,T,\theta] = -\beta^{-1}\,\ell n \operatorname*{Tr}_{\{n_{\underline{i}}\}} \exp\left\{-\beta\, H_{ad}\,[n_{\underline{i}},\underline{u}(\underline{\ell})]\right\} \tag{5}$$

is the free energy of the lattice gas system of adatoms for a fixed displacement configuration of the substrate atoms. In the low coverage limit, where there is no intrinsic ordering tendency due to adatom-adatom interaction, it is a reasonable approximation to calculate F_{ad} in mean field theory giving

$$F_{ad} = 1/2 \sum_{\underline{ij}} J_{\underline{ij}}\, \bar{n}_{\underline{i}}\, \bar{n}_{\underline{j}} - \sum_{\underline{i}} h_{\underline{i}}\, n_{\underline{i}} + T\sum_{\underline{i}} [\bar{n}_{\underline{i}}\ell n\bar{n}_{\underline{i}} + (1-\bar{n}_{\underline{i}})\,\ell n(1-\bar{n}_{\underline{i}})] \tag{6}$$

Here

$$h_{\underline{i}} = \sum_{\underline{\ell}} W[\underline{R}(\underline{i}) - \underline{R}(\underline{\ell}) - \underline{u}(\underline{\ell})]$$

and $\bar{n}_{\underline{i}}$, the mean occupation number of site $\underline{i}$, is to be determined from the condition

$$\frac{\partial F}{\partial \bar{n}_{\underline{i}}} = \mu \tag{7}$$

Solving (5), (6) and (7) together leads to an effective lattice Hamiltonian H_{eff} with temperature and coverage-dependent parameters. The details of this procedure are described elsewhere [23]. The effect of the adsorbate depends crucially on the symmetry of the adsorption site. Hydrogen is known to adsorb on the bridge site of W(001) [3]. For the commensurate c(2x2) phase on W(001), the major effect of H adsorption at low coverage is to increase the transition temperature and switch the displacement vector from <1,1> to <1,0> direction above a critical coverage [22,23]. On the other hand, oxygen which adsorbs on 4-fold symmetric hollow sites, has a drastically different effect. It tends to suppress the c(2x2) phase in favor of a (2x1) phase [24].

5. Soliton Picture for the Commensurate-Incommensurate Transition

On the hydrogen chemisorbed surface H/W(001) as well as clean Mo(001), a commensurate-incommensurate transition also occurs. This transition is far less well understood than the (1x1) to c(2x2) reconstruction. Various mechanisms have been proposed for the C - IC transition [25], [26], [27]. To date,experimental data are not conclusive enough to favour any particular model. FASOLINO, SANTORO AND TOSATTI [27] propose that the incommensurate phase arises not because of long-range interactions,but rather from the coupling of two vibrational modes; one perpendicular to surface,and one in the plane of surface. This coupling exists only when one moves away from symmetric (commensurate) wavevectors for the reconstruction. This is a special case of the general mechanism proposed by HEINE and MCDONNEL for the incommensurate phase in ferroelectrics [28]. However, in both references 27 and 28, the incommensurate state is represented by a sinusoidal state with a single incommensurate wavevector $\underline{q}$. This is known to overestimate the energy of the incommensurate state, and leads to a first order C - IC transition [29]. For other C - IC transitions occuring in physisorption systems,as well as the transition metal dichalcogenides, it is known that the incommensurate state near the C - IC transition is formed from a periodic array of anti-phase domain walls or "solitons", separating commensurate regions [30]. WILLIS [31] has suggested that similar situation occurs for the C - IC transition on H/W(001).

In this section, we demonstrate how one can derive a soliton Hamiltonian for the study of C - IC transition on H/W(001) and Mo(001). Let the two modes that are coupled be vibrations along the $\hat{y}$ and $\hat{z}$ directions. Keeping only these degrees of freedom, the starting lattice Hamiltonian has the form

$$H = \sum_{\underline{\ell}} \left\{ \frac{1}{2} A_y u_y^2(\underline{\ell}) + \frac{1}{2} A_z u_z^2(\underline{\ell}) + \frac{1}{4} B_y u_y^4(\underline{\ell}) + \frac{1}{4} B_z u_z^4(\underline{\ell}) + \frac{1}{4} B_{yz} u_y^2(\underline{\ell}) u_z^2(\underline{\ell}) \right\}$$

$$+ \frac{1}{2} \sum_{<\underline{\ell},\underline{\ell}'>_y} \left\{ J_\ell [u_y(\underline{\ell}) - u_y(\underline{\ell}')]^2 + J_z [u_z(\underline{\ell}) - u_z(\underline{\ell}')]^2 \right\}$$

$$+ \frac{1}{2} \sum_{<\underline{\ell},\underline{\ell}'>_x} \left\{ J_t [u_y(\underline{\ell}) - u_y(\underline{\ell}')]^2 + J_z [u_z(\underline{\ell}) - u_z(\underline{\ell}')]^2 \right\}$$

$$+ \sum_{<\underline{\ell},\underline{\ell}'>_y} \frac{J}{2} \left\{ u_y(\underline{\ell}) u_z(\underline{\ell}+a\hat{y}) - u_y(\underline{\ell}+a\hat{y}) u_z(\underline{\ell}) \right\} \tag{8}$$

Here, the sum over $\{\underline{\ell},\underline{\ell}'\}$ is over nearest neighbours only. $\underline{u}(\underline{\ell})$ is related to the real displacement vector by an alternating phase factor as before. Now, we change from $\underline{u}(\underline{\ell})$ to the phase and amplitude variables

$$u_y(\underline{\ell}) = \rho_1(\underline{\ell}) \cos \phi(\underline{\ell})$$

$$u_z(\underline{\ell}) = \rho_2(\underline{\ell}) \cos \{\phi(\underline{\ell}) + \delta(\underline{\ell})\} \tag{9}$$

In the commensurate phase $\phi(\underline{\ell})=0$, $\delta(\underline{\ell})={}^{\pi}/2$ and $\rho_1(\underline{\ell}) = \rho_o = -A_y/B_y$. Near the C - IC transition, amplitude fluctuations are costly in energy,and we can use the fixed amplitude approximation for the incommensurate phase

$$\rho_1(\underline{\ell}) = \rho_o = -\frac{A_y}{B_y} \quad \text{and}$$

$$\rho_2(\underline{\ell}) = \lambda\rho_o \quad . \tag{10}$$

with λ to be determined below. Also to maximize the coupling between y-z modes, $\delta(\underline{\ell})$ should remain at the constant value ${}^{\pi}/2$. Thus,the only spatially varying quantity is the phase angle $\phi(\underline{\ell})$. If we now go to the continuum limit, ϕ becomes a function of x and y. Keeping only the lowest order gradient terms, the following Hamiltonian results

$$H = \int d\underline{r} \ \{v(\phi) + \frac{1}{2}f_x(\phi) \ (\frac{\partial\phi}{\partial x})^2 + \ \frac{1}{2}f_y(\phi) \ (\frac{\partial\phi}{\partial y})^2 - J\lambda\rho_0^2 \ \frac{\partial\phi}{\partial y}\} \quad \text{where} \tag{11}$$

$$v(\phi) = v_0 \ (1-\cos 2\phi) - B_0(1-\cos 4\phi)$$

$$v_0 = \frac{1}{4}\rho_0^2(-\frac{1}{2}A_y + A_z\lambda^2 + \frac{1}{2}B_z\lambda^4\rho_0^2)$$

$$B_0 = (\rho_0^4/32) \ (B_y + B_z\lambda^4 - B_{yz}\lambda^2)$$

$$f_y(\phi) = 2J_\ell\rho_0^2 \ (1+\zeta \cos^2\psi)$$

$$f_x(\phi) = 2J_t\rho_0^2 \ (1+\zeta \cos^2\phi)$$

$$\zeta = \lambda^2 \ J_z/J_\ell - 1 \tag{12}$$

The Hamiltonian in (11) is a typical 2D anisotropic soliton model. In the x-direction only commensurate terms appear. The linear term in $\frac{\partial\phi}{\partial y}$ provides the driving force for the incommensurate phase. It differs slightly from the standard Hamiltonian for C - IC transition through the ϕ dependence of the elastic coefficients $f(\phi)$ and $f_y(\phi)$.

At T = 0, we only need to find the ground state configuration $\phi(x,y)$ that minimizes the total energy. Obviously $\partial_x\phi=0$. Minimization of the energy leads to the solution

$$V(\phi) = f_y(\phi) \ (\frac{\partial\phi}{\partial y})^2 + C \tag{13}$$

Now we consider a single soliton or anti-phase domain wall where the phase angle ϕ varies from 0 to π. For this case, the integration constant C = 0. The energy of the soliton per unit length is

$$\varepsilon_s = \varepsilon_s^0 - \pi \ J\lambda\rho_0^2 \qquad \text{where} \tag{14}$$

$$\varepsilon_s^0 = \int_0^\pi d\,\phi \ (2f_y(\phi) \ V(\phi) \ )^{\frac{1}{2}} \tag{15}$$

The C - IC phase transition occurs at $\varepsilon_s = 0$, the value of λ is determined by minimization of the soliton energy. As can be easily verified, the present soliton solution provides a lower energy than the simple one-mode approximation, in which the phase $\phi(y)$ is taken to be just the simple form $\phi(y) = qy$. This is true except in a small region around $\rho_0=0$ where one expects the "phase modulation only" assumption to be a poor description. The physical picture of the incommensurate phase here consists of domains of commensurate region with displacements in the plane of the surface separated by solitons (anti-phase domain walls), in which there are mixed parallel and perpendicular displacements. This is shown schematically in Fig 1.

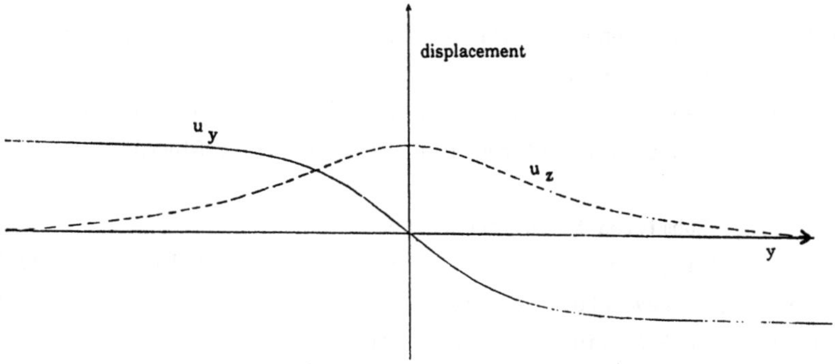

Fig 1: Displacement pattern in an anti-phase domain wall of the incommensurate phase of H/W(001)

At finite temperatures, it is the free energy rather than the energy of the soliton that needs to be considered. Since there are many configurations for which a soliton line can bend or deform, it leads to a large entropy and a lower free energy at finite temperatures. Thus,an increase in temperature always favors the incommensurate phase. We have indicated in a recent work [29] how the commensuration potential v(ϕ) can be renormalized to yield the transition criterion at finite temperatures. Actually, for the two-dimensional soliton lattice, a liquid phase separates the commensurate and incommensurate phases when dislocations are properly included in the theory [30]. However, it is expected that as far as short to medium range order is concerned,such as that studied in typical LEED experiment, the liquid phase differs little from the solid phase,and the C - IC boundary determined without consideration of dislocations still provides a useful description.

References

1. L.D. Roelofs and P.J. Estrup, Surface Sci. 125, 51 (1983)
2. T.F. Felter, R.A. Barker and P.J. Estrup, Phys. Rev. Letters 38, 1138 (1977); M.K. Debe and D.A. King, Surface Sci. 81, 193 (1979)
3. R.A. Barker and P.J. Estrup, J. Chem. Phys. 74, 201 (1980); D.A. King and G. Thomas, Surface Sci. 92, 201 (1980)
4. I. Stensgaard, L.C. Feldman and P.J. Silverman, Phys. Rev. Letters 42, 247 (1979); I. Stensgaard and L.C. Feldman, Surface Sci. 87, 410 (1979)
5. M.R. Barnes and R.F. Willis, Phys. Rev. Letters 41, 1727 (1978)
6. R. Jaeger and D. Menzel, Surface Sci. 100, 561 (1980)
7. J.E. Inglesfield, J. Phys. C11, L69 (1978); 12, 149 (1979)
8. K. Terakura, I. Terakura and N. Hamada, Surface Sci. 103, 103 (1981)
9. A. Freeman, this volume
10. M.I. Holmes and T. Gustafson, Phys. Rev. Letters 47, 443 (1981)
11. W. Mcmillan, Phys. Rev. B16, 643 (1977)

12. E. Tosatti in Korpacz Winter School of Theoretical Physics, Ed.A Pekalski and J. Pryzstawa (Springer, Berlin 1978)
13. A. Fasolino, G. Santoro and E. Tosatti, Phys. Rev. Letters 44, 1684 (1980); Proceedings of International conference on Phonon Physics, Bloomington, Indiana (1981)
14. G.Y. Hu and S.C. Ying, Submitted to Surface Science
15. S.C. Ying and L.D. Roelofs, Surface Sci. 125, 218 (1983); L.D. Roelofs, G.Y. Hu and S.C. Ying, Phys. Rev. B28, 6369 (1983)
16. G.Y. Hu, L.D. Roelofs and S.C. Ying, to be published
17. J. Wendelken and G.C. Wang, Proceedings of 30th National Symposium, and private communications
18. A.D. Bruce, Advances in Physics 29, 111 (1980)
19. S. Ohnishi, A.J. Freeman, E. Wimmer, Phys. Rev. B29, 5267 (1984)
20. G.Y. Hu and S.C. Ying, to be published
21. P. Estrup, private communication
22. K.H. Lau and S.C. Ying, Phys. Rev. Letters 44, 1222 (1980)
23. L.D. Roelofs and S.C. Ying, to appear in Surface Science
24. P. Estrup, L.D. Roelofs and S.C. Ying, to be published
25. K.H. Lau and S.C. Ying in: Ordering in Two Dimensions, Ed. S.K. Sinha (North-Holland, New York 1980)
26. T. Inaoha and A. Yoshimori, Surface Sci. 115, 301 (1982)
27. A. Fasolino, G. Santoro and E. Tosatti, Surface Sci. 125, 317 (1983)
28. V. Heine and J.D.C. McConnel, Phys. Rev. Letters 46, 1092 (1981)
29. S.C. Ying and G.Y. Hu, Proceedings of 1st International conference on the structure of solid surfaces, Berkeley, 1984
30. W. Mcmillan, V.L. Pokrovsky and A.L. Talapov, Sov. Phys. JETP 51, (1) 134 (1980)
31. R.F. Willis in Many-Body Phenomena at Surface, Eds. Langreth and Suhl, p. 297 (Academic Press, 1984)

Contributions

Charge Density Waves in Layered Compounds

P.C. Klipstein

Cavendish Laboratory, University of Cambridge, Madingley Road
Cambridge, United Kingdom

The Transition Metal Dichalcogenide compounds (eg. TaS_2) have a layered structure where each layer consists of a hexagonally packed plane of metal atoms sandwiched between two of chalcogen atoms [1]. In the metallic compounds, the Fermi surface consists of cylindrical electron or hole pockets with axes perpendicular to the layer [2]. This "two-dimensional" electronic character leads in most cases to a Periodic Lattice Distortion (PLD) below some temperature T_d [3]. The distortion is screened by the electron gas, giving rise to an associated Charge Density Wave (CDW).

The simplest example of a PLD-CDW is a linear atomic chain with lattice constant a and one valence electron per atom, resulting in a half-full conduction band. Dimerising the chain creates a new Brillouin Zone half the size of the original one and a gap, Δ, at the Fermi surface (F.S.) lowering the electronic energy. If the lattice distortion energy is less than this gain in electronic energy, a second order transition to the distorted (dimerised) state will occur when $kT < \Delta$[4]. In 3D the principle is the same, but not all states near the F.S. are involved. The driving force for the transition depends on the electron-phonon coupling, $g(\bar{q})$, the stiffness of the lattice in the absence of electron-phonon coupling, $\hbar\omega(\bar{q})$, and the proportion of occupied electron states whose energy is lowered by the distortion with wave-vector $\bar{q}$, a measure of which is given by the free electron susceptibility $\chi^{o}(\bar{q})$ defined in reference 3. A PLD will occur in a simple non-interacting electron picture if $4g^2/\hbar\omega > 1/\chi$ [5]. $\chi^{o}(\bar{q})$ is plotted in figure 1. The transition criterion, including exchange and correlation effects, is discussed in references 5 and 6.

A CDW may be commensurate (CCDW) or incommensurate (ICDW) depending on the size of the F.S. For a CCDW of fixed amplitude, the condensation energy depends on the phase. For an ICDW it does not. Thus a CCDW may gain extra "Lock-In" energy by adjusting its phase to the most beneficial value [7]. This may lead in some cases to a CCDW ground state, even though the best "nesting" wave-vector across the F.S. would favour an ICDW.

For an ICDW, the displacement may be written $u = u_o \sin(qx + \phi(x))$, where $u = u_o \sin qx$ is a CCDW and $\phi < qx$. There are two ways that ϕ may be expected to vary with x [9]. First, ϕ may vary in proportion to x and the CDW is simply sinusoidal. Second, ϕ may vary "stepwise" with regions of CCDW separated by "discommensurations" where the phase changes rapidly, such that the phase change over a large distance containing many discommensurations is the same as in the simple sinusoidal case (see figure 2). This situation is favoured by the gain in lock-in energy over the commensurate regions, and has a lower free energy than the sinusoidal ICDW. Note that the phase change at a discommensuration is $2\pi/n$ when the wavelength of the CCDW regions is na. McMillan shows that for a CCDW ground state and ICDW nesting of the F.S. a second order ICDW $\rightarrow$ CCDW transition will occur on cooling to some temperature T_o when the concentration of discommensurations decreases smoothly to zero [9]. In this case, the wave-vector of the ICDW appears to change continuously with temperature. This behaviour is observed in X-ray diffraction experiments on 2H-$TaSe_2$ [10] where the presence of

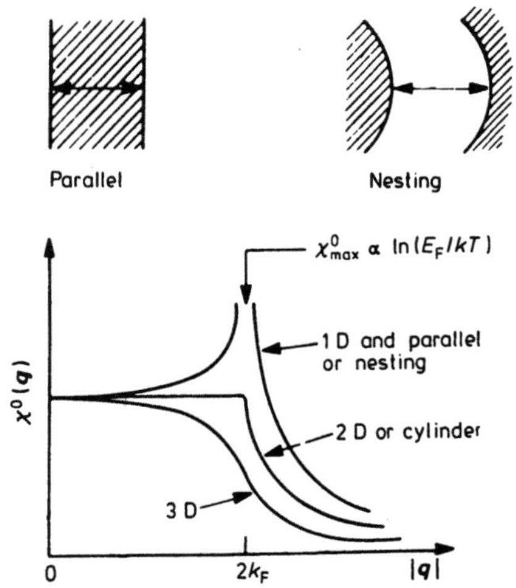

FIGURE 1 Free electron susceptibility, $\chi^o(\bar{q})$, plotted against q for a 1D, a circular cylindrical and a spherical Fermi surface (Ref 3)

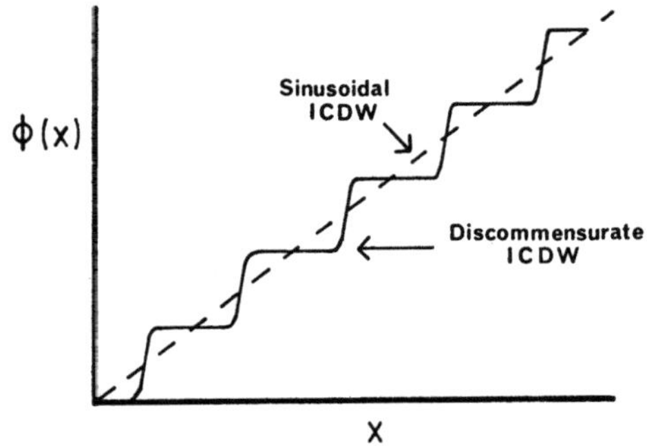

FIGURE 2 Two forms for $\phi(x)$ in an incommensurate CDW

discommensurations has elegantly been confirmed by direct imaging in electon microscopy [11,12] Hysteresis effects and a lowering of symmetry in the "striped" ICDW phase of this material (which exists only on warming from the CCDW phase) may be directly related to the repulsive interaction between discommensurations. A phase transition,which is observed at 280K only on warming in another material 1T-TaS_2 [13,14,15] shows the existence of a new phase between the CCDW to ICDW transition at ∿200K and this transition at 280K. This phase has recently been shown to be associated with a resistance anisotropy within the layer [14,15,16], which indicates symmetry lowering and may provide a new example of discommensuration phenomena in a layered compound.

REFERENCES
1. J.A. Wilson and A.D. Yoffe: Adv. Phys. 18, 193 (1969)
2. J.A. Wilson, F.J. DiSalvo and S. Mahajan: Adv. Phys. 24, 117 (1975)
3. R.H. Friend and D. Jerome: J. Phys. C. 12, 1441, (1979)
4. G.A. Toombs: Phys, Rep. 40C, 181 (1978)
5. S.K. Chan and V. Heine: J. Phys. F. 3, 795 (1973)
6. P.B. Littlewood and V. Heine: J. Phys. C. 14, 2943 (1981)
7. P.A. Lee, T.M. Rice and P.W. Anderson: Sol. St. Comm. 14, 703 (1974)
8. W.L. McMillan: Phys. Rev. B12, 1187 (1975)
9. W.L. McMillan: Phys. Rev. B16, 643 (1977)
10. R.M. Fleming, D.E. Moncton, D.B. McWhan and F.J. DiSalvo: Phys. Rev. Lett. 45, 576 (1980)
11. K.K. Fung, S. McKernan, J.W. Steeds and J.A. Wilson: J. Phys. C. 14, 5417 (1981)
12. C.H. Chen, J.M. Gibson and R.M. Fleming: Phys. Rev. B. 26, 184 (1982)
13. O. Sezerman, A.M. Simpson and M.H. Jericho: Sol. St. Comm. 36, 737 (1980)
14. S.C. Bayliss, A.M. Ghorayeb and D.R.P. Guy: J. Phys. C. 17, L533 (1984)
15. H. Fujimoto and H. Ozaki: Sol. St. Comm. 49, 1117 (1984)
16. M. Nunez-Regueiro and C. Ayache: Private communication (1984)

Dynamics of Mismatched Overlayers

K.M. Martini, S. Burdick, M. El-Batanouny
Department of Physics, Boston University, Boston, MA 02215, USA
G. Kirczenow
Department of Physics, Simon Fraser University, Burnaby, B.C. V5A 1S6, Canada

The interest in the dynamics of surfaces and overlayers on surfaces has recently intensified. This is largely due to the development of new, more detail-yielding, experimental techniques [1] and the investigation of new materials. The complexities inherent in these systems have hindered the development of a clear theoretical understanding of their dynamic behavior. The simultaneous occurrence of nonlinear interaction laws, incommensurabilities, disorder due to pinning effects and the resulting complexity of the excitation spectra did not favor a simple analytical approach [2]. We present here the results of a molecular dynamics study. This numerical method, coupled with a simple, but very effective, Fourier transformation allowed us to investigate this complicated problem without further approximations.

We use the Frenkel-Kontorova model of a rigid, sinusoidal substrate potential with a harmonic-coupled chain overlayer, since it has been widely used to study and interpret surface phenomena. The potential energy in this model is

$$V=C/2\sum_i (x_i-x_{i+1})^2 + W/2\sum_i [1-\cos 2\pi([i-1]b+x_i)/a] \tag{1}$$

where a is the the period of the substrate potential, x_i is the deviation of the ith overlayer particle measured from its ideal position in the absence of the substrate, W the substrate potential depth, b the length of an unstretched overlayer bond, and C the spring constant. We introduce a quantity $SR=(2\pi/a)^2 W/8C$, the <u>strength ratio</u>, which will serve as a parameter of the relative importance of the substrate potential with respect to the overlayer spring stiffness. The natural mismatch between the overlayer and the substrate is described by the ratio a/b. The systems simulated are finite overlayer islands on a substrate of infinite extension. We treated a chain with free ends, in order to permit the system to assume its natural length in the presence of the substrate, and to study end-effects. (We have also studied systems with periodic boundary conditions, in order to exclude the endmode effects.)

We start the calculation by distributing the overlayer particles into some arbitrary locations, and then allow them to relax down to their T=0 equilibrium positions. Due to the free ends, the equilibrium length of the overlayer chain can differ markedly from its value in the absence of the substrate. (For the case studied mainly - a/b=9/8 and 128 overlayer particles - we found, as a function of SR, between 0 and 14 solitons in the ground state equilibrium configuration.) After equilibrium is reached, a small random displacement ($<10^{-3}$a) is superimposed on the position of each particle to generate the initial conditions for our molecular dynamics simulation. We set the velocity of each atom equal to zero initially, and integrate Newton's equations of motion numerically to find the subsequent motion. The vibrational spectrum is obtained directly from the Fourier transformation

$$G(k,w)=\int dt \sum_j u_j(t) \exp[-i(jbk + wt)] \tag{2}$$

where $u_j(t)$ is the displacement of atom j from its equilibrium position. We identify modes with the local maxima of $[G(k,w)]^2$.

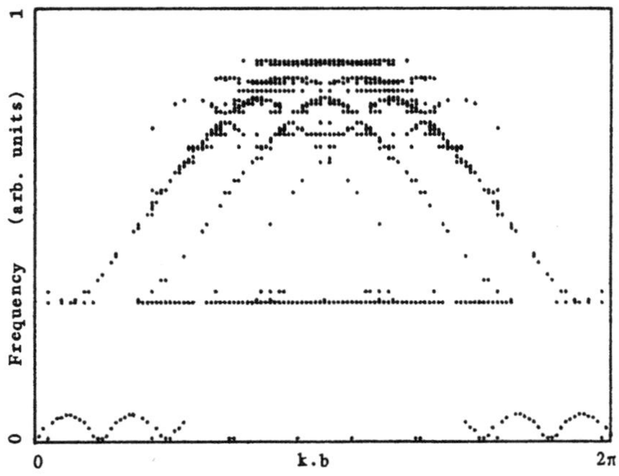

Fig. 1 :

Dispersion relation of the incommensurate overlayer as a function of the renormalized wave vector k (mapped onto the extended Brillouin zone of the unstretched chain) with a/b=1.125 and SR=0.1

The first phase of the calculation not only provided the respective equilibrium position of the system under investigation, but also the energy of that configuration. This information, as a function of the parameters SR and natural mismatch, was compiled into the energy phase diagrams. These showed that the number of solitons in the ground state can vary as a function of SR (between 0 and a maximum number, determined by the natural misfit and the island size). We could also continue each soliton arrangement into the metastable state region. Furthermore, we found that the solitons in the overlayer chain can be in a disordered arrangement if SR is too large, creating a multitude of additional metastable states. The disorder-order transition is driven by the soliton-soliton interaction,and it depends on the local soliton density, the natural mismatch and SR. At small SR values , a periodic soliton array is formed. The ends of the free chain also repel solitons and can indeed be represented as part of a soliton [3].

The vibrational features can be divided into three major parts. Figure 1 shows a typical dispersion relation. The dispersion of the lowest-lying frequency modes originates from the soliton-soliton interactions. The solitons behave like a chain of coupled quasiparticles,and we can describe the results in terms of the extended Brillouin zone scheme corresponding to their spacing. We verified this relation between soliton density and modulation wave vector in all cases investigated. The minimum frequency goes to zero at the disorder-order transition of the soliton system (this is a local effect governed by the local soliton density [3]). A similar transition was described for a more restricted case by AUBRY [4]. This branch loses its dispersion at high SR values (the solitons do not overlap any more and behave like Einstein oscillators). The second major part in the dispersion relation can be atttributed to the phonon-like modes. They are confined to a main branch and several zonefolded ones. The zonefolding wave vector is the same as the wave vector of the soliton Brillouin zone (i.e., the inverse of the soliton spacing and not related to the natural mismatch) [3]. The third component is due to the endmodes. They show up as flat bands in the dispersion relation because of their strong spacial localization. A prominent example can be seen at the lower frequencies of the phonon branch.

1. M.G. Brusdeylins, R.B. Doak, J.P. Toennies: Phys.Rev.Lett. 16, 937 (1981); J.M. Szeftel, S. Lehwald, H. Ibach, T.S. Rahman, J.E. Black, D.L. Mills: Phys. Rev. Lett. 51, 268 (1983)
2. A.D. Novaco: Phys. Rev. B 22, 1645 (1980)
3. K.M. Martini, S.Burdick, M. El-Batanouny and G. Kirczenow: Phys.Rev. B30, 492 (1984) and to be published
4. M. Peyrard and S. Aubry: J.Phys. C. 16, 1593 (1983) and refrences therein.

Part IV

Electronic States

Electronic Structure of Surfaces, Interfaces and Superlattices

A.J. Freeman

Department of Physics and Astronomy,and Materials Research Center
Northwestern University, Evanston, IL 60201, USA

1. Introduction

The existence of this Course at the International School of Material Science and Technology on "Dynamical Phenomena at Surfaces, Interfaces, and Superlattices" is, in itself, a strong recognition of the growing importance of this area of research. Clearly, a vast amount of detailed information has been obtained during the past decade, which is due to the development and refinement of sensitive experimental techniques and to highly precise theoretical/computational methods for determining theoretically the phenomena associated with these structures. This advance in research has been spurred by the enormous technological relevance of surface and interface phenomena in areas such as microstructure electronics, catalysis, and corrosion. Associated with this has been the recognition that one of the most important developments in the last decade lies in the preparation of synthetic structures on the sub-micron level. The excitement surrounding this development lies in the promise of producing materials with desired properties to specification, of permitting new scientific phenomena to be investigated and novel device applications to be made on artifical materials not found in nature. In other words, what we are witnessing today is the development of materials engineering on the microstructure scale[1].

Examples of the microstructure domain include thin film junctions (sandwiches) of two dissimilar metals which are either magnetic or superconducting, etc., and modulated structures or superlattices (both crystal and amorphous) consisting of alternating layers of materials A and B, i.e., ABABA--. In these structures, one has essentially only interface phenomena whereas for free surfaces one deals with a largely unexplored frontier field,in which new insights into the electronic structure are expected because (i) the existence and role of surface states and (ii) the reduced coordination and symmetry lead to important property differences with respect to bulk systems.

In these lectures we present some indication of the important role that theory can play in elucidating the phenomena associated with the electronic structure at surface, interfaces, and superlattices. Because space is limited and this subject is vast, only a few examples can be given.

2. Surfaces and Overlayers

2.1. Introduction

It has been possible recently to extend well-known bulk methods of calculation to treat surfaces. Thus, first-principles energy-band studies for surfaces have demonstrated a fair degree of sophistication in tackling a number of complex problems involving the electronic structure of free surfaces (including magnetic properties, surface reconstruction and relaxation), chemisorption bonding of atomic absorbates, and interface phenomena. A principal aim of such calculations is to achieve self-consistency between the calculated electronic charge density and the input potential. Such effects are especially important for systems with sizeable charge transfers between the atomic constituents.

While the electronic properties of semiconductor and simple metal surfaces have been the subject of extensive study, it is only recently that attempts have been made to deal theoretically with the additional complexity arising from the localized d-electrons in noble and transition metals. For example, a fairly recent review article [2] needed about one page only to describe the few realistic calculations for transition metal surfaces which had been reported at that time, and none of these was self-consistent. It is precisely the d-band electrons, however, which play such an important role in such interesting and important surface phenomena as surface reconstructions, differences between surface and bulk magnetic properties, and catalytic activity (especially the group VIII transition metals and their alloys). While the importance of the d-band electrons in these phenomena is generally accepted, the details of their influence are less clear.

Recently we have developed a highly precise, and unified method for calculating surface structures, the full-potential linearized augmented plane wave (FLAPW) method [3]. This is a unified method in that it can easily treat not only simple metals and semiconductors but also transition metal surfaces. It is also unified, in the sense that it is capable of treating molecular adsorption on surfaces and also the extreme limit of the isolated molecule and the clean surface. Furthermore, as will be demonstrated, through the calculation of the total energy of different atomic configurations, the method can be used to determine the preferred structure of systems of interest.

Our focus is on the treatment of extended surfaces with or without ordered overlayers of adsorbed atoms or molecules. Probably the most successful structural model uses a thin-slab to simulate both surface and bulk effects on an equal footing. Calculations in this model are performed for a finite thickness slab, which is infinitely periodic in the plane parallel to the surface. Typically, slabs 5-13 atomic-layers thick are used. To ensure accurate results, the slab should be thick enough so that the electronic structure in the interior of the slab resembles closely the expected bulk structure.

2.2. Local-Spin-Density-Functional (LSDF) Theory

Major progress in the last fifteen years has been made in the theoretical understanding of ground state properties of many-electron systems. This achievement is based largely on the success of LSDF theory [4-7]. LSDF theory introduces an effective one-electron potential which is a functional of the charge density, $\rho(\vec{r})$, and the spin-density, $n(\vec{r}) = \rho^{\uparrow}(\vec{r}) - \rho^{\downarrow}(\vec{r})$:

$$V^{\uparrow}_{eff}(\vec{r}) = V^{\uparrow}_{eff}\,[\rho(\vec{r}),\, n(\vec{r})]$$
$$V^{\downarrow}_{eff}(\vec{r}) = V^{\downarrow}_{eff}\,[\rho(\vec{r}),\, n(\vec{r})] \qquad (2.1)$$

(For paramagnetic systems, $n(\vec{r}) = 0$ and $V^{\uparrow}_{eff} = V^{\downarrow}_{eff}$.) These lead to one-particle Schrodinger-like equations for spin-up and spin-down one-particle states (in Rydberg atomic units)

$$[-\nabla^2 + V^{\uparrow(\downarrow)}_{eff}(\vec{r}) - \varepsilon_i^{\uparrow(\downarrow)}]\, \psi_i^{\uparrow(\downarrow)}(\vec{r}) = 0 \qquad (2.2)$$

In LSDF theory the one-particle energies, $\varepsilon_i^{\uparrow(\downarrow)}$, and states, $\psi_i^{\uparrow(\downarrow)}$, have essentially no physical meaning other than specifying the charge- and spin-densities

$$\rho^{\uparrow(\downarrow)}(\vec{r}) = \sum_i \psi_i^{\uparrow(\downarrow)*}(\vec{r})\, \psi_i^{\uparrow(\downarrow)}(\vec{r})\,\Theta(E_F - \varepsilon_i^{\uparrow(\downarrow)}) \qquad (2.3)$$

where E_F is the Fermi energy and the step function $(E_F - \varepsilon_i^{\uparrow(\downarrow)})$ ensures that the summation is over occupied states only. The essential feature of (2.2) and (2.3) is that the potential $V_{eff}^{\uparrow(\downarrow)}$ must be self-consistent with the charge- and spin-densities. Thus, these equations must be solved by iteration until self-consistency is achieved.

The effective potential can be expressed as the sum of two terms:

$$V_{eff}^{\uparrow(\downarrow)}(\vec{r}) = V_{coul}[\rho(\vec{r})] + V_{xc}^{\uparrow(\downarrow)}[\rho(\vec{r}),n(\vec{r})] \tag{2.4}$$

where V_{coul} is the classical Hartree potential due to the charge distribution of all the electrons and nucleii in the system. $V_{xc}^{\uparrow(\downarrow)}$ is the exchange-correlation potential and includes all many-body effects on ground state properties. The functional form for $V_{xc}^{\uparrow(\downarrow)}$ is usually determined by calculations for the homogeneous electron gas at different densities and spin-densities [6,7]. In the course of iterating (2.2-2.4), large changes occur from one iteration to the next in V_{coul} due to sizeable charge rearrangements as self-consistency is approached. For low symmetry systems like surfaces, these must be strongly damped for the iterative process to converge. Typically, this is done using some sort of attenuated feedback method.

LSDF theory has been used successfully to understand many bulk ground state properties. For example, MORUZZI et al. [8], have calculated ground state properties for elements in the periodic table through the 4d series, and the agreement with experiment is typically about 10% for such properties as binding energies, lattice parameters and bulk modulus. Similarly, the theory of itinerant electron magnetism has been considerably advanced in recent years by the success of *ab initio* self-consistent band theory calculations in providing a quantitative understanding of the bulk ground state properties to the ferromagnetic transition metals iron, cobalt, and nickel (indeed, the calculations [8], predict that these are the only elemental ferromagnets). These calculations have been remarkably successful in obtaining good agreement with such experimental quantities as magnetization, neutron form factors, hyperfine field, lattice parameter, bulk modulus, cohesive energies, and Fermi surface properties. This is particularly impressive considering that all many-electron effects are included only through an effective one-electron local potential. This achievement is a major confirmation of the utility LSDF theory, which provides the formal justification for using the single-particle picture (2.2)-(2.4) to determine ground state properties.

While the ground state properties are now quantitatively understood on this basis, this is not true, unfortunately, for the elementary excitations of many-electron systems as measured, for example, by photoemission. Spin-density-functional theory provides the formal justification for using single-particle band structures and charge densities to obtain ground state properties only.

2.3. Methodology and Approach

In its local approximation, density-functional theory leads to Schroedinger-like one-particle equations (Kohn-Sham equations) containing an effective potential energy operator, which is determined by the self-consistent charge distribution. Thus, the local density-functional one-particle equations have to be solved iteratively.

One of the most accurate and powerful schemes to solve the LSDF one-particle equations for the film geometry is the all-electron full-potential linearized-augmented-plane-wave (FLAPW) method [3]. The basic idea in this variational method is the partition of real space into three different regions, namely spheres around the nuclei, vacuum regions on both sides of the film, and the remaining interstitial region. In each of these regions the "natural" form of the variational basis functions is adopted, i.e. plane waves in the interstitial region, a product of radial functions and spherical harmonics inside the spheres, and in the vacuum a product of functions which depend only on the coordiate normal to the film and 2-dimensional plane waves. Each of these basis functions is continuous in value and derivative across the various boundaries. This is possible because inside the spheres (and analogously in the vacuum) two radial functions for each ℓ-value are used, namely the solution of the radial Schröedinger equation for the current potential and its energy derivative.

In the FLAPW method, no shape approximations are made to the charge density and the potential. Both the charge density and the effective one-electron potential are represented by the same analytical expansions, i.e. a Fourier representation in the interstitial region, an expansion in spherical harmonics inside the spheres, and a 2-dimensional Fourier series in each vacuum plane parallel to the surface. The generality of the potential requires a method to solve Poisson's equation for a density and potential without shape approximations. This is achieved by a new scheme, which goes beyond the Ewald method [9]. The key idea in this new scheme is the observation that the potential outside a sphere does not depend on the actual charge density inside the sphere, but only on the multipole moments of the sphere. Now, Poisson's equation is solved straight-forwardly when the charge density is given in a Fourier representation. Because of the sharp structure of the charge density in the core region (including the nuclear charge), a Fourier expansion of the total density would be extremely slowly convergent. However, since the potential outside the sphere depends on the charge inside only through the multipole moments, the true charge density can be replaced by a smooth density, which has a rapidly converging Fourier series and the same multipole moments as the true density. With this replacement of the density inside the spheres, we have a Fourier expansion of a charge density which gives the correct potential outside and also on the sphere boundaries. To find the potential inside the spheres, we are faced in a final step with a standard boundary-value problem of classical electrostatics, which can be solved from the original charge densities inside the spheres and the potential on the sphere boundaries by a Green's function method.

Thus, the FLAPW method allows a fully self-consistent solution of the LSFD one-particle equations for the film geometry and yields charge densities and spin densities close to the LSFD limit. Besides the total charge density, the key quantity in density-functional theory is the total energy corresponding to the ground state charge density. Recently, we have presented a new scheme [10] to calculate accurate and stable all-electron density-functional total energies and have applied it within the FLAPW method. The capability of total energy calculations for various geometrical arrangements provides us with a powerful theoretical tool to study the energetics and, at least in principle, the dynamics of surfaces and overlayers.

In the thin film geometry and the LSDF approach, the wave functions for each state are solutions of the one-particle equations

$$[-\nabla^2 + V_{eff}(\vec{r})]\ \psi_i(\vec{k},\vec{r}) = \varepsilon_i(\vec{k})\ \psi_i(\vec{k},\vec{r}) \tag{3.1}$$

where $\vec{k}$ is a vector of the two-dimensional first Brillouin zone and i is a band index. The effective potential, V_{eff}, is given as the sum of the electrostatic Coulomb potential, related to the charge density by Poisson's equation, and the local exchange-correlation potential as obtained from many-body theory.

In the FLAPW method for thin films, the electronic charge density is represented in each of the three spacial regions by the "natural" representation, namely

$$\rho(\vec{r}) = \begin{cases} \sum_j \rho_j \exp(i\vec{G}_j \cdot \vec{r}) & \text{for } \vec{r} \in \text{interstitial} \quad (3.2a) \\ \sum_L \rho_L(r)\, Y_L(\hat{r}) & \text{for } \vec{r} \in \text{sphere} \quad (3.2b) \\ \sum_q \rho_q(z) \exp(i\vec{K}_q \cdot \vec{r}) & \text{for } \vec{r} \in \text{vacuum} \quad (3.2c) \end{cases}$$

Here, $\vec{k}_j = \vec{k} + G_j$ where $\vec{G}_j$ is a vector of the three-dimensional reciprocal lattice.

The electrostatic potential is obtained from the electronic charge density and the nuclear charges by solving Poisson's equation using the technique described by [9] as implemented into the FLAPW method [3]. The exchange-correlation potential is calculated from the local electronic charge density by a least squares fitting technique, where the root-mean-square deviation is usually about 1 mRy in the interstitial region and better than 0.1 mRy inside the spheres and in the vacuum. Finally, the effective one-electron potential as the sum of the electrostatic Coulomb potential and the exchange-correlation potential is represented in the form completely analogous to the charge density as given by (3.2a-c).

The LSDF one-particle equations (3.1) are now solved iteratively. A starting density in the form (3.2) is constructed from a superposition of self-consistent atomic densities. From this density, the corresponding potential is calculated which defines the effective one-particle operator in (3.1). These solutions define now the film-wavefunctions $\psi_i(\vec{k})$ which in turn yield, according to Fermi-Dirac statistics, a new charge density

$$\rho'(r) = \sum_{\text{occ.}} \int_{BZ} \psi_i^*(\vec{k},\vec{r})\, \psi_i(\vec{k},\vec{r})\, d^2k \tag{3.3}$$

where the summation runs over all occupied states. The density of the core electrons is obtained by solving fully-relativistically a free-atom like problem using the current effective potential. This completes one iteration cycle. The new density is fed back and self-consistency is achieved when $\rho' = \rho$, i.e. when the output density is equal to the input density. Practically, self-consistency is assumed when the potentials corresponding to the input and output densities differ on the average by less than about 1 mRy.

Several examples of the power and applicability of this approach are given below.

2.4. Surface and Interface Magnetism

2.4.1. Magnetism of Ni(001) and Fe(001)

We first discuss the surface-induced changes in the magnetism of the classic ferromagnet, bcc iron which has a bulk magnetic moment of $2.12\mu_B$. We represent the surface by a 7-layer Fe(001) slab and use the von Barth-Hedin spin-polarized exchange-correlation potential [6]. The FLAPW result [11] of a surface-induced enhancement of the magnetic moment agrees with that of the earlier LCAO calculation by WANG and FREEMAN [12]. However, in this new calculation, we do not find a significant Friedel oscillation. The FLAPW results for the magnetic moments going from the bulk-like center of the 7-layer film to the surface are 2.27, 2.39, 2.35 and $2.96\mu_B$, i.e. the moment at the surface is increased by 30% compared to the center. A study of the density of states (DOS) decomposed into atomic (i.e. layer) and ℓ-like components gives insight into the mechanism of the surface-induced enhancement of the magnetism. The DOS in the central layer of the 7-layer Fe(001) film is very close to the DOS of bulk bcc Fe: the DOS shows a three-peak structure, so typical for the bcc structure with a pronounced minimum below the highest peak. The d-band for the majority spin has a small unoccupied part, whereas for the minority DOS the Fermi energy falls into the characteristic minimum, leaving about 30% of the minority d-band unoccupied. The resulting spin-imbalance is reflected in the large magnetic moment of bcc iron. The DOS for the surface layer is dramatically changed compared with the bulk-like DOS of the central layer. Due to the reduced symmetry and fewer number of nearest neighbors, the d-band is narrowed. As a consequence, the majority d-band is now almost completely filled. For the minority DOS we find a peak just at the Fermi energy, with its center of gravity slightly above E_F. The states leading to this peak in the surface DOS fall in the minimum of the bulk-like DOS and can be identified as surface states. These surface states are also present in the majority DOS and are shifted to lower energies by an exchange-splitting of about 2 eV. The net result of the surface-induced d-band narrowing and the occurrence of surface states is a larger spin-imbalance corresponding to an enhancement of the magnetic moment in the surface.

The spin density of the Fe(001) surface is dominated by positive spin densities with only small pockets of negative spin densities between the atoms [11]. In the bulk-like center, the shape of the spin density shows important non-spherical components of t_{2g} symmetry. Surprisingly, this bulk-like shape is found even for the iron atoms just one layer below the surface, and only the surface atoms exhibit a different shape of the spin density which is markedly more spherical. The polarization on the vacuum side of the surface atoms is found to be positive.

We have also carried out a similar study for the Ni(001) surface. Three important results of our highly accurate 7-layer calculations [13] shed new light onto the magnetism of the Ni(001) surface: (i) the magnetic moments in atomic-nearest neighbor volumes are (from the central layer to the surface) 0.56, 0.59, 0.60, 0.68 μ_B, i.e. we find an enhancement of the magnetic moment of ∿20% from the surface compared to bulk, (ii) no Friedel oscillation of the moment going from the surface to the center of the 7-layer film, and (iii) the majority-spin surface state $\bar{M}_3$ lies at 0.14 eV below the Fermi energy, i.e. there is no majority-spin d-hole at $\bar{M}$.

The decomposition of the majority and minority charges into ℓ-projected partial charges inside the atomic spheres gives insight into the mechanism of the surface-induced enhancement of the magnetic moments [13]. The dominant partial charge inside the Ni-spheres has d-character. For the majority spin this d-like charge is increased for the surface atom compared to the interior of the film, and the s,p-charge is decreased; in other words, for the atoms with a reduced number of nearest neighbors the charges become more d-like. For the minority spin, we find a similar trend, but here also the d-charge decreases in the surface leading to an increased d-moment. The majority d-band is completely filled, giving a rather small DOS at the E_F of 16.9 states/Ry-cell whereas the partially filled minority d-band gives a high DOS at the E_F of 175.8 states/Ry. The exchange-splitting varies from 0.19 eV for a state near the bottom of the d-band with s-admixture to 0.69 eV for a pure d-state about 1 eV below E_F. For the $\bar{M}_3$ surface states with the majority state just below E_F, the unoccupied minority state is found 0.78 eV higher in energy.

The spin density for the 7-layer Ni(001) slab exhibits [13] bulk-like features for all atoms except those at the surface. We observe a localized positive spin density inside atomic spheres with a radius of about 2 a.u. and a pronounced negative spin density in the interstitial region. The spin density inside the atomic spheres includes important nonspherical components. The dominant component of this spin density maps out the top part of the completely filled d-band, which has no occupied counterpart in the minority spin system. The shape of the spin density of the surface atoms is quite different from the bulk-like atoms in the interior of the film, particularly on the vacuum side where the spin density shows an egg-like shape. The spin density in the vacuum region is slightly negative.

It is remarkable that for both Fe and Ni the reduced number of nearest neighbors on the surface lead to an enhancement of the magnetic moment. This trend becomes even more obvious, and a simple picture emerges when we include linear chains [14] and free atoms in the consideration: going from bulk to a (001)-surface, then to a linear chain and finally to the free atom, the moments are 0.56, 0.68, 1.1, and 2.0 for Ni and 2.27, 2.96, 3.3, and 4.0 for Fe. Thus, as the dimensionality is decreased, the magnetic moments approach the values of the free atoms.

2.4.2. Interface Magnetism of Ag/Fe(001

We have recently carried out a set of self-consistent spin-polarized local spin density calculations for the Ag-covered Fe(001) system in order to obtain a better understanding of magnetism at vacuum-metal and metal-metal interfaces [15]. The results show that the charge density quickly heals within approximately one layer of the interface, but the spin density has a larger screening length. We find that there is an increase of the Fe magnetic moment at the Ag interface by about 10% compared to an increase of 30% for the clean surface. The single particle spectra (both valence and core) are modified at the interface; in particular, the Fe local density of states at the Ag interface is neither bulk- or surface-like, but rather has features of both.

One of the main motivations behind the present work was the Mössbauer measurements [16] of the hyperfine field (H_{hf}) in thin Fe films. These experiments showed an increase in magnitude of the (negative) H_{hf} at the Ag/Fe interface, which was interpreted as indicating an increased magnetic moment at the Fe surface. In our earlier study [11] of the clean Fe(001) surface, however, we found a decrease in H_{hf} due to a large positive valence contribution, although the magnetic moment increased. We have found that coating the Fe surface with Ag causes the Fe valence contribution at the interface to be negative again; this result then resolves the apparent contradiction between theory for the clean Fe surface and experiment for the Ag/Fe system. These results, however, point out that the contributions to the hyperfine field at the surface are different than those in the bulk, and that it is dangerous to apply the simple bulk interpretations of the data to surface work. Furthermore, the first principle results we have presented here not only give a simple and consistent explanation of this and other available experimental data [16-18] on the Fe(001) and Ag/Fe(001) systems, but also demonstrate quite clearly that surfaces and interfaces are different, and individually interesting systems.

2.4.3. Magnetism of the Ni/Cu Interface

One of the best studied interfaces between magnetic and non-magnetic metals is that of the Ni/Cu system. Because of the good match of their lattice constants (a(Ni) = 3.526 Å and a(Cu) = 3.615 Å), overlayers and interfaces are readily accessible experimentally. However, there are still open questions about the magnetism of Ni overlayers on a Cu surface. Interface phenomena also play a key role in layered coherent modulated structures (superlattices) for which the Ni/Cu system is an interesting prototype.

We have carried out the first determination [19-21] of the electronic structure and magnetism of Ni overlayers on Cu(001). We used our linearized augmented plane wave (LAPW) thin film method [22,23] to obtain accurate self-consistent spin-polarized semi-relativistic energy band solutions for Ni overlayers on a Cu(001) substrate, consisting of a 5-layer Cu(001) slab plus one or two p(1x1) layers of Ni on either side, referred to as Ni/Cu and 2Ni/Cu, respectively.

The spatial distribution of the spin density in both cases shows that the magnetization is localized in the Ni layers; the magnetization is essentially zero on the Cu layers. The vacuum and interstitial region are slightly polarized in the opposite direction, similar to that reported for the clean Ni(001) film. Examination of the layer-by-layer magnetic moments (contributed by the electrons inside the touching muffin-tin spheres), indicates that the magnetic moment of the surface Ni layer of the 2Ni/Cu film <u>increases</u> by about 28% to $0.69\mu_B$ compared to the "bulk" value[21]. The moment of the interface layer Ni(I) of the 2Ni/Cu film <u>decreases</u> by 21%, and the Ni layer of the Ni/Cu system <u>decreases</u> by 30% (to $0.39\ \mu_B$) compared to the calculated bulk value. From an orbital angular momentrum decomposition, we find that the contribution to the moments arises almost completely from the d-like component, similar to that in the clean Ni film.

Examination of our results leads to two observations which relate to the problem of surface magnetism. First, the surface and interface affect the total number of electrons, so that in general the surface atoms have fewer electrons and the Ni atoms in contact with the Cu substrate have more electrons due to charge transfer from Cu to Ni. Secondly, the change in the total number of electrons arises almost completely from the minority spin electrons, and this leads to the decrease of the moment with an increase of the total number of electrons.

One of the interesting results of our analysis is the correlation of the magnetic moment with the number of p-like electrons. At the surface,this dehybridization of the s,p, and d electrons acts to increase the magnetic moment and is related to the d-band narrowing seen there. This dehybridization is also related to the simultaneous removal of p-electrons from the muffin-tin (MT) spheres (where the d-electrons are mainly localized) as they spill out into the vacuum region. It is not surprising that the number of p-electrons in the MT region is correlated with the degree of dehybridization; in the free Ni atom, the p-orbitals are competely <u>unoccupied</u> . There is a remarkable correlation between the total s and p charge of both spins, $q_s + q_p$ (μ_s and μ_p are essentially zero) and the magnetic moment. Since Q^p_s is relatively unchanging ($\sim \pm$ 0.03 electrons) this is, indeed, a correlation with q_p. In the unsupported Ni monolayers, where the electrostatic shift mechanism is absent, this dehybridization accounts for the large increase of μ_d compared to the bulk value. In all cases, the total charge in each vacuum region (there are two per slab) is equal (to within $\sim$0.01 electrons) to the loss of p-electrons from the MT spheres.

In the 2Ni/Cu slabs, other effects are seen; charge-transfer (about 0.1 electron) into the d-bands of the interfacial Ni layer reduces the magnetic moment to $0.48\ \mu_B$ in this layer. Since this atom still has a coordination number of twelve, the dehybridization should not take place, as is also seen. By contrast, the surface Ni atoms of the 2Ni/Cu slab show some dehybridization, and could also have an upward electrostatic shift. Indeed, the Ni surface atom does show a $3p_{3/2}$ core-level shift of 0.26 eV to reduced binding energy relative to the Ni interface atom. Both of these effects are consistent with the increased moment of $0.69\mu_B$.

2.5. <u>Surface Multilayer Relaxation: W(001</u>

So far we have discussed the magnetism of metal surfaces and interfaces. Now we give an example which shows the power and precision of the total energy FLAPW approach - multilayer relaxation in W(001)[24].

It is well known that relaxation and reconstruction play a fundamental role in the physics and chemistry of surfaces, and thus have been the object of intense experimental efforts. Their observation in such diverse systems as semiconductors and clean and adsorbate covered metal surfaces indicates that these structural changes can be considered to be the rule rather than the exception. In some cases, it has been possible recently to present experimental evidence for multilayer relaxation effects. For the important class of transition metal surfaces, no quantitative study of any surface multilayer relaxation has been reported and little is known about the energetics of such a process. For well studied surfaces such as W(001), no multilayer relaxation has been considered, and even the extent of the relaxation of just the first layer has been a matter of controversy.

We have very recently determined [24] the energetics and the mechanism of the multilayer relaxation of the W(001) surface by the all-electron local density functional total energy approach. In agreement with recent experiments, we predict a contraction of the topmost layer by 5.7%. This is accompanied by an outward relaxation of the second and third layers leading to an increase of the second and third interlayer spacings by 2.4% and 1.2%, respectively. We also find, surprisingly, that the relaxation of the second and third interlayer spacings does not influence the equilibrium spacing between the two topmost layers, i.e. keeping the inner layers unrelaxed leads to practically the same equilibrium distance between the first and second layer as is found for the fully relaxed system. Thus, the equilibrium between the adjacent layers appears to be governed by highly screened local interactions. This decoupling of the relaxation for the topmost and the inner layers is the more remarkable, since 25% of the total relaxation energy of 0.06 eV originates from the relaxation of the inner layers.

These results provide a sensitive demonstration that the all-electron LDF total energy approach is capable of describing subtle aspects in the energetics of transition metal surfaces such as multilayer relaxation processes involving even high-Z atoms such as tungsten. Importantly, the results reveal that the equilibrium distance between the two topmost layers of the W(100) surface is unaffected by additional relaxation of the inner layers. This demonstrates that because of the highly effective screening in a transition metal, the basic interaction mechanism which drives the surface relaxation in W(001) is of such a local nature that essentially only interactions between adjacent layers are involved. This finding should be valid for a variety of transition metal surfaces similar to W(100).

3. Metallic Superlattices

3.1. Introduction

It has now been established that artifically made bimetallic superlattices have added a new dimension to research. In a parallel development, artificial single- or multiple-bimetallic sandwiches [25] are being investigated as a way of producing unusual properties of materials.

A key feature of property modification lies in the observation that when a metal A is deposited on a single crystal of an isostructural metal B, the initial layers of A form with a lattice parameter, a_o, between those of A and B[26]. Large property modifications have been reported for both bimetallic sandwiches and modulated structures. Stimulated by the observations of THALER et al.[27] of strongly enhanced ferromagnetic resonance in CuNi composition modulated alloys, considerable theoretical [28] and experimental [29] interest has focused on the properties of these systems [30].

BRODSKY and FREEMAN [25] demonstrated that exceptionally large increases in the magnetic susceptibility (indicating nearly-magnetic ordering) of thin films of Pd sandwiched between thicker Au films occurred at low T - presumably due to the tetragonal stretching of the Pd lattice constant by the Au. Striking effects in other sandwich systems have also been observed [31] recently.

In the work on modulated structures, several important parameters are modulation amplitudes and wavelengths, concentration, interface matching and orientation of the modulation. Most of the experimental work has been performed on composition modulated structures, in which the concentration varied in a sinusoidal fashion along the modulation direction. Our theoretical studies have been performed on coherent modulated structures (CMS) in which the composition of any given layer is exactly 100% of either metal A or metal B.

In this paper we cite results of our theoretical self-consistent local density (for paramagnetic structures) and local spin density (for ferromagnetic structures) energy band studies of a PdAu CMS oriented along (100) and NiCu CMS oriented along (111) and (100) for different wavelengths and compositions [28].

3.2. Theoretical Approach

In these studies, the electronic structures were determined using the spin-polarized self-consistent linear muffin-tin orbital (LMTO) band method [32]. The basis functions included s,p and d for all sites with f in the three center terms to improve convergence. The core states were recalculated in each iteration (and renormalized over the WS spheres) thus making studies of core-level shifts possible. Furthermore, the core levels were calculated relativistically, while valence states are semirelativistic (i.e. all relativistic terms except for spin-orbit coupling). This treatment of the core is important, as it gives an additional relativistic contribution to the Fermi contact hyperfine field. The potential is of local (spin) density type and the exchange correlation is included using the HEDIN-LUNDQVIST[33] formula for the non-spin-dependent part and the GUNNARSSON-LUNDQVIST[7] spin-dependent part is added in the spin-polarized calculations.

3.3. Results for NiCu and PdAu

3.3.1. CuNi

Extensive theoretical studies were performed[28] for CuNi CMS with the modulation along the (111) direction: one with a wavelength of six layers, 3Cu + 3Ni, and the other with twice that wavelength, 6 Cu + 6 Ni. Results for the former structure showed local density of states (DOS) broadening on interface sites as well as the reduction of the Ni moment. The major effect of the CMS is on the resulting magnetic moment [28]. For the 6 layer (111) CuNi CMS, the Ni interface layer (i.e. the Ni layer which has an adjacent Cu layer) has its moment reduced to 0.37 μ_B and the 'bulk' layer moment is reduced to 0.50 μ_B. For the 12 layer (111) CMS, the 'bulk' layer again has a moment of 0.50 μ_B and the interface layer has a moment of 0.39μ_B. (Recall that similar calculations give a moment of 0.54μ_B for fcc Ni.) These predictions which were in disagreement with the enhanced ferromagnetic resonance measurements [27], were later confirmed in several experiments[29].

In the calculations, the WS-radii are chosen proportional to the local distortions, which are assumed to be 1.25% in the 6 layer CMS and 2.5% in the 12 layer CMS. Thus, the charge differences found are exaggerated somewhat, since the Cu spheres are largest in the 12 layer CMS and so on. The charge transfers are, in both cases, very small (0.07-0.04 electron per atom) but the signs are opposite in the two cases. It is likely that large charge transfer will generate large electrostatic forces in the planes and counteract the effects of local distortion. The actual distortion is not known, but one would believe that a larger distortion than assumed for the 6 layer CMS and a smaller one than assumed for the 12 layer CMS minimizes the charge transfer in both cases. Thus about 2% distortion, or a little less, is inferred from these charge transfer arguments.

The ferromagnetic Ni moments in the CMS are all reduced compared to the fcc Ni value ($0.54\mu_B$). The small reduction of bulk-like Ni site moments follows from what is expected form the local distortion, while the large reduction of interface sites must be due to the proximity of Cu. The total Ni moment has 98% 3d character. The s- and p-electrons are itinerant also over the Cu region, and may be important in mediating long-range spin interactions, in particular those between separated Ni layers. In fcc Ni, the s- and p-moments are polarized opposite to the d-moment, so that effectively there is a negative spin-polarization in the interstitial region between the Ni atoms - as observed by neutron diffraction experiments. The s- and p-polarization is essentially the same for tetragonally distorted Ni[34], while for bulk-like Ni sites in the CMS, the situation is changed. This shows that the presence of Cu is important for the Ni s- and p-polarization, even though the Cu is several layers away, and thus there is an effective long-range s- and p-interaction.

To summarize the CuNi(111) results, one may conclude that the local distortion has an influence on the Ni-d moments (and DOS) whereas the s and p moments are affected by the modulation and presence of Cu. The range of interaction for the s-p moments is quite long; at least 6 Ni layers are affected by the CMS. On the other hand, the second layer of Ni has the d-band almost unaffected by the Cu hybridization. In line with this long-range coupling, we have discussed elsewhere [28] the possible role of Fermi surface nesting in promoting antiferromagnetic ordering in periodically diluted systems such as a CMS. For CuNi, we concluded that for a (111) modulation, the minimum wavelength for antiferromagnetic coupling was 5-6 layers. While it is possible to test out this possibility with actual spin-polarized calculations, the large unit cell involved for antiferromagnetic coupling makes such studies quite elaborate and formidable.

Our calculations for the CuNi CMS modulated in the (100) direction were designed to give information about the concentration and wavelength dependence. Here, a total of 8 layers in the CMS are considered for all combinations from Cu_7Ni_1 to Cu_4Ni_4. No local distortions are assumed.

One of the quantities of interest is the reduction of the Ni moments as the composition changes. In these (100) systems, each Ni has 4 other Ni as nearest neighbors within each layer, whereas each Ni has 6 nn at the same distance in the (111) CMS. This difference may lead to a more rapid decrease of the Ni moments in the (100) case as the Ni thickness is reduced. We especially wanted to see if one or two layers of Ni become "magnetically dead", as has been inferred from experiments on Ni overlayers on a Cu(100) surface [35], but shown to be dead in modulated structures [36].

The Ni interface moments are rather effectively reduced as the number of Ni layers is reduced; however, one single Ni-layer between seven Cu layers still has a non-neglible moment, $0.12\mu_B$. For two layers, the moment per atom is doubled, and for three layers there is another increase of about $0.12\mu_B$. In contrast to the (111) results, the "bulk-like" sites also have a sizable moment reduction comparable to that of the Ni-interface sites ($0.38\ \mu_B$ for Cu_5Ni_3 and $0.44\mu_B$ for Cu_4Ni_4). This may be interpreted as an effect of the different number of nearest neighbors (nn) in the two structures. In the (100) modulation, there are few nn within the layer, resulting in weak interaction, while the interaction between different layers can be relatively stronger than in the (111)-CMS due to an additional nn in the next layer.

3.3.2. PdAu

Our theoretical studies were made on two PdAu CMS modulated along (100), a Pd_2Au_2 structure (two layers each) and a Pd_3Au_3 structure (three layers each). This permits a separation between pure interface effects (which exist only in Pd_2Au_2) and the effects of bulk vs interface (which occur in Pd_3Au_3). The Au_2Pd_2 system has four atoms/unit cell and two inequivalent layers, whereas the Au_3Pd_3 has six atoms/unit cell and four inequivalent layers. A more realistic matching of the Pd and Au lattice constants is obtained than in the NiCu case studied earlier (where a simple average was used) by use of the Pd and Au elastic constants. (A more careful determination of the distortion condition is more important here than in CuNi because of the larger lattice difference involved.) This matching gives a 0.25% smaller x-y lattice constant than the simple average, and a smaller local volume change, now about 1%. Further, the WS-sphere radii are very close to those of the pure fcc cases.

The band structure results were obtained self-consistently at 144 k points and 108 k points in the irreducible Brillouin zone for the Au_2Pd_2 and Au_3Pd_3 CMS, respectively. For the Au_2Pd_2 CMS, the Pd DOS is due only to the interface states, which consists of strong hybridization with the Au layers. The Fermi energy lies at the high energy side of the DOS peak, giving total DOS at E_F of 2.01/eV atom, which is lower than that of pure Pd(2.40/eV atom). By contrast, E_F in the Pd_3Au_3 CMS lies at a peak in the total DOS. In this case, the partial DOS for both the interface (Pd_i) and 'bulk' (Pd_b) layers (2.58 and 3.38 states/eV atom, respectively) is higher than that for pure Pd metal. Whereas the Pd_i DOS bears little resemblance to that for pure Pd, the Pd_b DOS does indeed resemble that for fcc Pd. Although there are some differences between the two, their peak positions are at the same energies.

Charge transfer effects are small, amounting to 0.06 electrons from Pd to Au in both Pd_2Au_2 and for Pd_i and Au_i in Pd_3Au_3; by contrast, there is at best only a very small (0.02 electrons) transfer from Pd_b.

Finally, as an estimate of the possibility of magnetic ordering in these CMS, we have calculated the generalized Stoner factor, S using the local spin density Hamiltonian. For Pd_2Au_2 we find S = 0.6 i.e. well below the value for fcc Pd (S = 0.85). However, for Pd_3Au_3 we find an increase in S to 0.93 - which more than doubles the exchange enhancement factor of fcc Pd and suggests for these thin (few layers) Pd CMS that one may be approaching a ferromagnetic instability.

As we saw, the Pd_2Au_2 CMS is a distinctly different system from Pd_3Au_3 and a poor representation of the actual situation found for the PdAu sandwiches. By the same reasoning, based on the above results, we would expect somewhat thicker Pd layers to give a more realistic representation of the physical situation. We have therefore estimated the Stoner factor, which would result from making up a thicker film assuming the results found for Pd_3Au_3 to hold for the thicker films. For Pd_4Au_4 (i.e. 2 'bulk-like' and 2 interface layers) S is increased to 0.98 and for Pd_5Au_5 (3 'bulk' and 2 interface layers) S is 1.01. This prediction of a ferromagnetic instability for thicker sandwiches is being investigated using the all-electron energy band approach described above.

4. Acknowledgements

I am grateful to C.L. Fu, T. Jarlborg, H. Krakauer, S. Ohnishi, D.S. Wang, M. Weinert, E. Wimmer, and J.-H. Xu, with whom the work reported here was carried out. Part of this work was supported by the U.S. National Science Foundation (grant number DMR82-16543), and through the Northwestern University Materials Research Center (grant number DMR82-16972), and the U.S. Office of Naval Research (grant number N00014-81-K-0438).

References

[1] A.J. Freeman, J. Mag. Magn. Matls. 15-18 , 1070 (1980)
[2] J.A. Applebaum and D.R. Hamann, Rev. Mod. Phys. 48 , 479 (1976)
[3] E. Wimmer, H. Krakauer, M. Weinert, and A.J. Freeman, Phys. Rev. B 24 , 864 (1981)
[4] P. Hohenberg and W. Kohn, Phys. Rev. 136 , B864 (1964)
[5] W. Kohn and L. Sham, Phys. Rev. 140 , A1133 (1965)
[6] U. von Barth and L. Hedin, J. Phys. C5, 1629 (1972)
[7] O. Gunnarsson and B.I. Lundqvist, Phys. Rev. B 13 , 4274 (1976)
[8] V.L. Moruzzi, J.F. Janak and A.R. Williams, Calculated Electronic Properties of Metals (Pergamon Press, New York, 1978) and references therein.
[9] M. Weinert, J. Math. Phys. 22 , 2433 (1981)
[10] M. Weinert, W. Wimmer, and A.J. Freeman, Phys. Rev. B 26 , 4571 (1982)
[11] S. Ohnishi, A.J. Freeman, and M. Weinert, Phys. Rev. B 28 , 6741 (1983)
[12] C.S. Wang and A.J. Freeman, Phys. Rev. B 24 , 4364 (1981)
[13] E. Wimmer, A.J. Freeman, and H. Krakauer, Phys. Rev. B (1984) to appear
[14] M. Weinert and A.J. Freeman, J. Mag. Magn. Matls. 38 , 23 (1983)
[15] S. Ohnishi, M. Weinert and A.J. Freeman, Phys. Rev. B (1984) to appear
[16] J. Tyson, A.H. Owens, J.C. Walker, and G. Bayreuther, J. Appl. Phys. 52 , 2487 (1981)
[17] G. Bayreuther and G. Lugert, J. Mag. Magn. Matls. 35 , 50 (1983)
[18] A.M. Turner, Y.J. Cheng, and J.L. Erskine, Phys. Rev. Lett. 48 , 348 (1982); A.M. Turner and J.L. Erskine, Phys. Rev. B 28 , 5628 (1983)
[19] D.S. Wang, A.J. Freeman, and H. Krakauer, J. Appl. Phys. 52 , 2502 (1981)
[20] D.S. Wang, A.J. Freeman, and H. Krakauer, Phys. Rev. B 26, 1340 (1982)
[21] A.J. Freeman, D.S. Wang, and H. Krakauer, J. Appl. Phys. 53 , 1997 (1982); D.S. Wang, A.J. Freeman, and H. Krakauer, Phys. Rev. B 26 , 1340 (1982)
[22] H. Krakauer, M. Posternak, and A.J. Freeman, Phys. Rev. B 19 , 1706 (1979)
[23] M. Posternak, H. Krakauer, A.J. Freeman, and D.D. Koelling, Phys. Rev. B 21, 5601 (1980)
[24] C.L. Fu, S. Ohnishi, E. Wimmer, and A.J. Freeman, Phys. Rev. Lett. (submitted)
[25] M.B. Brodsky and A.J. Freeman, Phys. Rev. Lett. 45 ,133 (1980)
[26] J.E. Hilliard, in Modulated Structures-1979 , eds. J.M. Cowley, M.B. Salamon and B.J. Wuensch, AIP Conf. Proc. No. 53 (American Institute of Physics, New York, 1979); F.C. Frand and J.H. van der Merwe, Proc. Roy. Soc. London Ser. A 198 , 205 (1949); R.W. Vook and C.T. Horng, Phil. Mag. 33 , 843 (1976)
[27] B.T. Thaler, J.B. Ketterson, and J.E. Hilliard, Phys. Rev. Lett. 41 , 336 (1978)
[28] T. Jarlborg and A.J. Freeman, Phys. Rev. Lett. 45 , 653 (1980); T. Jarlborg and A.J. Freeman, J. Appl. Phys. 52 , 1622 (1981)
[29] E.M. Gyorgy, J.F. Dillion, Jr., D.B. McWhan, L.W. Rupp, Jr., L.R. Testardi and P.J. Flanders, Phys. Rev. Lett. 45 , 57 (1980); G.P. Felcher, J.W. Cable, J.Q. Zheng, J.B. Ketterson, and J.E. Hilliard, J. Mag. Magn. Matls. 21 , L198 (1980); J.Q. Zheng, C.M. Falco, J.B. Ketterson and I.K. Schuller, Appl. Phys. Lett. 38 , 424 (1981); E.M. Gyorgy, D.B. McWhan, J.F. Dillon, Jr., L.R. Walker, and J.V. Waszczak, Phys. Rev. B 25 , 6739 (1982)
[30] E.M. Gyorgy, J. Mag. Magn. Matl. 31-34 , 915 (1983)
[31] M.B. Brodsky, J. Appl. Phys. 52 , 1665 (1981)

[32] O.K. Andersen, Phys. Rev. B 12 , 3060 (1975); T. Jarlborg and G. Arbman, J. Phys. F 7 , 1635 (1977)
[33] L. Hedin, B.I. Lundqvist and S. Lundqvist, Solid State Commun. 9 , 537 (1971)
[34] T. Jarlborg and A.J. Freeman, J. Mag. Magn. Mat. 22 , 6 (1980)
[35] G. Bergmann, Phys. Rev. Lett. 41 , 2332 (1980)
[36] N.K. Flevaris, Ph.D. Thesis, June (1983)

Electrons and Phonons in One and Two Dimensions in Semiconductor Structures

M. Pepper

Cavendish Laboratory, Department of Physics, University of Cambridge Cambridge CB3 OHE, United Kingdom, and

G.E.C. Hirst Research Centre, Wembley, Middlesex, United Kingdom

In this article an outline is given of dimensionality effects, localisation and quantisation of Hall resistance in semiconductor structures. The Si inversion layer is used as the model system in view of the relative simplicity of interpretation of phenomena.

Some of the recent topics in semiconductor structures are described in this article. It is schematic and for a more complete description of events the reader is referred to the Surface Science series of proceedings of the two-dimensional conferences, also Ando et al[1].

The topics discussed are strong and weak localisation and interaction effects in 2D, Dimensionality transitions where the change of a controlling voltage and/or temperature causes transport to become 3D or 1D and the Quantum Hall Effect.

Strong Localisation

The inversion layer of the Si MOS device can be used to simulate an amorphous material in which strong localisation is produced by charges at the Si-SiO_2 interface[2,3].

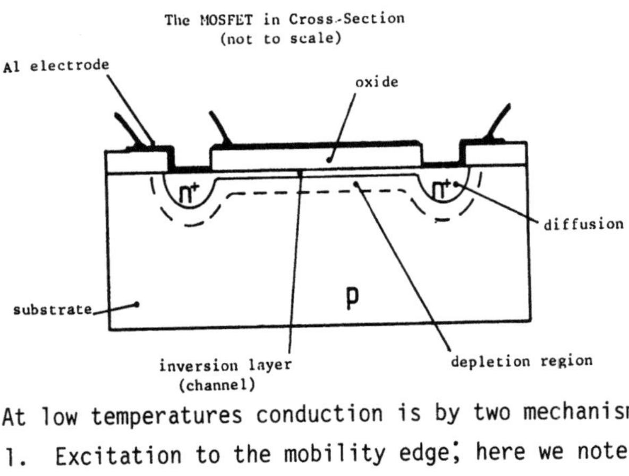

Figure 1.

At low temperatures conduction is by two mechanisms in parallel.

1. Excitation to the mobility edge; here we note that at temperatures above ~ 1K inelastic scattering cuts off the weak localisation and gives a sep-

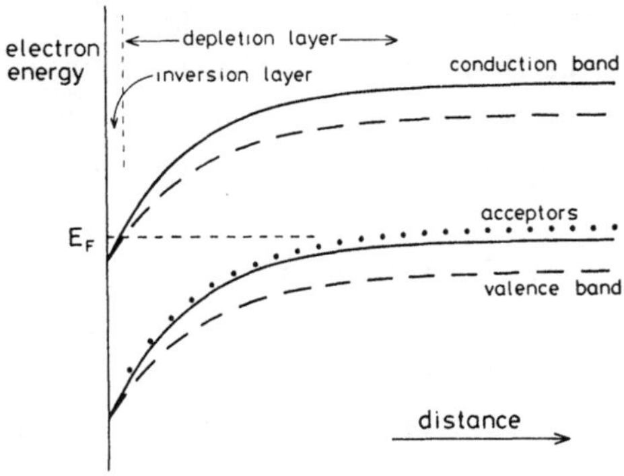

Fig. 2. - Formation of an n-type inversion layer at the surface of a p-type semiconductor. The broken curves show the effect of substrate bias.

aration between extended and localised states.[4] The conductivity σ can be written

$$\sigma = \sigma_{min} \exp(-W/kT)$$

$W = |E_c - E_f|$, is the separation between the mobility edge and the Fermi energy. σ_{min} is the minimum metallic conductivity which in 2D is $\approx 0.1e^2/\hbar$ ($\approx 3.10^{-5}\,\Omega^{-1}$).

2. Variable range hopping by electrons at the Fermi energy produces a temperature-dependence given by $\sigma = \sigma_0 \exp - C/T^{1/3}$. The pre-exponential factor σ_0 depends on the electron-phonon interaction and the various percolation paths through the system.[5] The constant $C = 3\alpha^2/\pi N(E_f)k$ here α^{-1} is the decay-constant of the wavefunction ($\psi \sim \exp -\alpha r$) k is Boltzmann's constant and $N(E_F)$ is the density of states at the Fermi energy.

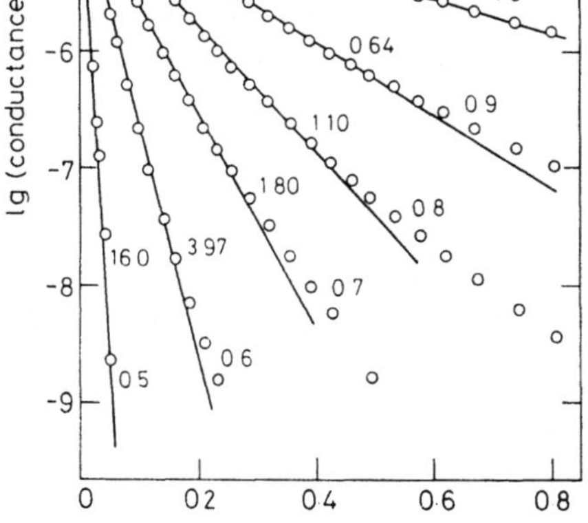

Fig. 3. - Plots of lg σ against 1/T for various gate voltages (positions of the Fermi level) in the temperature-range where conduction is dominated by excitation of carriers to the mobility edge. The activation energies are also shown. σ_m is about 2 x $10^{-5}\,\Omega^{-1}$ This device had a total density of localised states of about 1.8 X 10^{15} m^{-2} and a net oxide charge density (as reflected at the interface) less than 1 X 10^{15} positive charges m^{-2}. The oxide thickness was 60 nm.

The derivation assumes $N(E_F)$ is fairly constant over an energy range $\sim kT$ which is reasonable at low temperature. The theory is a one electron theory and in the presence of interactions both the temperature-dependence and σ_0 are altered. In 3D σ varies as $\exp -(B/T^{1/4})$ a law which has been widely observed.[4,5]

Figure 3 shows the temperature-dependence of a silicon inversion layer for different values of carrier concentration. The results show evidence for excitation to the mobility edge with a common pre-exponential factor, the deviation at higher temperatures is due to excitation to higher mobility states. At lower temperatures, points depart from the extrapolated dependence due to the onset of variable range hopping.

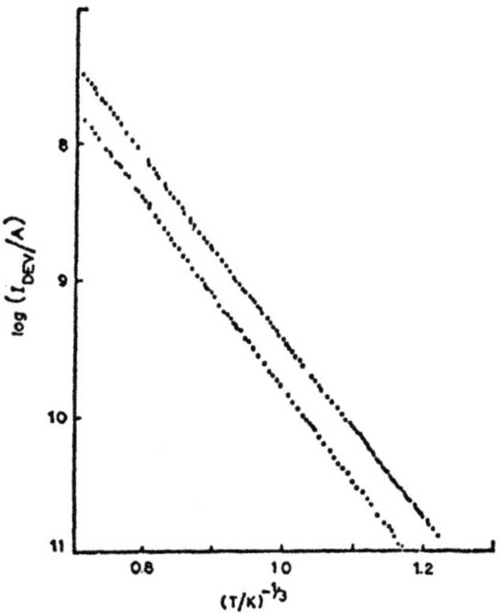

Fig. 4. - Log I is plottet against $(T^{-1/3})$ illustrating variable range hopping.

Figure 4 shows the temperature-dependence of variable range hopping obtained from a different sample. The excellent fit to the T^{113} plot (analysis suggested that log $\sigma \propto T^{-0.32}$) was not altered by a small temperature-dependence of σ_0 .

It is possible to investigate the density of localised states from analysis of the activation energy. Deep in the tail, this behaviour is exponential but, near E_c. $N(E_F)$ can be obtained from the relation

$$\frac{dn}{dW} \approx N(E_f)$$

where dn is the change in carrier concentration necessary to change the activation energy by dW. It was found that if the total number of localised carriers is less than $2.10^{11} cm^{-2}$ then $N(E_c)$ is near $\frac{3}{4}$ of the unperturbed value, in reasonable agreement with theoretical estimates.[2] However, if $n > 2.10^{11}$ then the obtained $N(E_F)$ is much greater than the unperturbed value. This cannot be so, and indicates the failure of the analysis, the cause of the failure may lie in the movement of E_c with increasing carrier concentration. A tendency for the electron-electron interaction to reinforce disorder in localising carriers is present in semiconductor impurity bands where disorder and the Hubbard bands are responsible for the metal-insulator transition.[6]

The role of substrate bias on strong localisation has been investigated by Fowler.[7] It was found that forcing electrons closer to the interface resulted in a decrease in the extent of strong localisation. The effects of substrate bias on the temperature-dependence are shown in Figures 5,6 and 7.[45]

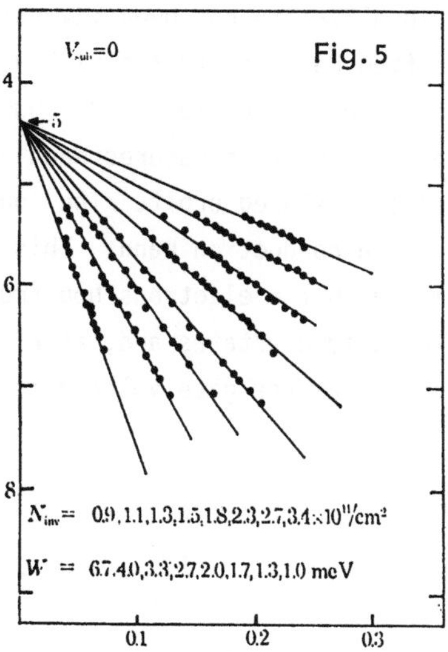

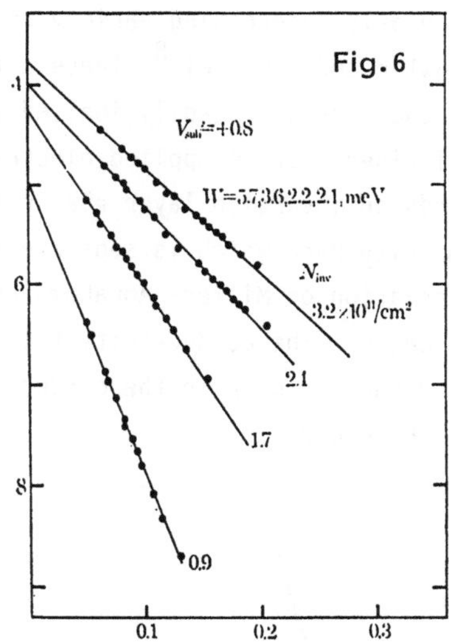

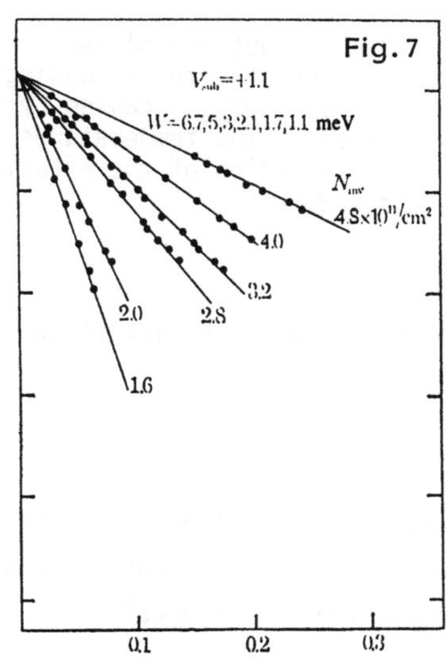

Fig. 5-7. - In these figures the progressive effect of a forward substrate bias on the plots of $\ln\sigma$ against $1/T$, when conduction is by excitation to E_c. The values of substrate bias and carrier concentration are indicated, 1 V on the gate changes the carrier concentration by 7.5×10^{10} electrons/cm^2; the substrate doping was 2×10^{15} acceptors/cm^3 and the net oxide charge was 1.70×10^{11}/cm^2.

It is seen that pulling carriers away from the interface causes an increase in localisation and an increase in the value of σ_{min}., A possible model for this is based on the potential fluctuations becoming longer-range as the electrons are pulled away from the interface. This prevents tunnelling through the fluctuations and so reduces delocalisation. The increase in σ_{min} can be explained along similar lines, when delocalisation occurs the electron wavelength is shorter than would otherwise be and σ_{min} is increased. The cause of the potential fluctuations at the Si-SiO_2 interface is not clear. However, if there are compensating charges present, much like the D^+, D^- states in chalceogenide glasses,then the potential fluctuations would vary with distance from the interface in the manner proposed.

A particularly interesting variety of strong localisation has been investigated by Fowler and Hartstein[8]. These authors drifted Na^+ ions to the Si-SiO_2 interface, the ions are readily ionised and drift through the SiO_2 at $\approx$ 100C under the influence of an applied electric field. At low temperatures each Na^+ ion binds an inversion layer electron in a pancake-shaped orbit. The ions form an impurity band which is separate from the main conduction band. This allows observation of Miller- Abrahams hopping,in which the electrons hop from centre to centre. The conductivity follows the density of states and, at a fixed temperature, a peak in the conductance is found as the gate voltage is increased - Figure 8.

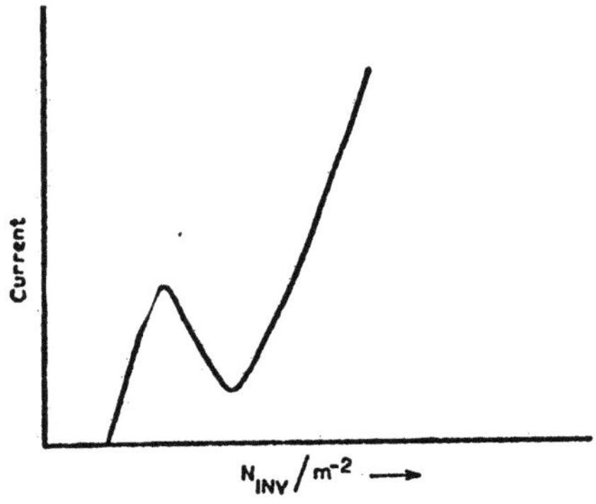

Fig. 8. - Schematic outline of the peak in device current with increasing carrier concentration N_{INV} caused by a Na^+ impurity band. The temperatures required for observation of the effect are below about 20 K.

This system has been of considerable importance in the investigation of 2D hopping.

Weak Localisation

Numerical calculations of the 2D conductance first indicated that all states were weakly localised in 2D and that increasing the sample size, and increasing the cut-off length in the localisation, reduced the conductance[9].

Subsequently many papers[10,11,12,13] showed that for $k_F l > 1$, where k_F is the Fermi wavevector and l is the mean free path for elastic scattering, the conductance is given by:

$$\sigma = \sigma_B - \frac{\alpha e^2}{\pi^2 \hbar} \ln (L/l)$$

Here α is a constant taking values of ½ or 1 depending on the extent of spin scattering, L is the specimen length at absolute zero, or the inelastic diffusion length L_{in} at finite temperature, σ_B is the Boltzmann conductance. The process giving rise to the logarithmic correction is coherent back-scattering (quantum interference), a process which is cut off by inelastic scattering. At low temperatures the dominant inelastic scattering process is due to electron-electron scattering rather than electron-phonon. If the rate of scattering $1/\tau_{e-e}$ varies as T^x then L_{IN} varies as $T^{x/2}$ and the conductivity correction $\delta\sigma$ is given by

$$\delta\sigma = \frac{\alpha x e^2}{2\pi^2 \hbar} \ln T$$

Another mechanism which gives a logarithmic correction to the conductivity arises from the electron-electron interaction in the presence of elastic scattering.[14,15] A minimum in the density of states occurs at E_F and the conductivity is reduced by a correction $\delta\sigma_I$

$$\delta\sigma_I = \frac{2e^2}{\pi^2 \hbar} (1-F) \ln (kT\tau/\hbar).$$

τ is the mean free time for elastic scattering. The factor F arises from the effects of screening on the interaction and for Si inversion layers, with a short screening distance, F $\approx$ 0.85- .95 over the range of carrier concentration used in the experiment. Other 2D systems such as Ga As have a larger screening length, and here F can be $\approx$ 0.5 with the result that both interactions and localisation affect the conductivity,[16] for this reason the Si inversion layer is most appropriate for the study of weak localisation. Theories of 2D weak localisation are not in agreement on the possible existence of a mobility edge. It is argued[10,11] that all states are expontentially localised in 2D and a 'quasi mobility edge' exists at finite temperatures, because inelastic scattering cuts off the localisation. On the other hand it has been suggested that a localisation edge separates states which are exponentially localised in the tail from those which fall off with distance r as $1/r^s$, s decreases to unity as $k_F l$ increases.[11,13,17] This gives a power-law dependence of conductance,

$$\sigma = \sigma_B (L_{IN}/l)^{-\gamma}$$

$$\gamma = 2/\pi k_f l$$

σ_B is the Boltzmann conductivity $e^2 k_f l/\pi\hbar$ approximating to a first order logarithmic correction.

In the regime of weak localisation where L_{IN} is relatively short, and transport is diffusive, the magneto-resistance is negative. The loss of momentum determinacy results in a decrease in the effective length scale and suppression of the logarithmic correction. Crudely, we can account for this by noting that the wavefunction will fall off with a characteristic distance L_c (L_c is the cyclotron radius $(e\hbar/B)^{\frac{1}{2}}$). The effective length L' can be regarded from a physical basis as

$$\frac{1}{L'} = \left(\frac{1}{L_c^4} + \frac{1}{L_{IN}^4}\right)^{1/4}$$

Arguments have been presented which suggest that in the non-magnetic case,[4] the various length scales can be combined in the form

$$\frac{1}{L'^2} = \frac{1}{L_1^2} + \frac{1}{L_2^2} + \cdots\cdots$$

The theory of the interaction process suggests that in 2D if the magnetic field lifts the spin degeneracy ($g\beta B > kT$) then a change in screening occurs converting the pre-logarithmic factor into (2-F) from (2-2F).[18]

The theory of the Hall effect indicates that there are pronounced differences between these two mechanisms.[14] In the case of weak localisation, the Hall mobility has the same logarithmic correction, i.e. there is no correction in the Hall constant R_H. On the other hand, the interaction model suggests that the Hall constant increases at twice the rate of the resistance R

$$\frac{\delta R_H}{R_H} = \frac{2\delta R}{R} = -\frac{2\delta\sigma}{\sigma}$$

At lower temperatures the heating of an electron gas by an applied field can be used as a thermometer of the electron temperature. As pointed out by Anderson et al[11] the rise in temperature T can be calculated as follows,

$$\sigma F^2 \tau_{e-p} = cT\Delta T$$

F is the electric field, τ_{e-p} is the phonon emission time, cT is the specific heat of the electron gas and T is the rise in temperature. If the temperature of the gas is much larger than the lattice temperature

$$\sigma F^2 \tau_{e-p} = cT^2$$

If τ_{e-p} varies as T^{-p} then $F \propto T^{1+p/2}$ and experimentally determined field-dependence of temperature can be used to explore the temperature-dependence of the electron-phonon relaxation time and phonon dimensionality; P = 2,3, or 4 if phonons are 2D or 3D in the dirty and clean limits respectively.

The maximum temperature of the localisation correction is determined by the temperature-dependence of scattering mechanisms. These can be electron-phonon or the temperature-dependence of the screening in elastic scattering. In general the negative magneto-resistance can be observed at higher temperatures than the logarithmic term, as other temperature-dependent scattering terms do not have a low field magneto- resistance

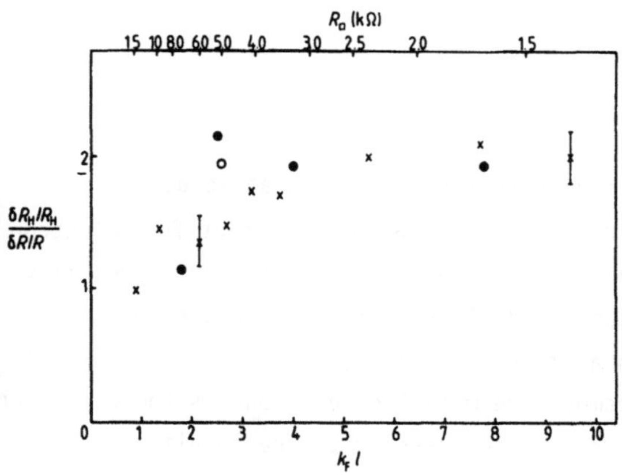

Fig. 9. - The Hall ratio plotted against $k_F \ell$ and R (k Ω). (o, ●) versus temperature (B = 1.31 Tesla), (x) versus field

Figure 9 illustrates the relationship $\frac{\delta R_H}{R_H} = 2\frac{\delta R}{R}$ observed when the magnetic field is sufficiently great to bring about the interaction correction. As seen, the ratio of 2 is maintained until $k_F \ell \approx 1$, when the weak corrections are lost.

More recent work has established that the interaction correction in the Hall effect is present down to very small values of magnetic field, which are insufficient to convert the (2-2F) term into (2-F).[19] It appears that screening affects the Hall coefficient in a different manner to the conductivity. Figure 10 clearly illustrates the supression of localisation, with loss of the logarithmic correction, by a magnetic field. When the correction is suppressed the magneto-resistance is still negative, being present until $L_c \sim \ell$. When $g\beta B > kT$ the magneto-resistance becomes positive and the logarithmic correction returns with the increasing importance of the interaction correction. It is noticeable that increasing the electron temperature suppresses the

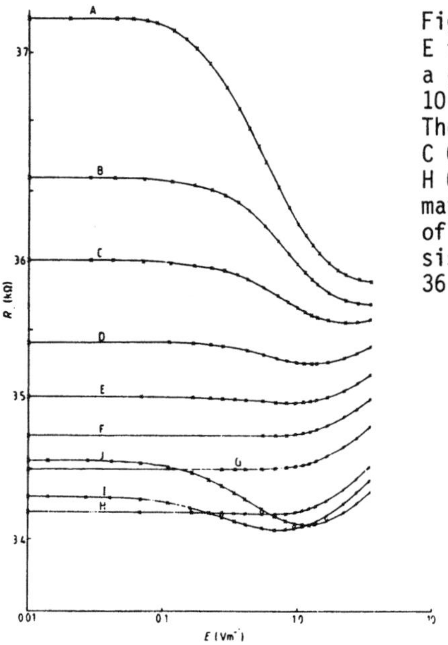

Fig. 10. - R (kΩ) is plotted against log E for various values of magnetic field with a constant carrier concentration of 3.8 X $10^{15}m^{-2}$ and a lattice temperature of ~50mK. The magnetic fields in T are: A 0; B 0.005; C 0.021; D 0.047; E 0.074; F 0.1; G 0.15; H 0.26; I 0.32; and J 0.52. The crosses mark the points used for the integration of the originally measured differential resistance. The elastic scattering length was 36 nm.

interaction correction and now the magneto-resistance is negative as the localisation correction is always present. This figure shows that for electron temperatures above 0.05K there is a range of magnetic field for which metallic behaviour is found. This would disappear at lower temperatures due to the onset of the interaction correction at much lower magnetic fields. The use of lower mobility samples would also remove the metallic behaviour, as the condition $g\beta B > kT$ would be present before the suppression of the localisation at $L_{in} < L_c$.

The presence of the weak localisation can be used to investigate electron-electron scattering. Hikami[20] and co-workers have shown that the magneto-conductance is given by

$$\Delta\sigma = (\alpha e^2/2\pi^2\hbar)[\psi(1/2 + 1/a\tau_i) - \psi(1/2 + 1/a\tau)]$$

ψ is the digamma function, τ_i and τ are the inelastic and elastic scattering times, α =1 and a =4eBD/$\hbar$ where D is the electron diffusion coefficient. By fitting experimental results to this formula (often the value of α is slightly decreased) τ_i can be found as a function of temperature and disorder. It is found that τ_i (due to electron-electron scattering) is given by

$$\frac{1}{\tau_i} = A_1 T + A_2 T^2$$

The T^2 term is the normal Landau-Baber scattering and the experimental value of A_2 is in reasonable agreement with the theoretical estimate of $k^2/\hbar E_F$. The T term arises from scattering, such that the small change in momentum is within the indeterminacy produced by elastic scattering. Again experiment[21] is in reasonable agreement with the theoretical estimate of $2k/\hbar k_F l$ Figures 11 and 12.

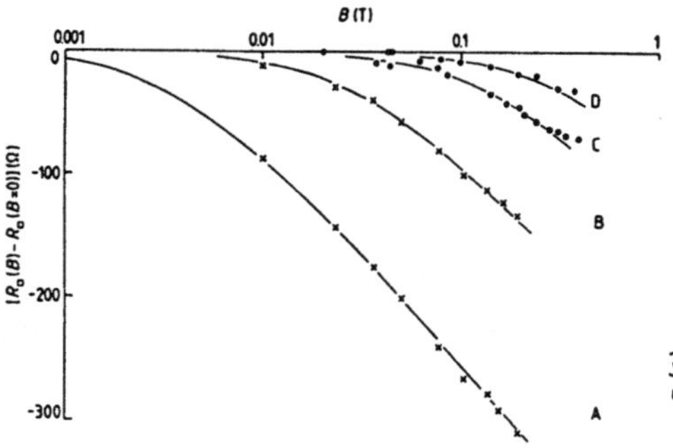

Fig. 11. - The change in resistance is plotted as a function of magnetic field for temperatures A=.11K, B= .55K, C= 1.2K and D= 2.1K.

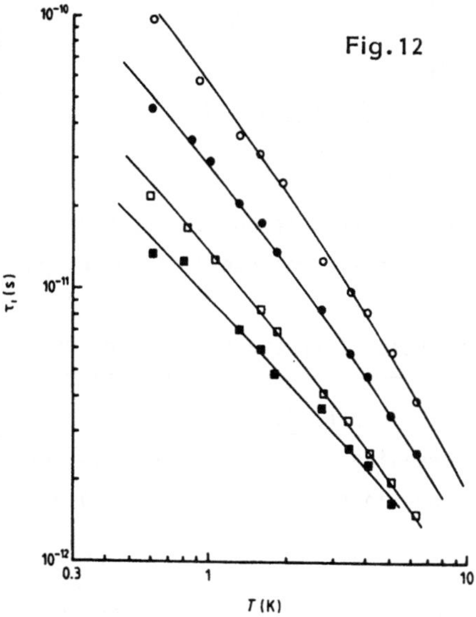

Fig.12

The inelastic scattering time τ_i, in seconds, is plotted against temperature. The lines are fitted curves $1/\tau_i = A_1 T + A_2 T^2$

Symbol	$n_{inv}(10^{16}\,m^{-2})$	$k_F l$	$A_1(10^{10}\,K^{-1}\,s^{-1})$	$A_2(10^{10}\,K^{-2}\,s^{-1})$
○	3.6	22	1.3	0.43
●	1.9	11	3.0	0.51
□	1.0	4.2	6.5	0.7
■	0.63	1.05	13.0	†

† The fitted value was smaller than the experimental error.

Similar effects are found with the emission of phonons by hot electrons[22]. Measurement of the temperature of the gas as a function of electric field allows extraction of the temperature-dependence of phonon emission. In the low disorder case, the energy relaxation time varies as T^{-4} but disorder, and indeterminacy, introduce an extra time which varies as T^{-2}, Figure 13.
The Silicon inversion layer is the simplest system for studying the quantum corrections to the conductivity. In other systems such as heterojunctions or InP inversion layers, separation of localisation and interaction effects

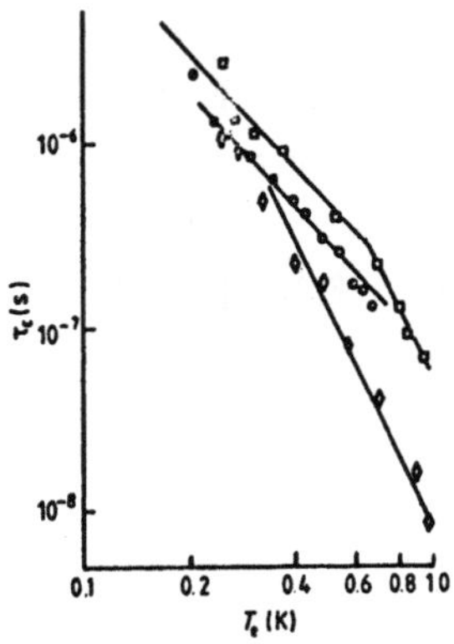

Fig. 13. - Energy relaxation time versus electron temperature for different values of the elastic mean free path , on log scales. The lines drawn show $\tau_{E-p} \propto T^{-4}$, $\tau_{E-p} \propto T^{-2}$
The values of $k_F \ell$ are 2.8 (□), 1.4 (○) and 6.5 (◇)

can be difficult, and the spin-orbit coupling[23] can play a role in reducing the negative magneto-resistance, particularly at low values of magnetic field.

Dimensionality transitions in the conductivity.

a) 3D to 2D

Decreasing the channel thickness of a GaAs FET, Figure 14, results in a change in the form of the density of states, Figure 15.

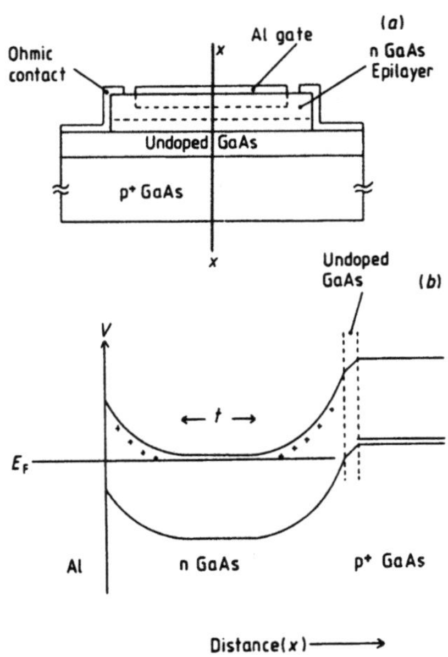

Fig. 14. - The Ga As FEt used in these experiments and an outline of the potential distribution.

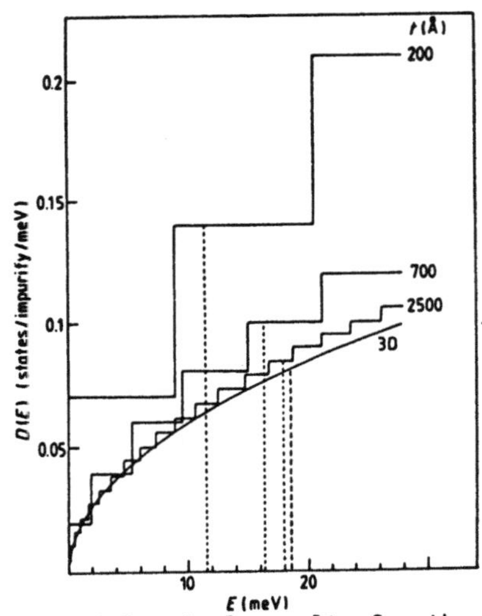

Fig. 15. - Semiclassical results for the density of states D(E)(in units of states/channel impurity/meV) as a function of channel thickness for the N_D 2 X $10^{17} cm^{-3}$ device. The parabola shows the 3D D(E). The case with = 2500 A is not applicable to the device investigated in this work. The vertical dotted lines show E_F for each thickness.

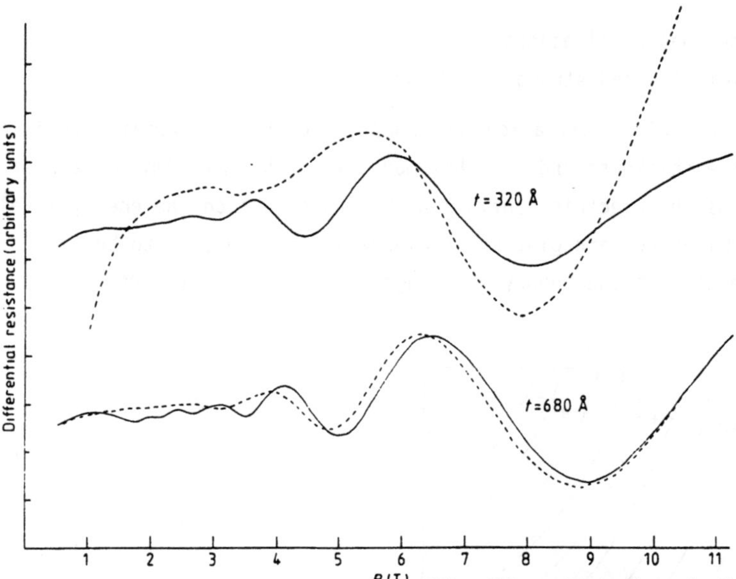

Fig. 16. - Shubnikov-de Haas oscillations in the differential of resistance for B applied perpendicular (Full curve) and parallel (Broken curve), $N_D = 2 \times 10^{17}$ cm^{-3}.

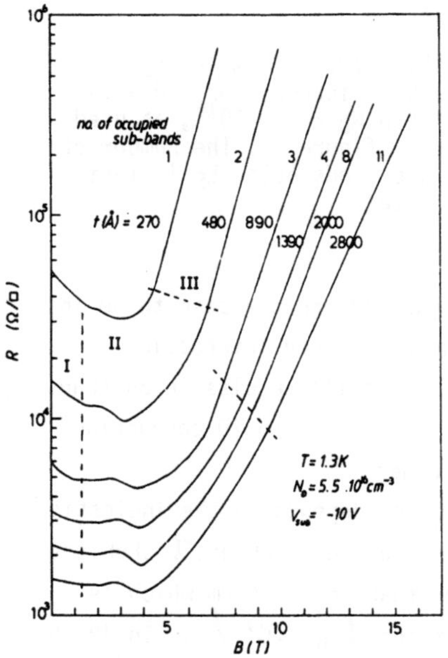

Fig. 17.

Measurement of the periodicity of Shubnikov-de Haas oscillations gives the Fermi energy, E_F, and this can be related to the thickness of the conducting channel. Figures 16,17 show the magnetic oscillations with both parallel and transverse field for different thickness.

The magnetic field induced 3 regimes of behaviour

1. Negative magneto-resistance and the suppression of quantum interference.

2. Shubnikov-de Haas oscillations.

3. Magnetic field-induced strong localisation.

As expected, the low field oscillations are suppressed when the orbit size is comparable with the thickness and, in this mode, when the oscillations appear they are shifted as the electric confinement has an effect on the energy levels. The values of Fermi energy are plotted in Figure 18 and are seen to be in reasonable agreement with the theory of Berggren for non-interacting electrons.[24]

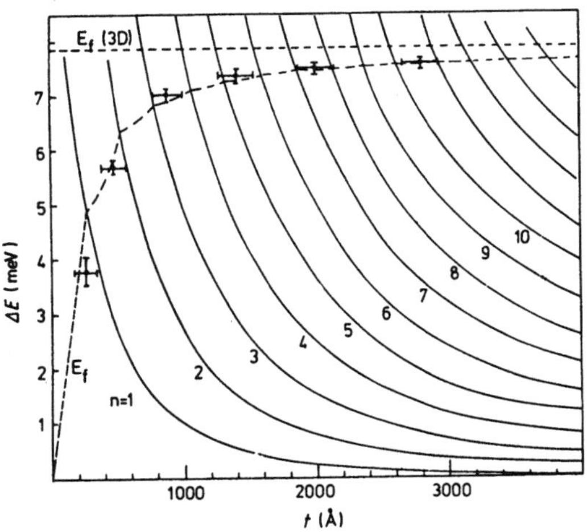

Fig. 18. - Fermi energy versus thickness for a device with N_D = 5.5 10^{16}, as used for figure 17. The number of sub-bands below E_F is indicated.

The values of Fermi energy are plotted in Figure 18 and are seen to be in reasonable agreement with the theory of Berggren for non-interacting electrons.[24] Initially, decreasing the thickness still results in 3D transport, but as the number of sub-bands falls, a transition to 2D conduction occurs and then E_F drops with increasing thickness.

The dimensionality of the localisation correction depends on the inelastic diffusion length, L_{IN}. If L_{IN} exceeds the sample thickness t then the correction becomes 2D, and logarithmic rather than the 3D form which is $\delta\sigma \propto (1/l - 1/L_{IN})$ and is not present for $L_{IN} \gg l$. In 3D the magneto-conductance varies as $1/L_c$, ($B^{\frac{1}{2}}$), and, like 2D varies as B^2 when $L_c \gg L_{IN}$. Figure 19 illustrates the magneto-conductance of a Ga As FET at 4.2K and 1.2K.

As seen, 3D behaviour is present at the widest value, and decreasing the temperature has a pronounced effect on the dimensionality transition.[25] When the negative field is parallel, little effect is found at low values, at high values it is similar to the transverse mode.

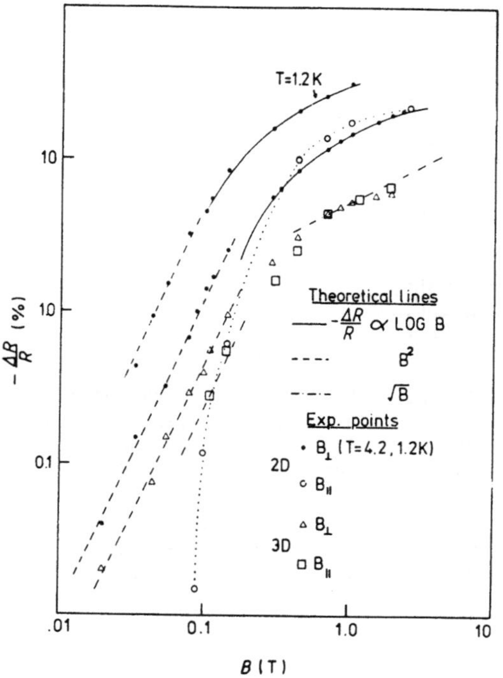

Fig. 19. - Negative magnetoresistance arising from the suppression of quantum interference as a function of thickness and dimensionality.

b) <u>2D to 1D</u>

All states in 1D are localised but, as in 2D, inelastic scattering cuts off the localisation and can produce quasi-extended states. Coherent backscattering produces a reduction in the conductivity given by:

$$\delta\sigma = \frac{e^2}{\hbar} L_{IN}$$

1D behaviour can be seen in quite wide samples as the criterion for localisation is that $L_{IN} > a$ where a is the specimen width. The localisation length, L_0, in the case of an Si inversion layer with a valley degeneracy of 2 (100 surface) is

$$L_0 = a k_F l / \pi$$

Thouless has proposed a general expression for the Diffusion coefficient D in one-dimensional transport.

$$D = D_0 \left(1 - \left[2(D_0 \tau_i)^{1/2} / L_0\right] \tanh L_0 / 2(D_0 \tau_i)^{1/2}\right)$$

D_0 is the "Boltzmann" diffusion coefficient. For short τ_i this formula gives the correction factor proportional to L_{IN} but for long inelastic times $D \approx L_0^2/12\tau_i$, ie $\sigma \approx N(E_F) L_0^2/12\tau_i$, $N(E_F)$ is the appropriate 2D expression. This conductivity mechanism corresponds to an electron inelastically scattered by other electrons, and each time a scattering event

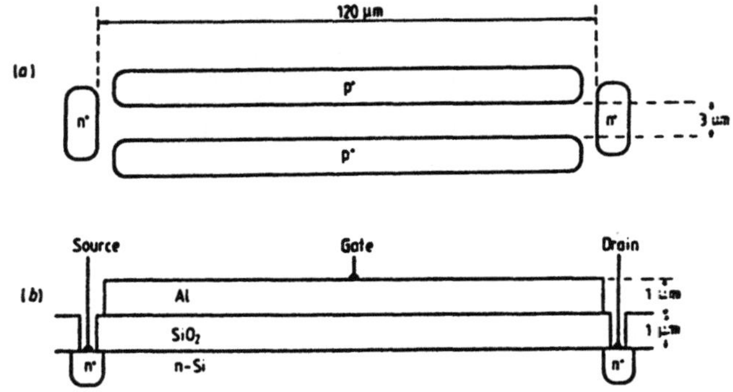

Fig. 20. - Schematic illustration of the accumulation-layer device used in this work. (a) is a plan of the n^+ source and drain diffusions and the p^+ regions on either side of the channel. (b) is a vertical section showing the 1μ m oxide layer and the 120μ m X 3μ m gate aluminium.

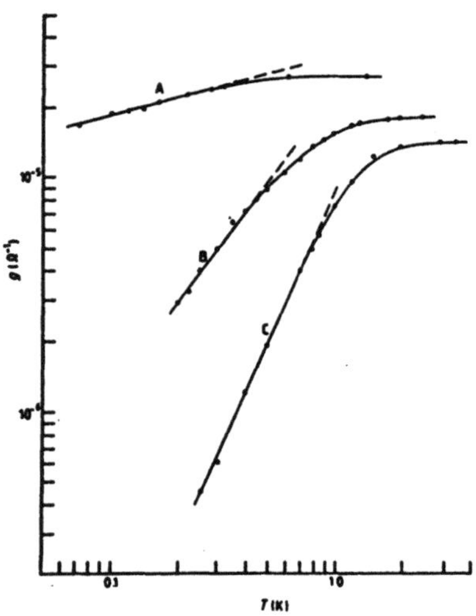

Fig. 21. - Conductance g as a function of temperature T at V_P = 0 V for three values of gate voltage V_G: (A) 40 V; (B) 30 V; (C) 25 V.

occurs (τ_i seconds) the electron jumps a distance L_0. The type of device used in this type of work[27] is shown in Figure 20, application of a bias voltage to the p type regions results in a decrease in channel width.

The temperature-dependence of conduction is illustrated in Figure 21, analysis of these results shows that a strong temperature-dependence sets in when $L_{in} \geq a$ The interaction correction in these experiments was always 2D but enhanced by the rapid decrease in diffusivity with increasing temperature. At higher temperatures such that $L_{in} < a$ the conductance shows the normal 2D logarithmic correction; this can be extremely small as $k_f l$ can be large. Analysis of the field-dependence of heating showed that when 1D behaviour was present, the electron- phonon relaxation time was three orders of magnitude greater than 2D systems with the same $k_f l$ and at the same temperatures. The explanation for this may lie in the reduction in emission rate by a reduction in overlap of wavefunctions or the reduction in inter sub-band scattering caused by the lateral quantisation.[28]

Finally, oscillations have been found in the conductance when the channel is narrow. Resonant tunnelling has been suggested as the cause when states are strongly localised[29] and the 1D sub-bands when not[30]. These explanations have been supported by magneto-resistance measurements; on the other hand, oscillations have been found in a narrow channel with a high impurity concentration present, which are not affected by a magnetic field. Here Coulomb effects and the electron glass have been invoked as an explanation.[31]

THE QUANTUM HALL EFFECT

The application of a high magnetic field in a transverse direction to the 2D electron gas results in the formation of Landau levels. Pushing the Fermi energy through the levels gives rise to an oscillating conductance - the Shubnikov - de Haas effect. Many experiments have shown that states in the tails of Landau levels are strongly localised, and the conductance is thermally activated.[32,33] Two different geometries of devices are used in these studies, the Corbino disc for measurement of σ_{xx} , and the Hall device for measurement of ρ_{xx}, ρ_{xy}.

In strong fields such that $\omega\tau > 1$ ω is the cyclotron frequency eB/m , Landau levels are formed, and the number of states in each level is $g_s g_v eB/\hbar$, where g_s and g_v are the spin and valley degeneracies $g_v=1$ for Ga As and 2 for Si (100) surface although this is lifted to 1 at high fields. When E_F is in the gap between Landau levels, the Hall conductance σ_{xy} is given by

$$\sigma_{xy} = -ne/B + A\,\sigma_{xx}/\omega\tau\,, \quad A \approx 1.$$

When states at E_F are localised σ_{xx} can be very small and $\sigma_{xy} = -ne/B$, which becomes Ne^2/h where N is the number of Landau levels below E_F. This relation was thought valid only when exactly N levels are full, i.e. not one electron more or less. The surprising feature of the experiments is that the quantised value of σ_{xy} is found when σ_{xx} is very small, even though E_F is in the tails of the levels,[34] Figure 22.

The quantisation is very accurate and is of interest as a possible standard for the determination of the fine structure constant. The physical origin of this effect lies in the confinement of electrons to equipotentials through the sample.[35,36] The smallest meaningful flux between equipotentials is the flux quantum h/e which results in a quantisation of area between equipotentials. If a region does not carry current, due to localisation, then no voltage is dropped and an enhancement of the field occurs across regions which do

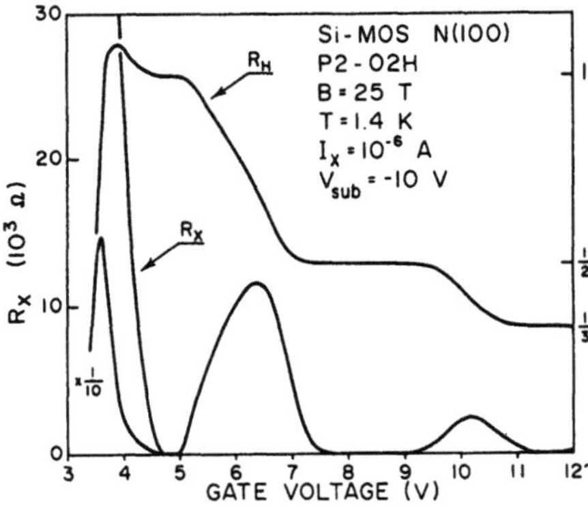

Fig. 22. - R_H and R_X plotted against V_G in a field of 25 Tesla, from reference 39.

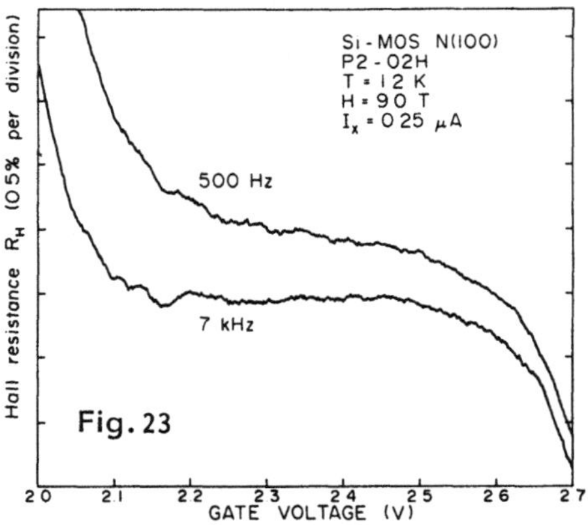

Fig. 23. - The frequency-induced creation of a quantised plateau when all states below E_F are localised.

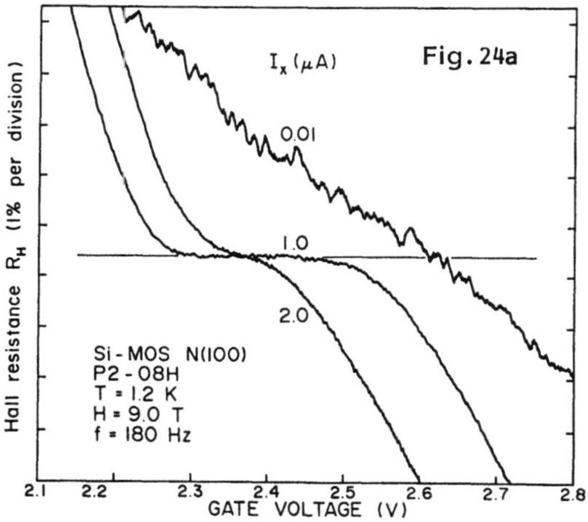

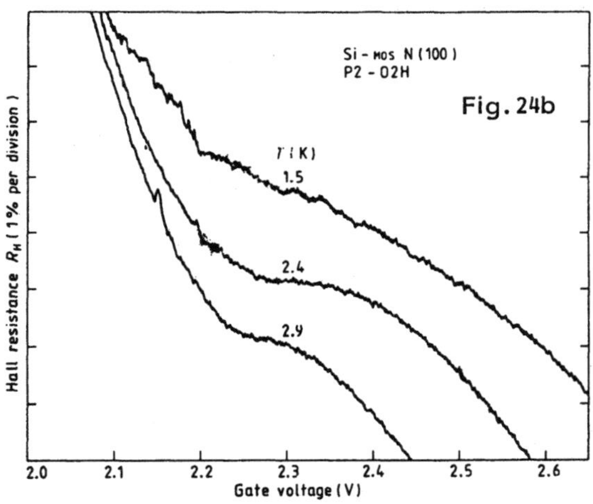

Fig. 24. - a) Increasing the sample temperature by the current results in plateau formation when states below E_F are localised. b) Similar results to a) but the ambient temperature is changed.

carry the current. This results in the quantised Hall resistance when states at E_F are localised and σ_{xx} is very small.[34]
The conditions for the observation of the Q.H.E. are extended states below E_F and localised states at E_F. When states below E_F are localised but with

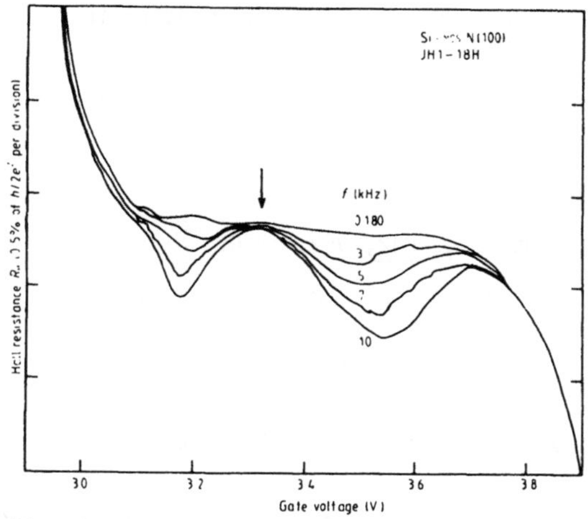

Fig. 25. - The destruction of a plateau by frequency is illustrated. The quantum value occurs when E_F is between Landau levels.

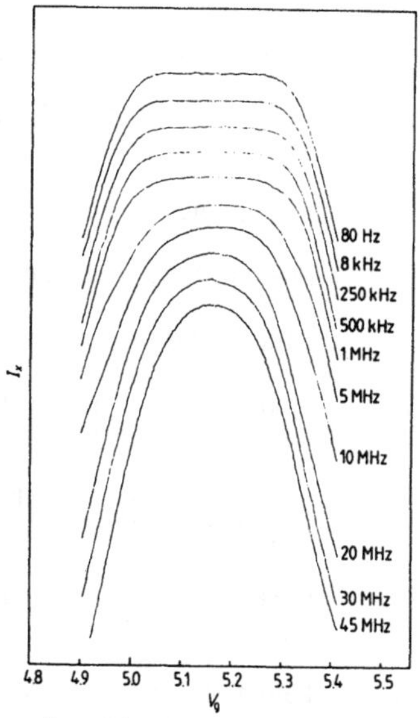

Fig. 26. - Detail of the i = 4 step (ground Landau level) of the quantised Hall resistance in a 400 μ m long, 50μm wide Hall geometry sample at various frequencies (sample P2H9). The magnetic field = 8T; temperature = 1.2 K and electric field E_{sd} = 75 mVcm^{-1} (RMS)

large localisation lengths, an increase in frequency can result in an effective delocalisation and the appearance of the quantised plateau. Figures 23, 24.

At very high frequencies, states in the tails of the Landau levels behave as if delocalised, and the plateau turns into a point,[37,38,39] Figures 25,26. Using Ga As - Ga Al As heterojunctions with very high values of mobility Tsui and colleagues[40,41] have found a new many-body effect.[42] Gaps in the Landau levels and plateaus of quantised resistance appear at fractional values of filling factor, $\frac{1}{3}$, $\frac{2}{3}$, ... At present there is considerable theoretical and experimental interest in this effect, which grows as the effects of disorder and localisation are reduced. Fractional effects have been observed at very high frequencies where the drift path of the electron is sufficiently small for the effects of disorder to be minimised.[43,44]

REFERENCES

1. Ando T, Fowler A.B. and Stern F, Rev. Mod. Phys. 54, 437, 1982.

2. Mott N.F., Pepper M,, Pollitt S., Wallis R.H. and Adkins C.J., Proc. Roy. Soc. A 345, 169, 1975.

3. Fang F.F. and Fowler A.B., 1968, Phys. Rev. 174, 823, 1968.

4. Uren M.J., Davies R.A. Kaveh M and Pepper M., J. Phys. C. 14, 5737.

5. Mott N.F. and Davis E.A., Electronic Processes in Non-Crystalline Materials, 2nd Edition, Oxford 1979.

6. Mott N.F. 'Metal-Insulator Transitions, Taylor and Francis, London 1974.

7. Fowler A.B., Phys. Rev. Lett. 34, 15, 1975.

8. Fowler A.B. and Hartstein A, Phil Mag. B42, 949, 1980.

9. Licciardello D.C. and Thouless D.J., J. Phys. C 11, 925, 1978.

10. Abrahams E., Anderson P.W., Liciardello D.C., Ramakrishnan T.V., Phys. Rev. Lett. 42, 673, 1979.

11. Anderson P.W., Abrahams E and Ramakrishnan T.V., Phys. Rev. Lett. 43, 718, 1979.

12. Gorkov L.P., Larkin A.J., and Khmelnitskii D.E., J.E.T.P. Lett, 30, 228, 1979.

13. Kaveh M. and Mott N.F., J. Phys. C. 14, L177, 1981.

14. Altshuler B.L., Khmelnitskii D., Larkin A.J. and Lee P.A., Phys. Rev. 109 1492, 1980.

15. Kaveh M. and Mott N.F. J. Phys. C. 14, L183, 1981.

16. Poole D.A. Pepper M. and Hughes A., J. Phys. C 15, L1137, 198

17. Haydock R.V., Phil Mag. B43 , 203, 1981.

18. Kawabata A. Surf. Sci. 113, 527, 1982.

19. Davies R.A. and Pepper M., J. Phys. C 16, L361, 1983.

20. Hikami S., Larkin A.J. and Nagaoka Y. Prog. Theor. Phys. 63, 707, 1980.

21. Davis R.A. and Pepper M., J. Phys. C. 16, L353, 1983.

22. Payne M. Davis R.A., Inkson J.C. and Pepper M., J. Phys. C. 16, L291, 1983.

23. Poole D.A., Pepper M. and Hughes A., J. Phys. C. 15, L1137, 1982

24. Poole D.A., Berggren K.F., Hill G., and Myron H.W., J. Phys. C. L21, 1982.

25. Poole D.A., Pepper M. and Myron H.W., Physica 117B, 697, 1983.

26. Thouless D.J., Solid State Comm. 34, 683, 1980.

27. Dean C.C. and Pepper M., J. Phys. C. 15, L1287, 1982.

28. Dean C.C. and Pepper M., J. Phys. C., in the press.

29. Fowler A.B., Hartstein A., and Webb R.A., Phys. Rev. Lett., 48, 196, 1982.

30. Wheeler R.G., Choi K.K. and Wisnieff R., Proc[1] 5th Int. Conference on Electronic Properties of 2D systems, Oxford 1983, to be published in Surface Science.

31. Pepper M., Uren M.J., and Oakley R.E., J. Phys C 12, L897, 1979.

32. Kawaji S., Surf. Sci. 73, 46, 1978.

33. Pepper M., Phil Mag. 37b, 83, 1978.

34. von Klitzing., Dorda G., and Pepper M., Phys. Rev. Lett., 45, 494, 1980.

35. Laughlin R.B., Phys. Rev. B23, 5632, 1981.

36. Kazarinov R.F. and Luryi S. Phys. Rev., 25,7626, 1982.

37. Pepper M., and Wakabayashi J., J. Phys. C. 15, L861, 1982.

38. Pepper M., and Wakabayashi J., J. Phys. C. 16, L1113, 1983.

39. Wakabayshi J., Myron H.W., and Pepper M., Physica 117B, 691, 1983.

40. Tsui D.C., Stormer H.L., and Gossard A.C., Phys. Rev. Lett. 48, 1559, 1982.

41. Stormer H.L., Chang A.M., Tsui D.C., Hwang J.C.M., Gossard A.C. and Weigmann W., Phys. Rev. Lett. 50, 1395, 1983.

42. Laughlin R.B., Phys. Rev. Lett. 50, 1395, 1983.

43. Long A.P., Myron H.W., and Pepper M., J. Phys. C 17, L433, 1984.

44. McFadden C., Long A.P., Myron H.W., Pepper M., Andrews D and Davies G.J., J. Phys. C 17, L439, 1984.

45. Pepper M. Proc. Roy. Soc. Lond.A. 353, 225, 1977

RHEED and Photoemission Studies of Semiconductors Grown in-situ by MBE

P.K. Larsen

Philips Research Laboratories, NL-5600 JA Eindhoven, The Netherlands

B.A. Joyce and P.J. Dobson

Philips Research Laboratories, Redhill, Surrey RH1 5HA, United Kingdom

1 Overview

The electronic and crystallographic structure of semiconductor surfaces and interfaces is important not only fundamentally but also practically for the emerging range of low-dimensionality device structures [1]. The growth technique of molecular beam epitaxy (MBE) based on ultra high vacuum has been used successfully in both areas and it has been demonstrated that film properties can be related to surface conditions during growth [2].

In general, semiconductor surfaces are reconstructed (i.e. have reduced symmetry from the ideal bulk termination), and when prepared by MBE the reconstruction is strongly dependent on growth conditions. By combining MBE with surface-sensitive techniques such as photoemission spectroscopy (PES) and reflection high energy electron diffraction (RHEED), it is possible to study surfaces and interfaces which have been formed in-situ. Growth dynamics can also be followed directly. For most compound semiconductors (e.g. GaAs) there is no viable alternative to MBE for producing clean, damage--free, reconstructed surfaces other than by cleavage, which is restricted to a single orientation, usually (110). Techniques such as ion bombardment and annealing invariably introduce crystal damage and a surface region of unknown and variable stoichiometry. It may also be advantageous to produce surfaces of the elemental semiconductors Si and Ge by MBE, since there is evidence that this results in lower energy reconstructions than those formed by other preparation techniques such as thermal annealing or ion bombardment and annealing.

In this chapter (in section 2) we will first discuss the experimental techniques used, MBE, RHEED and angle-resolved PES (ARPES) with excitation by synchrotron radiation. The important feature of this combination of techniques is the compatibility of the measurement methods (RHEED and ARPES) with the growth process. In section 3 we demonstrate how RHEED can be used to evaluate the basic growth mechanism for both elemental and III-V compounds, and to give some indications of surface disorder and interface perfection. RHEED also enables us to determine surface symmetry, while photoemission from core levels allows us to establish the relationship between surface stoichiometry and reconstruction. In section 4 we show how the various spectral features in ARPES measurements from reconstructed surfaces can be identified, and discuss the determination of bulk valence bands and surface electronic structure. ARPES can also be used to measure semiconductor heterojunction band edge discontinuities. In the last section of this chapter two applications of surface-sensitive core level photoemission are treated. From line shape and intensity analysis, certain aspects of the surface chemistry can be established. The technique also allows determination of the compositional abruptness of a junction.

2 Basis of Experimental Approach

2.1 Molecular Beam Epitaxy

MBE is simply a refined form of vacuum evaporation, in which directed neutral thermal atomic and molecular beams impinge on a heated substrate under UHV conditions. The apparatus to achieve this with the necessary degree of control has become rather complex, and the elements of a system suitable for growth of III-V compound films are illustrated schematically in fig. 1. (Methods used for type IV elements are somewhat different, and we consider them later). It is based on a two or three-chamber stainless steel UHV system, usually ion- or cryo-pumped, and incorporates large areas of liquid-nitrogen-cooled panels. The provision of an interlock system, to enable the substrate to be introduced into the growth chamber without breaking the vacuum, is essential if films having high quality electrical and optical properties are to be prepared.

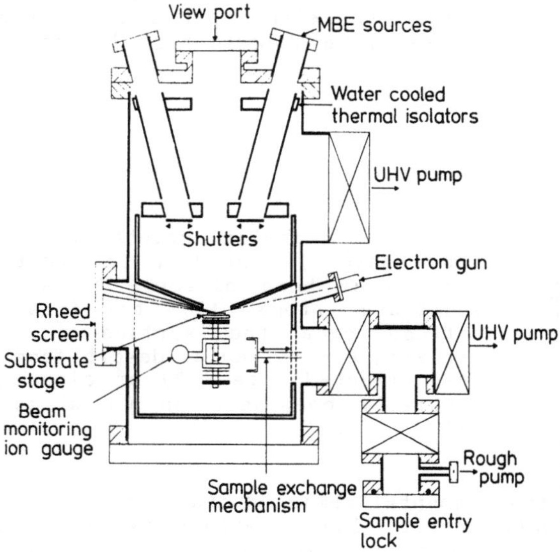

Fig. 1. Schematic diagram of an MBE system, showing cells, substrate rotation stage, RHEED, vacuum interlock arrangement and cryopanelling

The beams are formed in heated Knudsen effusion cells, and individual cells are thermally isolated by a succession of heat shields. The flux is regulated by a shutter operating in front of each orifice. A single microcomputer can be used to control all of the cell temperatures, and all of the shutters, to achieve any growth sequence. Fluxes are usually monitored by an ion gauge, which can be rotated in and out of the beams.

To provide the basis for high quality films, a substrate surface free of crystallographic and other defects and clean on an atomic scale (≤ 0.01 monolayer (ML) of impurities) must be prepared. The first stage usually involves chemical etching, which leaves the surface covered with a protective volatile oxide. This is subsequently removed in the vacuum system by heating. For III-V semiconductors it is necessary to carry out the heating in a beam of the Group V element to avoid surface dissociation. To preserve the clean surface at high temperatures (e.g. above about 300°C for GaAs) it is necessary to maintain the Group V element flux.

Source materials are usually elemental, and all of the Group III elements produce monoatomic beams. Group V sources are rather more complex, however, both in terms of the species produced and the methods used to produce them. When evaporation takes place directly from the elements, fluxes consist entirely of tetra-atomic molecules (P_4, As_4 and Sb_4). Dimers can be produced from the elements by using a two-zone Knudsen cell, in which a tetramer flux is formed conventionally and passed through an optically baffled high temperature stage, which can be designed to produce a complete conversion of tetramers into dimers [3]. Growth is initiated by bringing the substrate to the appropriate temperature (typically in the range 500-750°C) in a beam of the Group V element and then opening the shutter of the Group III element source. Growth rates are in the range 0.1-10 μmhr^{-1}, corresponding to beam fluxes from $\sim 5\times10^{13}$ to $\sim 5\times10^{15}$ atoms $cm^{-2}s^{-1}$. The Group V element flux is typically 3-5 times greater than that of the Group III element, but it is the latter which determines the growth rate. We show in fig. 2 a schematic model of this process.

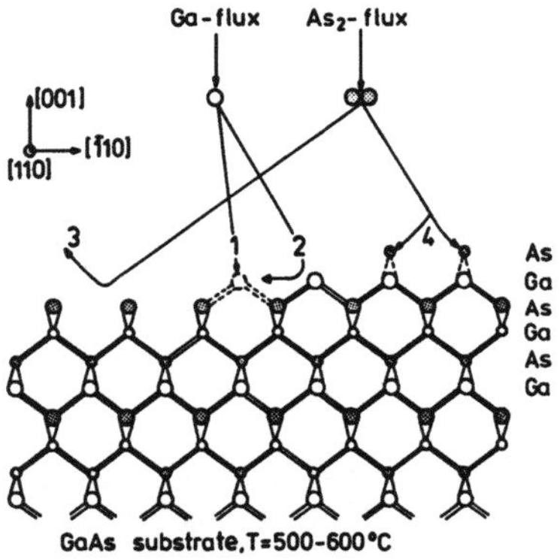

Fig. 2. Schematic growth model for GaAs(001). Incident Ga atoms can either bond directly to As-substrate atoms (1) or first diffuse over the surface (2). Arriving As_2 molecules desorb from As-terminated sites (3) or dissociatively chemisorb on Ga atoms (4)

The procedures used for the elemental semiconductors, Si and Ge, are somewhat different. Surfaces are cleaned either by chemical etching followed by high temperature annealing in the vacuum chamber, or by ion bombardment followed by annealing (low-temperature annealing). Ge can be grown using a Knudsen cell source, but for Si the vapour pressure is so low that electron beam evaporation is used [4].

2.2 Surface Reconstruction and RHEED

For a number of semiconductor surfaces, it is possible to prepare the same basic crystallographic orientation with different structures. If the bulk positions were maintained at the surface, there would be no reconstruction, and the surface would be said to be unreconstructed, having the 1x1 structure. However, in general with semiconductors, there will be a reduction in symmetry at the surface which results in a surface unit cell having larger periodicity than the unreconstructed form. This is described by a notation introduced by Wood [5], whereby the new primitive cell is expressed in terms of the dimensions of the unreconstructed unit cell. For example, the (001) surface of GaAs has several reconstructions, of which the most stable is the 2x4 structure. This means that along two orthogonal ⟨110⟩ directions, there is a two-fold and a four-fold increase in the periodicity, respectively.

The surface reciprocal lattice can be determined by electron diffraction techniques, using either normally incident low-energy electrons (of the order of 100 eV), usually referred to as LEED, or higher energy electrons (5-50 keV) at grazing angles (1°-3°) (RHEED). The forward scattering geometry of RHEED makes it ideally compatible with MBE since the molecular beams impinge normally to the substrate surface, therefore diffraction measurements can be taken during growth. This would not be possible with LEED. In fig. 3a we illustrate the experimental geometry of RHEED. An important aspect is the appearance of streaks normal to the shadow edge instead of spots. The simplest explanation is to consider that the penetration of the electron beam in the RHEED geometry is restricted to the uppermost layer of the crystal surface [6]. If we start with the assumption that only the first layer contributes to diffracted intensities, we can simplify the usual point reciprocal lattice picture, since the reciprocal lattice points are now drawn out into rods [7]. This can be thought of as a relaxation of a Laue condition, and the spacing between reciprocal lattice rods can also be related to the spacing between rows of atoms in the surface layer. The diffraction pattern can be visualized as an intersection of the Ewald sphere - whose radius is large at high energies - with the set of reciprocal lattice rods as illustrated in fig. 3b. In practice, this is a rather oversimplified model, since the penetration depth is not completely restricted to the first layer, and a

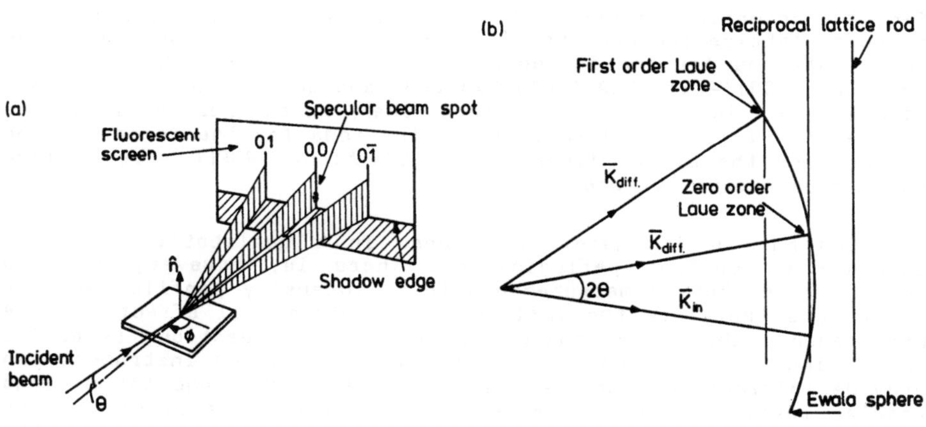

Fig. 3. (a) Schematic diagram of RHEED geometry showing incident beam at an angle θ to the surface plane, azimuthal angle φ.
(b) Ewald sphere-construction appropriate for a qualitative explanation of the formation of streaks in RHEED patterns. The wave vector K_{in} is very large compared with the inter-rod spacing.

modulated reciprocal lattice rod model is probably more appropriate. From the distance between the streaks on the fluorescent screen, the distance between this and the substrate (the camera length) and the energy of the incident electrons, the surface periodicity perpendicular to the beam azimuth can be evaluated from the Ewald sphere construction. A determination of the surface symmetry, therefore, in general requires measurements along various azimuths. In fig. 4 we show real and reciprocal space sections of three typical reconstructions, the corresponding surface Brillouin zones (SBZ) together with zero-order Laue zone RHEED patterns from them at three different azimuths.

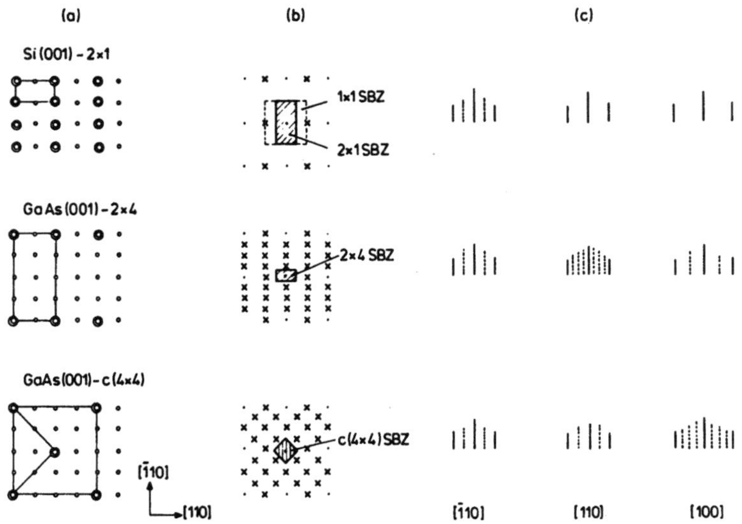

Fig. 4. Schematic diagram of the real space geometry (a), the reciprocal lattice rod geometry (b) and expected zero-order Laue zone RHEED patterns for the incident beam along the azimuths indicated (c) for Si(001)-2x1, GaAs(001)-2x4 and GaAs(001)-c(4x4). The double circles (a) indicate the surface real space periodicity, the crosses (b) refer to additional lattice rods for the reconstructed surface, and the broken lines (c) to additional RHEED streaks from the reconstructed surface

RHEED has so far mainly been used in a qualitative way with emphasis on surface periodicities. There is, however, a trend towards quantitative measurements (e.g. intensity profile measurements) and therefore, the influence of instrumental effects on the measurements should be known. This relates to questions regarding the maximum distance on the surface over which the instrument can provide information (the so-called transfer width) and the minimum angular resolution. The instrumental limitations are caused by the finite size of the electron source, the angular spread of the electron beam, etc. These questions have been discussed recently, with emphasis on LEED, by Comsa [8] and by Lu and Lagally [9]. The instrument response for RHEED instruments has been treated by Norton [10] and by van Hove et al. [11]. These authors indicate for incident electrons in the range 10-40 keV a transfer width along the path of the beam of several thousands of Angstroms, while the width perpendicular to the beam is much smaller, of the order of at most a few hundred Å, similar to typical values for LEED. As pointed out by Lu and Lagally [9] the most appropriate parameter to describe the resolving power of a diffraction instrument is the minimum angular resolution rather than the transfer width.

2.3 Angular Resolved Photoemission

The principles and various techniques of photoemission from solids have been treated extensively in the literature [12-14], and the growing field of inverse photoemission has recently been reviewed by Smith [15]. Here, some of the experimental aspects of obtaining surface relevant information from photoemission measurements on reconstructed semiconductors will be discussed, with emphasis on angle-resolved (or directional) photoemission from states in the valence band region, and on photoemission from core levels.

In fig. 5a the energy level diagram of a heteropolar semiconductor is shown schematically. The absorption of a photon of energy $h\nu$ can lead to different optical transitions, as indicated. Transitions 1 and 3 are from bulk valence band and bulk core level states, respectively. At the surface and also at semiconductor-semiconductor and semiconductor-metal interfaces, the energy states are modified, and localized surface (or interface) states appear. The binding energies of the core levels for surface atoms might shift, and we therefore have specific transitions, nos. 2 and 4, related to the surface (or interface) region. In a photoemission spectrum,

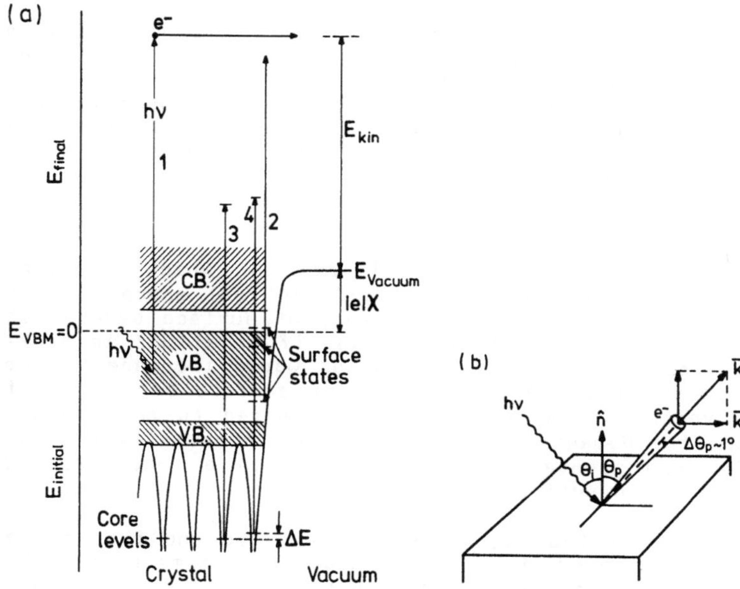

Fig. 5. (a) Schematic energy level diagram for a heteropolar semiconductor, illustrating the photoemission process. The various transitions, labelled 1-4, are discussed in the text.
(b) Geometrical arrangement of an angle-resolved photoemission experiment. θ_i is the photon incident angle, θ_p the polar angle of emission.

emission features due to the surface region may overlap with bulk features. We discuss in section 4 the identification and separation of surface state emission from bulk emission.

In the single particle picture of photoemission, assuming direct transitions, there is conservation of energy, $E_f = E_i + h\nu$, and momentum (in the reduced zone scheme), $\bar{k}_f = \bar{k}_i$. Electrons escaping into vacuum without being inelastically scattered have a kinetic energy $E_{kin} = E_f - |e|\chi$, where $|e|\chi$ is the electron threshold, that is the vacuum level measured with respect to the top of the valence bands ($E_{VBM} \equiv 0$). In an ARPES experiment only photoelectrons having a well-defined direction of propagation, given by the polar angle θ_p (see fig. 5b) and azimuthal angle φ, are measured. From θ_p, φ and E_{kin} the momentum of the electron in vacuum is determined, $\bar{P} = \hbar k$, and therefore the component of the wave vector parallel to the surface is given by

$$k'_{\parallel} = (2mE_{kin}/\hbar^2)^{1/2}\sin\theta_p. \qquad (1)$$

This quantity is conserved as the photoelectron escapes into vacuum and therefore

$$\bar{k}_{\parallel} = \bar{k}'_{\parallel} + \bar{g}_{\parallel} \qquad (2)$$

where k is the wave vector in the crystal and $\bar{g}_{\parallel}$ is a surface reciprocal lattice vector. The component of the wave vector perpendicular to the surface, $\bar{k}_{\perp}$, is however not conserved by the escape into vacuum.

For two-dimensional band structures $E_i = E_i(\bar{k}_{\parallel})$ it is possible from measurements of the energy position of peaks as a function of the polar angle to plot directly the energy dispersions. The initial energy E_i can simply be calculated from the measured kinetic energy from $E_i = E_{kin} + |e|\chi - h\nu$ and $\bar{k}_{\parallel}$ is determined by Eqs. (1) and (2). This is the case for surface band structures, where $E_i(\bar{k}_{\parallel})$ has the symmetry of the surface Brillouin zone :

$$E_i(\bar{k}_{\parallel}) = E_i(\bar{k}_{\parallel} + \bar{g}_{\parallel}). \qquad (3)$$

Consider as an example the GaAs(001)-2x4 SBZ which is eight times smaller than the unreconstructed SBZ (in real space the 2x4 unit cell is 8 times larger) and the number of surface bands must therefore have increased considerably, e.g. if a 1x1 SBZ is backfolded into an nxm SBZ, the number of bands would increase by the factor nxm. Because of the larger number of surface bands and of the smaller $\bar{g}_{\parallel}$ vectors of a reconstructed compared to an unreconstructed surface the demands for the energy and especially the angular resolution are greater for measurements on a reconstructed than an unreconstructed surface.

We have already seen that $\bar{k}_{\perp}$ cannot be measured and this means that three-dimensional band structures cannot be determined directly. If however the final state is known, e.g. from a band structure calculation, or if certain assumptions about it can be made, then it is possible to evaluate the bulk band structure $E_i(\bar{k}) = E_i(\bar{k}_{\parallel}, \bar{k}_{\perp})$. We show in section 4 how bulk bands can be plotted for GaAs and Ge. The use of tunable radiation, i.e. monochromatized synchrotron radiation, is essential for such measurements, since a change of the photon energy means a different optical transition with different $\bar{k}$-vector (and hence $\bar{k}_{\perp}$).

The intensity of a peak in a photoemission spectrum in the valence band region provides information on the initial states, but in a quite complex way, so that intensity measurements are mainly made to establish symmetries and orbital character of the states. A realistic calculation of the photoemission intensity requires a knowledge of the incident radiation (polarization), the dielectric character of the semiconductor, the initial and final states involved in the optical transition, and the factors affecting the propagation of the photoelectrons to the detector. In section 4 we compare experimental results for GaAs(001) surface state emission with calculations [16] based on Holland's theory of angle-resolved photoemission from localized orbitals at solid surfaces [17].

If we now turn our attention to studies of core levels, and especially surface core levels (see the review by Eastman et al. [18]), the main emphasis is on intensity and line shape measurements. To obtain a maximum surface sensitivity, the photon energy should be chosen to give a final energy in the range 30-100 eV since this corresponds to the range of minimum electron escape depths λ. For GaAs, Ge and Si, $\lambda_{min} \sim 6$ Å [19]. The use of tunable synchrotron radiation, combined with a good total energy resolution (monochromator plus electron energy analyzer), typically of the order of 0.2 eV, is necessary in line shape studies from which the surface atom binding energy-shifts may be resolved. These shifts depend on the surface atom-bonding geometry. Surface to bulk and anion to cation core level intensity ratios provide information on the surface stoichiometry. We note that for these measurements there is no need for any angular resolution, since the (localized) core level binding energies do not depend on $\bar{k}$. For photoelectron diffraction studies using excitation from core levels, this is not the case, however. Finally, core level measurements play an important role in photoemission determinations of band bending and of valence band discontinuities. We shall discuss the various core level measurements in sections 4 and 5.

3 The Growth Process

The experimental arrangement of MBE is unique among epitaxial thin film preparative methods in that it enables the growth process to be studied in-situ in several ways. In particular, two of these, modulated molecular beam-mass spectrometry (MBMS) and RHEED allow direct measurements of the kinetics of surface reactions and dynamics of growth, respectively. The kinetic measurements are outside the scope of this article, but have been reviewed by Foxon and Joyce [20]. Here we will concentrate on growth dynamic aspects, surface reconstruction in relation to surface stoichiometry and surface disorder effects, illustrated principally with reference to GaAs(001).

3.1 Growth Dynamics

It is possible to explore some features of the growth dynamics of MBE by monitoring temporal variations in the intensity of various features in the RHEED pattern. It has been found [21] that damped oscillations in the intensity of both the specular and diffracted beams occur immediately after initiation of growth. A typical example for the specular beam is shown in fig. 6. The period of oscillation corresponds exactly to the growth of a single monolayer, i.e. a complete layer of Ga and As atoms (= $a_0/2$, in the [001] direction) and is consequently independent of the azimuth of the incident beam and of the particular diffraction feature being measured. The amplitude, however, is strongly dependent on both of these parameters. For evaluation of growth dynamics,most of the information is contained in the specular beam, so we limit the discussion of growth dynamics to this feature.

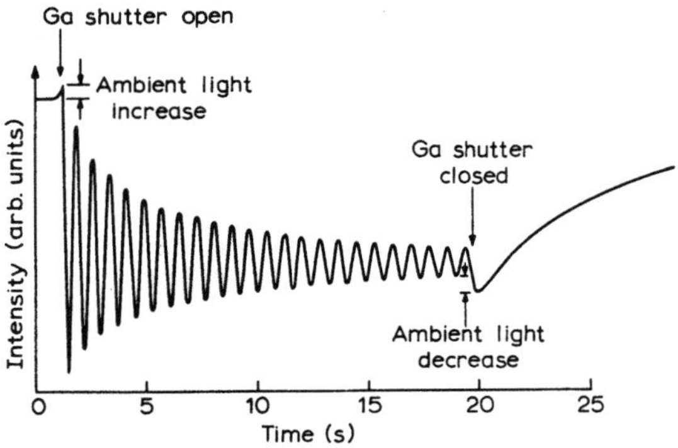

Fig. 6. Intensity oscillations of the specular beam in the RHEED pattern from a GaAs(001)-2x4 reconstructed surface, [110] azimuth. The period exactly corresponds to the growth rate of a single Ga+As layer and the amplitude gradually decreases

The observation of oscillatory effects in thin film growth studies [22-25] is usually associated with a layer-by-layer growth process, i.e. two-dimensional nucleation. In the Gilmer-Weeks theory [26] the equilibrium surface is smooth, and at the start of growth random two-dimensional clusters are generated. The layer is complete when all the clusters have coalesced, and all the holes are filled. If we now equate changes in intensity of the specular beam in the RHEED pattern with changes in surface roughness, the equilibrium surface is smooth, corresponding to high reflectivity. At the commencement of growth, two-dimensional clusters are formed at random positions on the crystal surface, leading to a decrease in the reflectivity. This decrease can be predicted from a simple optical analogue, since the Broglie wavelength of the electrons is ~ 0.12 Å while the bi-layer step height is ≃ 2.8Å, so diffuse reflectivity will result. In fig. 7 we illustrate the formation of a single layer. Nucleation is not restricted to a single layer, but can reoccur before the preceding layer is complete. In the early stages, however, one layer is likely to be almost complete before the next layer starts, so the reflectivity will increase as the surface again becomes smooth on the atomic scale, but with subsequent roughening as the next layer develops. This repetitive process will cause the oscillations in reflectivity gradually to be damped as the surface become statistically distributed over several incomplete layers. When growth is terminated and the layer is maintained at its growth temperature in a beam of arsenic, the reflectivity recovers to its original value, i.e. the surface becomes smoother.

The oscillation period of the specular intensity of the RHEED beam provides a continuous and absolute growth rate monitor with atomic layer precision. For GaAs(001), growth occurs through bi-layers, but for elemental semiconductors, where single atomic layer growth is possible, RHEED shows that this does in fact occur [27]. The amplitude of the intensity oscillations is, however, a complicated function of azimuth and temperature [28]. For growth of GaAs(001) under conditions for which the (2x4) reconstruction is observed, the amplitude of the oscillations is greater for the beam incident along the [110] than along the [$\bar{1}$10] direction. This would suggest that most of the steps which develop on the surface are along the [$\bar{1}$10] direction, as shown in fig. 7, i.e. they cause maximum diffuse scattering when their longer edges are normal to the incident beam. The temperature dependence may be related to a possible change from two to three-dimensional nucleation with increasing temperature to changes in surface reconstruction.

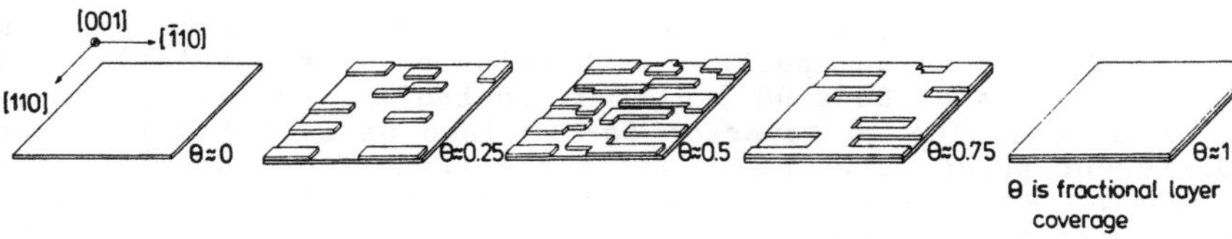

Fig. 7. Real space representation of the formation of a single complete layer, deduced from the oscillations of intensity in the specular beam of the RHEED pattern.

An alternative theoretical approach to growth dynamics has been formulated by Lent and Cohen [29]. Their treatment involves analysis of diffracted beam intensities and profiles from stepped surfaces, but is essentially a layer-by-layer growth model similar to that shown in fig. 7.

The results discussed in this section are important not only for the understanding of the growth process but also in the preparation of smooth and well-defined surfaces for subsequent studies by RHEED or photoemission at room temperature. After completion of growth, in order to obtain the smoothest surface, the substrate should be maintained at temperature in a flux of arsenic. For studies of heterojunctions, the ability to obtain exact control of the thickness of the grown overlayer is essential.

3.2 Surface Stoichiometry and Reconstruction

For polar III-V surfaces, e.g. the (001), (111) and ($\bar{1}\bar{1}\bar{1}$) orientations, the various surface reconstructions observed are related to the surface stoichiometry. In the high symmetry polar direction, the crystal can be considered to be built up of alternating layers of Group III or Group V element atoms (see fig. 2). Ideally, the surface would therefore be terminated with either anions or cations. It is observed, however, especially for the (001) orientation, that several different structures can be formed. With surfaces prepared by MBE, for example, the reconstructions depend on the relative and absolute fluxes of the incident beams and on the substrate temperature.

It is in principle possible to determine the surface stoichiometry either from Auger electron spectroscopy (AES) or from photo-

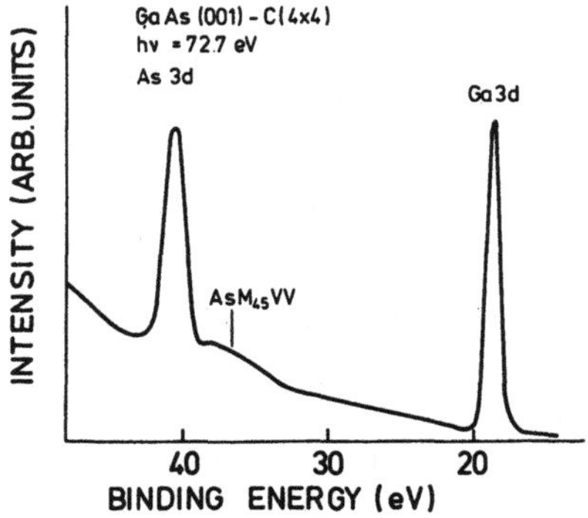

Fig. 8. As(3d) and Ga(3d) core level photoemission spectrum at $h\nu$ = 72.7 eV for a GaAs(001)-c(4x4) surface. The broad feature (labelled) is related to an As Auger transition

emission core-level intensity measurements. As an example, we show in fig. 8a photoemission spectrum with emission from the As(3d) and Ga(3d) core levels of GaAs(001)-c(4x4). In table 1 we give a compilation of data of the As coverage for various GaAs(001) structures from different authors. We have listed the reconstructions in order of increasing As coverage, and note that Chiang et al. [33] have measured a similar sequence from Ga to As core level intensity ratios. There are some contradictions in these results, which arise from two basic reasons. The first is an intrinsic problem, since it has been clearly demonstrated [34, 35] that a range of stoichiometry can exist for a fixed reconstruction. The other resides in the nature of the measurements themselves and in their interpretation. Methods used for data evaluation are based on layer-by-layer models involving electron escape depths, for which a wide range of values have been used, and also they do not take into account possible photoelectron diffraction effects. Consequently, the actual As coverages given in table 1 should be treated with extreme caution; they merely demonstrate a trend.

Table 1. Surface layer As coverage for various GaAs(001) reconstructions (in units of monolayers)

	c(8x2) 4x2	4x6	1x6	c(2x8) 2x4	c(4x4)
Auger data [30]	0.22	0.27	0.52	0.61	0.86
Core level data [31]	0.52	0.31	0.42	0.89	1.00
Core level data [32]					1.2-1.75

3.3 Surface Morphology Effects

In addition to providing information on surface symmetry, RHEED can be used as a diagnostic technique for the evaluation of surface morphology, i.e. the topography and the extent of ordering of the surface [36].

Two-dimensional disorder effects can be analyzed from the streak shape. Any lack of perfect ordering in a particular direction which restricts the average size of the ordered regions leads to the reciprocal lattice rod becoming two-dimensional, forming a solid el-

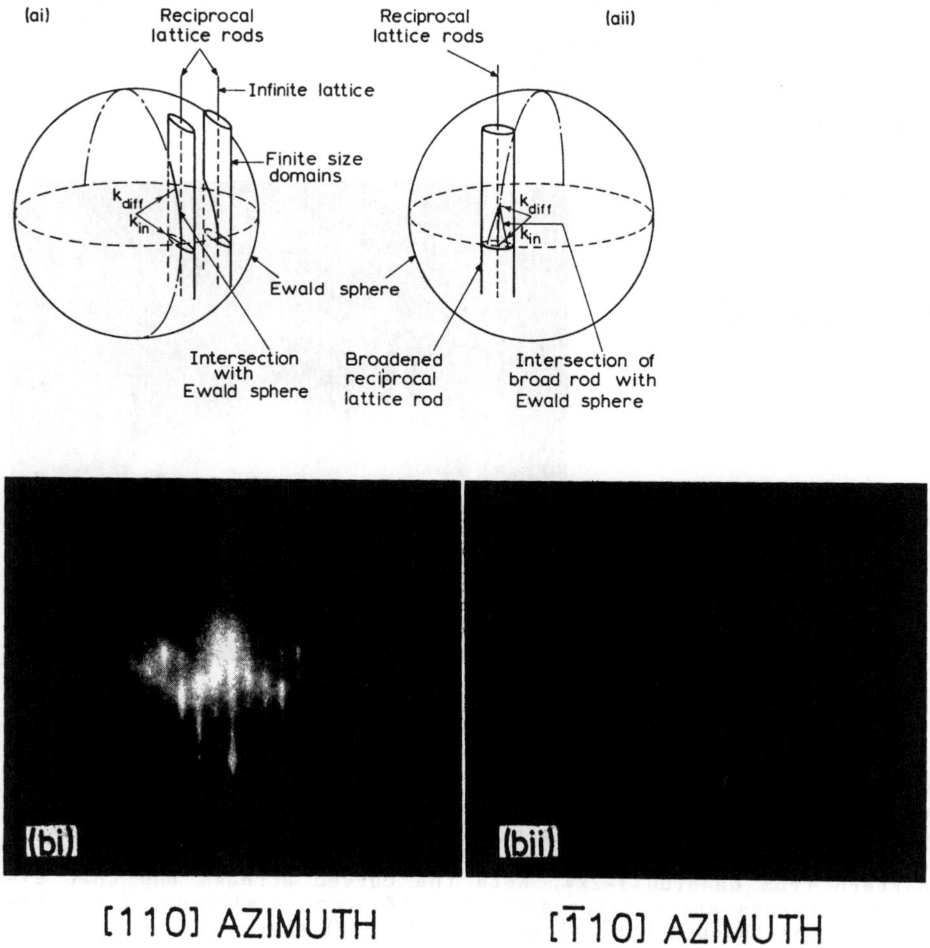

Fig. 9. (ai) and (aii): Reciprocal lattice-Ewald sphere construction showing lengthening and broadening of diffraction streaks resulting from lack of perfect ordering in a specific direction (bi) and (bii): RHEED patterns illustrating this effect, from GaAs(001)-2x4; [110] and [$\bar{1}$10] azimuths respectively.

lipsoidal cylinder. This point is illustrated in fig. 9a, where we show the effects along two orthogonal azimuths for the case when the domains are extensive in one direction and restricted in the perpendicular direction. With the beam parallel to the "short" domain side (fig. 9(ai)), the streaks are long and narrow, which corresponds to the diffraction pattern in fig. 9(bi). When the beam is parallel to the "long" domain side, the situation depicted in fig. 9(aii) prevails, which produces short and broad streaks, corresponding to those in fig. 9(bii). The domain structure giving rise to these has been discussed in detail elsewhere [37].

The presence of two-dimensional surface disorder, or domains, implies that they will be separated by one-dimensional (or antiphase) boundaries, which give rise to curved streaks, provided the

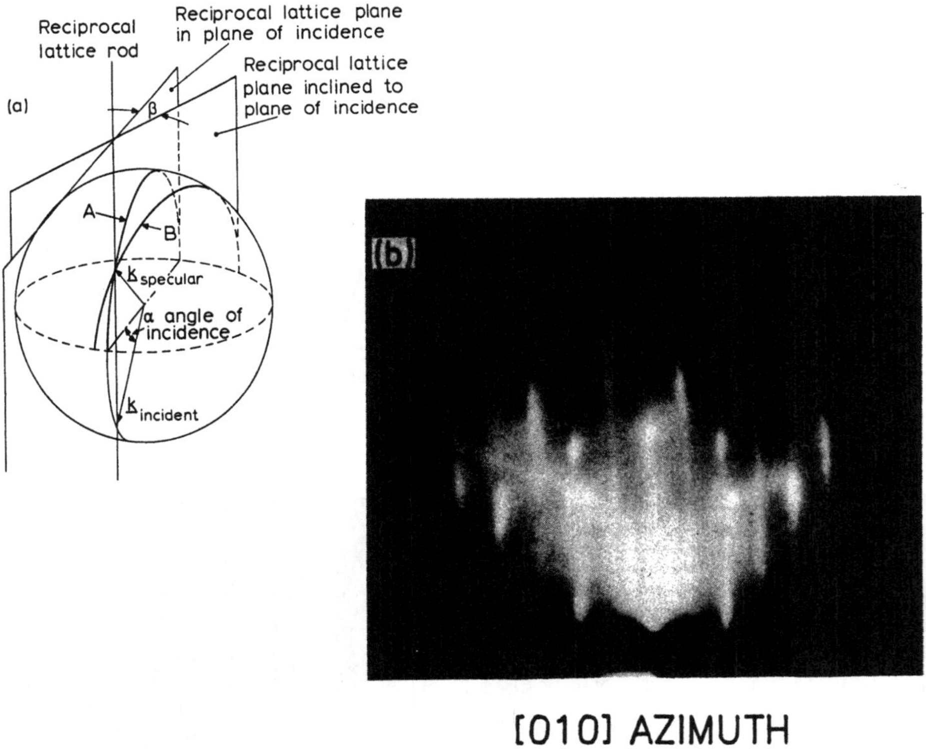

Fig.10. (a): Reciprocal lattice-Ewald sphere construction, showing the origin of curved streaks. A and B define the streaks originating from the reciprocal lattice plane in the plane of incidence and inclined at an angle β, respectively, (b): [010] azimuth RHEED pattern from GaAs(001)-2x4. Note the curved streaks and that extended streaks are only present in half-order positions, i.e. they do not pass through the origin of reciprocal space.

appropriate azimuth is used [38, 39]. Such boundaries give planes in reciprocal space, and when the incident electron beam is parallel to the boundaries, the Ewald sphere is tangential to the reciprocal-lattice planes, and there is only an increase in background intensity. For any other azimuth, however, the planes and the Ewald sphere intersect to give curved streaks, as shown in fig. 10(a), producing the type of RHEED pattern illustrated by fig. 10(b). This example was taken in the [010] azimuth from a GaAs(001)-2x4 reconstructed surface which contained one-dimensional disorder boundaries along $[\bar{1}10]$ [21].

At least four topographic features can be identified from RHEED patterns, but it is important to realize that in general, the length and width of streaks are not necessarily related to topography. The simplest case is surface roughness or asperities, on such a scale that the glancing incidence beam produces a transmission diffraction pattern, in which there are no streaks but only sharp diffraction spots. The asperities need to be <1000 Å thick (in the beam direction) to allow a transmission pattern to be formed at typical RHEED energies ($\sim$20 keV). The effect of step arrays on RHEED and

LEED patterns has been analyzed in detail by Henzler [40] and Hottier et al. [41], so we will not consider it further here, beyond the general comment that steps are revealed by splitting of the diffracted beams, in a manner characteristic of their orientation and spacing. Finally, the presence of facets shows up as additional streaks in the pattern, formed by diffraction from the facet planes, and so they are not normal to the shadow edge [36]. The streaks tend therefore to produce arrowhead-like features.

4 Electronic Surface Structure Determination

The surface electronic structure is directly related to the surface crystallographic structure, and therefore, in principle, each reconstruction results in a surface with a specific electronic structure. In this section, we discuss the experimental evaluation of the electronic structure, and compare the results for the specific case of GaAs(001)-(2x4) with theoretical model calculations. We also discuss how semiconductor heterojunction band edge discontinuities can be determined.

4.1 Identification of Surface and Bulk States

Although there are various ways of distinguishing between emission from bulk and surface states, the use of MBE to produce different structures of the same bulk crystal face provides a unique opportunity of making this distinction. This is demonstrated in fig. 11, which shows three spectra from different GaAs(001) surface structures measured under identical conditions. The spectra correspond to the 2x4, c(4x4) and the Pb-induced 1x3 reconstruction. The latter was obtained with a Pb coverage of ~ 0.1 ML deposited on the 2x4 surface. For initial energies < -2.7 eV, the spectra are qualitatively very similar, and this region is dominated by bulk emission. The peaks X_3 and Σ_1^{min} occur at the energies of the X_3 and Σ_1^{min} critical points, and are observed over a wide range of experimental conditions because of the flatness of the energy bands and the resulting high density of states, in conjunction with surface umklapp processes. The peak labelled 1 represents a transition from a valence band to a free-electron-like final state, which will be discussed in section 4.2.

When we consider energies higher than -2.7 eV, however, there are few similarities in the spectra shown in fig. 11. Energy positions of peaks and shoulders from the 2x4 and c(4x4) reconstruc-

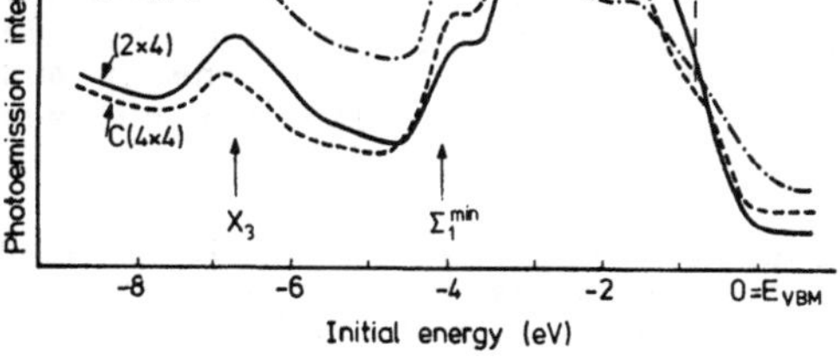

Fig. 11. Angle-resolved photoelectron spectra taken at a polar angle of 27° in the [010] azimuth for three GaAs(001) surface reconstructions

tions are quite different, and the emission is strongly quenched from the Pb-overlayer structure. This strongly indicates that the prominent emission features, e.g. S_2 and S_3 for the 2x4 structure, are surface state features. The experimental conditions correspond to a value of $\bar{k}_{\parallel}$ close to the K symmetry point for the 1x1 SBZ. If we project the bulk valence band of GaAs onto the (001) surface, the features S_2 and S_3 are situated in a band gap of this projected bulk band structure [37]. This is further support for their assignment as surfaces states. In principle, emission from bulk states could be observed in band gaps by means of surface umklapp processes, which are different for different reconstructions, but they could not be expected to be the origin of major peaks.

Another technique used to distinguish valence band and surface states, where the tunability of synchrotron radiation plays a decisive part, relies on keeping $\bar{k}_{\parallel}$ fixed at some point in the SBZ by choosing appropriate values of $h\nu$ and θ_p. Dispersive peaks must depend on $k_{\perp}$ and can be attributed to bulk interband transitions, while stationary features can be ascribed to surface state emission (apart from a few non-dispersive bulk critical points). Using this procedure the states S_2 and S_3 are again identified as surface states.

4.2 Bulk Valence Bands

In section 2.3 it was pointed out that the determination of three--dimensional band structures from ARPES measurements cannot be made if the final state is unknown. It is, however, possible to explain the presence of a number of bulk emission features, like peak 1 in fig. 11 by a direct transition model using free electron-like final states. This description for semiconductors was first given for the GaAs(110) cleavage surface [42] and later extended to the MBE grown GaAs(001) [33, 43] and Ge(001) [27] surfaces. Figure 12a shows normal emission spectra ($\theta_p = 0$, i.e. $\bar{k}_{\parallel} = 0$) as a function of the photon energy for GaAs($00\bar{1}$)-2x4. The features 1 and 3 disperse with $h\nu$ and therefore must be bulk transitions, while the stationary features S_1 and S_2 can be identified as surface states. We notice that assuming the direct transition model to be valid, both the initial and final energies are determined for each $h\nu$. A free-electron-like final state can be described by

$$E_f(\bar{k}) = \hbar^2(\bar{k}+\bar{G})^2/2m + E_0 \qquad (4)$$

where E_0 is the inner potential (~ -7.7 eV for GaAs [43]). With $\bar{k}_{\parallel} = 0$ and $\bar{G} = (2\pi/a)(0,0,2) = 2\Gamma X$, we find the parabola shown in fig. 12b. The measured values of E_f therefore determine $k_{\perp}$ and since $\bar{k}$ is conserved for a direct transition, the initial energy dispersion $E_i(\bar{k})$ can be determined as shown in fig. 12b. This procedure can also be applied to the case of non-normal emission [43].

The argument can be inverted in the sense that if the valence band structure is known, we can predict the energy positions E_i for the transitions from the valence bands for each photon energy and polar angle, and ensure that they do not overlap with the surface state transition to be investigated.

It should finally be noted that the observation of dispersing bulk transitions which can be interpreted using a free-electron-like final state is probably evidence for a rather perfect surface.

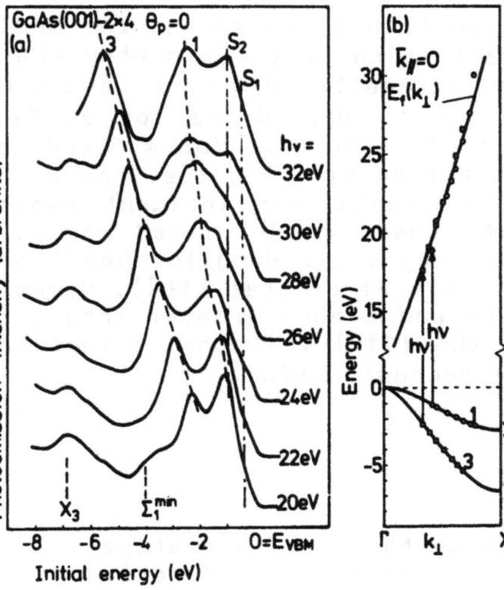

Fig. 12. (a): Photoelectron-energy spectra for GaAs(001)-2x4 taken at normal emission as a function of the photon energy. (b): Initial and final state energies versus $k_\perp$ for the transitions '1' (circles) and '3' (squares)

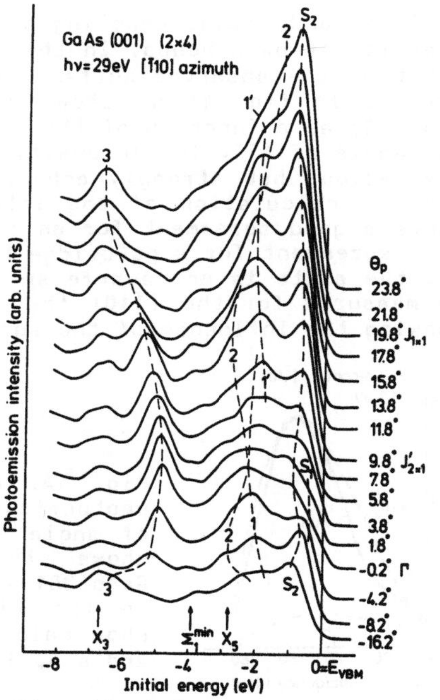

Fig. 13. Polar angle dependence for GaAs(001)-2x4 in the [110] azimuth

This is supported by the fact that for the GaAs(110) cleavage surface, the observation of direct transitions is related to the quality of the cleave; for bad cleaves, density-of-states peaks dominate the spectra [44].

4.3 Surface States: Energy Bands and Symmetries

The surface band structure $E_i = E_i(\bar{k}_{\parallel})$ can be directly determined by measuring the initial energy position of surface state transitions as a function of the polar angle and hence $\bar{k}_{\parallel}$. Figure 13 shows a typical set of spectra for GaAs(001)-2x4. Transitions 1, 2 and 3 are primary cone-emission features from the upper three valence bands to the quasi-free-electron final state, which agree with calculated dispersions to within 0.2 eV.

There is a very prominent surface state S_2, whose initial energy has a dispersion of 0.3 eV, with minimum values of $E_i = -1$ eV at normal emission (Γ-point) and at $\theta_p = \pm$ 18°, and a maximum value of $E_i = -0.7$ eV for $\theta_p = \pm$ 9°. The $\bar{k}_{\parallel}$ values for θ_p = 9° and 18° correspond to the J' and Γ points, respectively, for the 2x1 SBZ in an extended zone scheme, while the energy dispersion of S_2 displays two-fold symmetry along the $[\bar{1}10]$ azimuth of GaAs(001)-2x4. At a near-normal emission another surface state S_1, appears as a shoulder on S_2.

The surface states for covalently bonded semiconductors can be related to the disruption of surface bonds, the formation of new surface bonds (e.g. dangling bonds and bridge bonds) and the modification of back bonds. In the determination of the symmetry character of the observed surface states, intensity measurements can be useful. In fig. 14 we show the intensity dependence for surface state S_2 as a function of the angle of incidence θ_i for a fixed polar angle of 15°. The intensity is small at $\theta_i = 0°$, i.e. for s-polarization, but strongly enhanced for p-polarized incident radiation. A calculation of the intensity based on localized orbitals shows a good agreement for an initial sp_z (i.e. dangling bond) and poor agreement for a p_x(bridge bond) initial orbital [16]). However, the state is not a pure sp_z state and a much greater intensity is measured for the [$\bar{1}$10] than for the [110] azimuthal direction, showing the influence of the surface reconstruction.

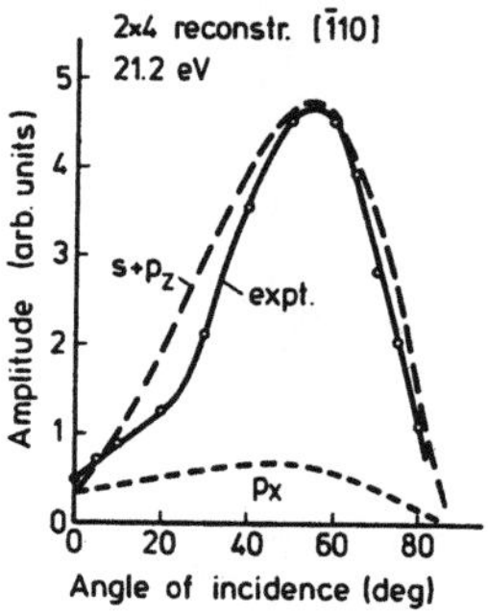

Fig. 14. Experimental and theoretical photoemission intensities as a function of angle of incidence. The full curve shows the measured intensity for GaAs(001)-2x4 at $h\nu$=21.2 eV and θ_p=15° in the [110] azimuth. Broken curves show calculated intensities for an sp_z and a p_x initial orbital

A complete surface band structure determination involves measurements along the symmetry lines of the SBZ, and an unequivocal evaluation requires measurements to be made at least at the symmetry points of the SBZ over a wide range of photon energies and also along equivalent symmetry lines in an extended zone scheme, since the probability of the transition can vary with both energy and momentum.

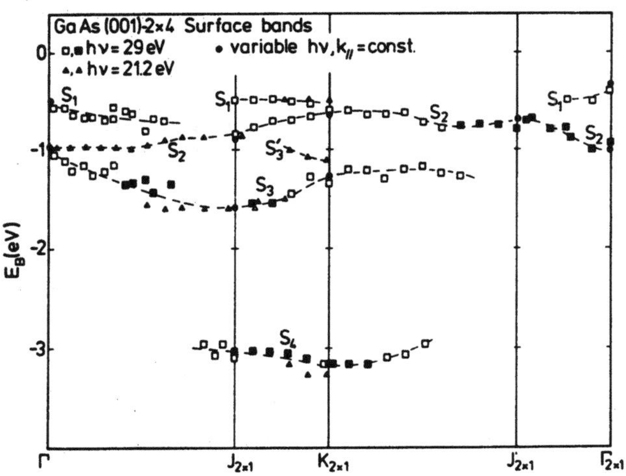

Fig. 15. Experimental surface energy bands for GaAs(001)-2x4 along the symmetry lines of the 2x1 SBZ

In fig. 15 we show the experimental surface band structure of GaAs(001)-2x4 within the framework of the 2x1 SBZ so that it can be compared with calculations. The closed symbols refer to strong surface state emission peaks in the spectra, whereas the open symbols denote the approximative energy positions of weak shoulders. In general, the surface state emission is found to be highly directional and particularly strong for off-normal directions, while energy dispersions of the surface energy bands are rather weak.

A calculation of the surface band structure using the scattering theoretical method on the basis of an empirical tight-binding description of the GaAs bulk crystal has been carried out for an aplanar As-As dimer model of the As terminated GaAs(001) surface having a 2x1 surface reconstruction. The results of this calculation are shown in fig. 16. The principal effect of the reconstruction as compared to a calculation for the unreconstructed surface is the introduction of the dimer-related state D_i at $\sim$-3.5 eV. This state is in very good agreement with the observed surface state S_4 (see fig. 15) and gives strong support for the dimer model. The dangling bond states near the top of the valence bands are found both theoretically and experimentally. More bands are observed experimentally than are calculated, but this is largely because the measurements were carried out for the 2x4 reconstruction while the calculations were done for a 2x1 structure.

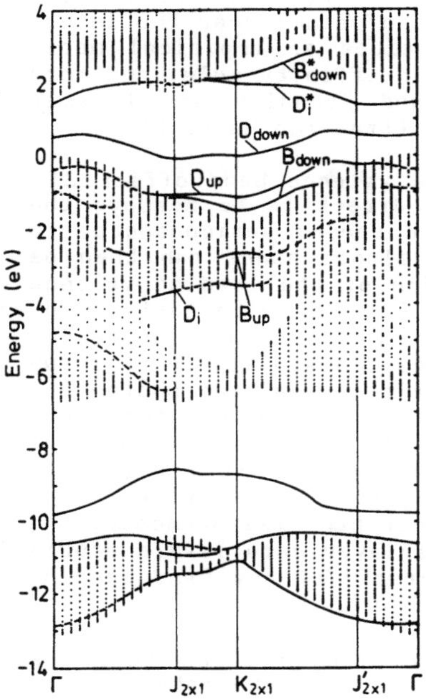

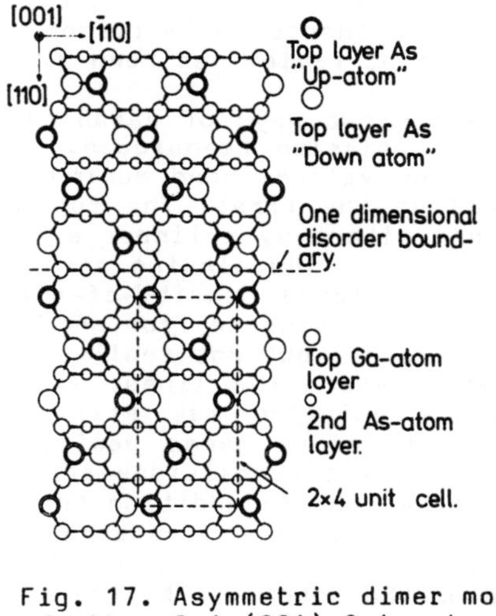

Fig. 17. Asymmetric dimer model of the GaAs(001)-2x4 structure and a model of a one-dimensional disorder boundary

Fig. 16. Surface band structure and projected bulk bands for a 2x1 asymmetric dimer model of the As-terminated GaAs(001) surface. The labeling "up" and "down" refers to the up and down atom of the dimer. The bands labeled D, D_i and B represent states derived from the dangling bonds, the dimer bond and the back bonds, respectively

There are further arguments for an asymmetric dimer model of As terminated surfaces of GaAs(001) from surface total energy calculations [45]. The occurrence of different dimers or the way they are connected may be the cause of the four-fold periodicity and in fig. 17 we show a possible model of the 2x4 unit cell on the basis of a single type of asymmetric dimer. To make the connection with the RHEED observations discussed in section 3.3, the figure also shows a one-dimensional disorder-boundary, which might be the origin of the curved streaks observed in RHEED diffraction patterns from this surface.

4.4 Semiconductor heterojunctions

Semiconductor heterojunctions are of extreme technological significance not only for many types of conventional devices based on compound semiconductors, but especially for new device structures involving confined electron states [1].

The most important characteristic of a semiconductor heterointerface is its energy band diagram, for which theoretical treatments generally assume a compositional transition extending over only one or two atomic layers. There is an abrupt change in energy gap at the interface which produces discontinuities in the conduction and valence band edges. The magnitude of the discontinuity is assumed to be an intrinsic property of the semiconductor pair, independent of doping and hence of Fermi level considerations, but possibly dependent on crystallographic orientation and other factors which influence the exact arrangement of atoms near the interface. The main theoretical treatments include an empirical model due to Anderson [46], known as the electron affinity rule, and more detailed treatments by Harrison [47] and Frensley and Kroemer [48]. A recent review has been given by Pollman and Mazur [49].

There are two ways of establishing the valence band offset ΔE_V using photoemission techniques. The most direct is to measure the shift in the valence band maximum when the second semiconductor is deposited on the first. The top of the valence bands can then be determined either by a linear extrapolation of the leading valence band edge (i.e. low binding energy edge) [50], or alternatively by using a calculated density-of-states [51]. The latter approach, which neglects emission from surface states, is applicable at X-ray photon energies, but not really in the direct-transition photon energy region where the linear extrapolation approach is used. Surface state emission may still complicate the determination of E_{VBM}. It therefore seems desirable to choose conditions ($h\nu, \theta_p$) for which direct transitions are occurring from states at energies close to the VBM. In addition to the valence band measurements, the energy positions of the substrate and overlayer core levels should be measured to establish and possibly correct for any band bending induced on the substrate by the overlayer [50].

The second technique used to determine ΔE_V assumes that the binding energy of the core levels, both for the substrate and the overlayer material, are well known. From the measured shift between a substrate and an overlayer core level, the valence band offset can then be directly determined [52].

A great number of experimental studies of band discontinuities of semiconductor heterojunctions have been carried out (see ref. 50).

It has apparently been established for GaAs(001)/Ge(001) that the band discontinuity can be independent of the Fermi level position [53], and that the Anderson electron affinity rule fails [54]. The rule predicts that $\Delta E_c^{Ge-GaAs} = \Delta\chi^{Ge-GaAs}$ (where ΔE_c = conduction band discontinuity and $\Delta\chi$ = electron affinity difference) but $\Delta\chi^{Ge-GaAs}$ (measured) was 0.01 eV, cf. $\Delta E_c^{Ge-GaAs}$ (measured) = -0.54 eV. Discrepancies seem to exist, however, between results obtained at different X-ray photon energies. For Ge overlayers on GaAs, Waldrop et al. [55] using 1486.6 eV report values of ΔE_v from 0.46 eV to 0.67 eV, depending on surface orientation and reconstruction, while values obtained using $h\nu$ =130 eV are in the range from 0.35 - 0.45 eV [50,56-57] and are independent of the surface reconstruction [53]. The differences might be caused by systematic experimental errors or by differences in the methods of preparation.

5 Core Level Photoemission

We have in the preceding sections indicated the use of core level photoemission in obtaining information on the surface stoichiometry and the importance of this technique in the determination of heterojunction band discontinuities. In this section we discuss, using specific examples, two other aspects of surface-sensitive core level photoemission.

5.1 Surface Chemistry

The core level binding energy for a surface atom is in general different from that for the same atom in the bulk. The As(3d) core level spectrum from a GaAs(001)-2x4 surface, measured with a photon energy corresponding to a small escape depth ~6Å, is shown in the upper part of fig. 18. The line shape can be resolved into a bulk spin-orbit split component B and a surface component S_L, which is shifted towards lower binding energies by ~ 0.32-0.38 eV [35,58]. We believe this level originates from the intrinsic As atom terminating the polar surface. The surface core level shift can in principle be caused by a charge transfer between bulk and surface atoms which changes the electrostatic potential at the core, long-range Madelung potential effects and final-state effects (screening of the core hole). Since in bulk GaAs there is an electron charge transfer towards the As atoms from the Ga atoms [59], one would expect As at the surface to have less charge than in the bulk. If we therefore invoke charge transfer to explain the shift, we would expect to see a higher surface binding energy, whereas we observe a lower one. This is a shift in the same direction as observed for Ge(001)-2x1 [60]. It appears, therefore, that an explanation of S_L requires a Madelung potential calculation (we disregard final state effects) which has not yet been carried out.

The As(3d) line shape for the c(4x4) reconstruction, shown in the middle panel of fig. 18, is clearly more complicated than that from the 2x4 structure. In addition to the B and S_L features, there is also a surface core level S_H, shifted by ~0.5 eV towards higher binding energies. This can be explained in terms of charges on surface As atoms, and may be illustrated by comparing the c(4x4) spectrum with the one shown in the bottom panel of fig. 18, which was obtained from a 7-monolayer arsenic film deposited on a GaAs substrate. The lower full line is the (small) contribution from the substrate, while the broken line is the contribution from

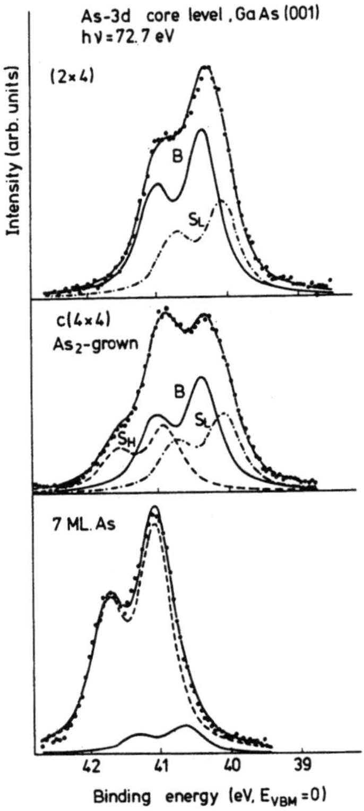

Fig. 18. As(3d) core level photoemission spectra for GaAs(001)-2x4 (upper panel), c(4x4) (middle planel) and a 7 ML amorphous arsenic film on a GaAs(001)-2x4 substrate (lower panel)

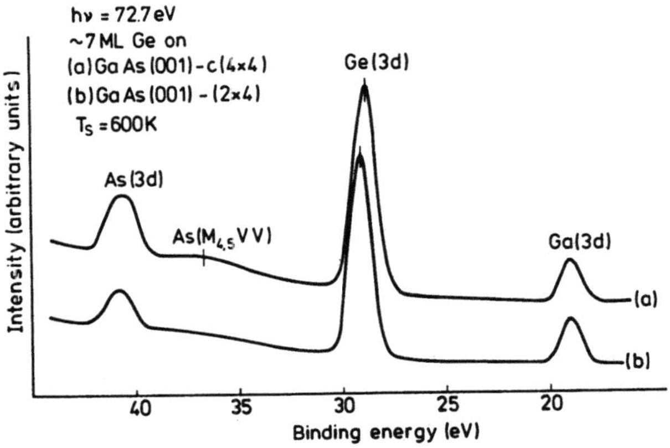

Fig. 19. As(3d), Ge(3d) and Ga(3d) core level spectra for 7 ML of Ge deposited at 600 K on (a) GaAs(001)-c(4x4) and (b) GaAs(001)-2x4

the amorphous overlayer in which the As is trigonally bonded. This is shifted ~0.6 eV towards higher binding energy relative to the position for As in bulk GaAs, which is very similar to that observed between B and S_H for the c(4x4) structure. We would suggest therefore that the surface As atoms which are responsible for this level are bonded to other As atoms, the bonds therefore being more covalent in character than bonds to Ga atoms in bulk GaAs [32,61]. There are several independent observations which help to confirm this model: (i) there is no involvement of Ga atoms since the Ga(3d) core level spectra from c(4x4) and 2x4 surfaces are identical, (ii) intensity ratios of the Ga(3d) and As(3d) levels indicate excess As (i.e. more than monolayer coverage) on the c(4x4) surface, and (iii) a study of As thermal desorption shows that the c(4x4) structure persists over a broad range of surface stoichiometry. A surface model derived from trigonally bonded chemisorbed As atoms was also found to be consistent with RHEED data [32].

5.2 Heterojunctions

It is of considerable importance to relate the microscopic properties of semiconductor heterojunctions to the preparation conditions and for this purpose surface-sensitive core level photoemission is an excellent tool. This is illustrated in fig. 19 where we show spectra of the 3d core levels of Ge, Ga and As from 7 ML Ge films grown by MBE on two different GaAs(001) reconstructions [27]. The Ga(3d) line shape is independent of reconstruction, but the energy positions of the Ge(3d) peaks differ by ~0.2 eV and the As(3d) line width is broader and the area under the peak is greater for the c(4x4) than for the 2x4 structure. This indicates that the Ge overlayer for the c(4x4) reconstruction reacts with the substrate As atoms.

Normalized Ga(3d) and As(3d) peak areas as a function of Ge coverage for films grown at 300 K and 600 K are shown in fig. 20. For deposition on a 2x4 surface an exponential decay is observed for both temperatures, while for growth on a c(4x4) reconstructed surface the decay of both the Ga and the As peaks, but especially the latter, is slower. For layer-by-layer growth an exponential decay of the substrate core level emission should be observed, with the exponent determined by the electron escape depth corresponding to the appropriate final energy of the photoelectrons. The important conclusion from these results is therefore that an abrupt interface is obtained with a 2x4 reconstructed surface, but not with a c(4x4). The difference arises because the 2x4 surface is an intrinsic GaAs(001) structure, while the c(4x4) surface is a chemisorption structure containing an excess of trigonally bonded arsenic. This excess arsenic must be distributed within the interface region and/or be segregated at the surface of the growing Ge film.

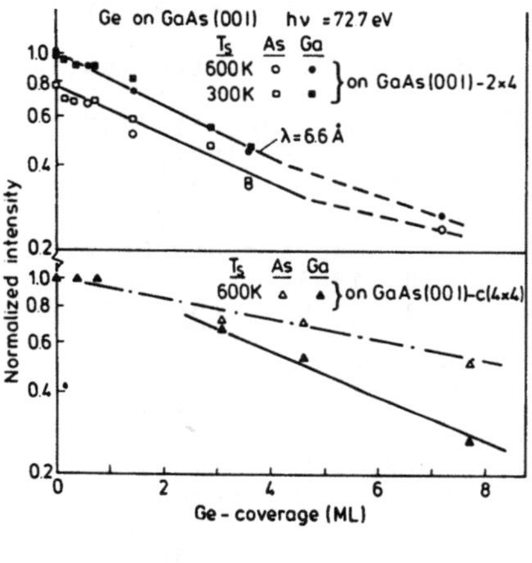

Fig. 20. Normalized Ga(3d) and As(3d) core level emission intensities as a function of the thickness of a Ge overlayer deposited at two temperatures on two GaAs(001) reconstructions

Acknowledgement

The authors are indebted to J.H. Neave and J.F. van der Veen for their valuable collaboration. We want to thank LURE, Université de Paris-Sud et CNRS (Orsay, France) for providing excellent synchrotron radiation facilities.

References

1 A.C. Gossard: Treatise on Materials Science and Technology, Vol. 24, p. 13 (Academic Press, New York 1983)
2 B.A. Joyce: Proc. 2nd Int.Conf. on MBE and Clean Surface Techniques, p. 9 (Tokyo 1982)
3 J.H. Neave, P. Blood and B.A. Joyce: Appl.Phys.Lett. 36, 311 (1980)

4 Y. Ota: Thin Solid Films 106, 3 (1983)
5 E.A. Wood: J.Appl.Phys. 35. 1306 (1964)
6 K. Britze and G. Meyer-Ehmsen: Surf.Sci. 67, 358 (1977)
7 E. Bauer, in "Techniques of Metals Research, Vol. II", R.F. Bunshah, Ed. (Interscience 1969)
8 G. Comsa: Surf.Sci 81, 57 (1979)
9 T.-M. Lu and M.G. Lagally: Surf. Sci. 99, 695 (1980)
10 N.G. Norton: Thesis, Imperial College of Science and Technol. (London, 1983)
11 J.M. van Hove, P. Pukite, P.I. Cohen and C.S. Lent: J.Vac.Sci. Technol. A1, 609 (1983)
12 Photoemission and the Electronic Properties of Surfaces, Eds. B. Feuerbacher, B. Fitton and R.F. Willis (Wiley, 1978)
13 Photoemission in Solids I + II, Eds. M. Cardona and L. Ley (Springer, 1978)
14 E.W. Plummer and W. Eberhardt: Adv.Chem.Phys. 49, 533 (Wiley, 1982)
15 N.V. Smith: Vacuum 33, 803 (1983)
16 P.K. Larsen, J.H. Neave and B.A. Joyce: J. Phys.C (Solid State Physics) 14, 167 (1981)
17 B.W. Holland: Surf.Sci. 68, 490 (1977)
18 D.E. Eastman, F.J. Himpsel and J.F. van der Veen: J.Vac.Sci. Technol. 20, 609 (1982)
19 W. Mönch: Thin Solid Films 104, 285 (1983)
20 C.T. Foxon and B.A. Joyce, in "Current Topics in Materials Science", Vol. 7,p. 1,E. Kaldis, Ed. (North Holland, 1981)
21 J.H. Neave, B.A. Joyce, P.J. Dobson and N. Norton: Appl.Phys. A31, 1, (1983)
22 V. Bostanov, R. Ronssinova, and E. Budevski: J. Electrochem. Soc. 119, 1346 (1972)
23 H. Dabringans and H.J. Meyer: J. Crystal Growth 16, 31 (1972)
24 Y. Namba and R.W. Vook: Thin Solid Films 82, 165 (1981)
25 K.D. Gronwald and M. Henzler: Surf.Sci 117, 180 (1982)
26 J.D. Weeks and G.H. Gilmer: Adv.Chem.Phys. 40, 157 (1979)
27 J.H. Neave, P.K. Larsen, B.A. Joyce, J.P. Gowers and J.F. van der Veen: J.Vac.Sci.Technol. B1, 668 (1983)
28 J.H. Neave, P.J. Dobson and B.A. Joyce: Appl.Phys. A34, 179 (1984)
29 C.S. Lent and P.I. Cohen: Surf.Sci. 139, 121 (1984)
30 P. Drathen, W.Ranke and K. Jacobi: Surf.Sci. 77, L162 (1978)
31 R.Z. Bachrach, R.S. Bauer, P. Chiradia and G.V. Hansson; J.Vac.Sci.Technol. 18, 797 (1981): 19, 335 (1981)
32 P.K. Larsen, J.H. Neave, J.F. van der Veen, P.J. Dobson and B.A. Joyce: Phys.Rev. B27, 4966 (1983)
33 T.-C. Chiang, R. Ludeke, M. Aono, G. Landgren, F.J. Himpsel and D.E. Eastman: Phys.Rev. B27, 4770 (1983)
34 J. Massies, P. Etienne, F. Dezaly and N.T. Link: Surf.Sci. 29 121 (1981)
35 J.F. van der Veen, P.K. Larsen, J.H. Neave and B.A. Joyce: Solid State Commun. 49, 659 (1984)
36 B.A. Joyce, J.H. Neave, P.J. Dobson and P.K. Larsen: Phys.Rev. B29, 814 (1984)
37 P.K. Larsen, J.F. van der Veen, A. Mazur, J. Pollman, J.H. Neave and B.A. Joyce: Phys.Rev. B26, 3222 (1982)
38 P. Delecluse and A. Masson: Surf.Sci. 100, 423 (1980)
39 P.J. Dobson, J.H. Neave and B.A. Joyce: Surf.Sci. 119, L339 (1982)
40 M. Henzler: Appl.Phys. 9, 11 (1976)

41 F. Hottier, J.B.Theeten, A. Masson and J.L. Domange: Surf.Sci. 65, 563 (1977)
42 T.-C. Chiang, J.A. Knapp, M. Aono and D.E. Eastman, Phys.Rev. B21, 3513 (1980)
43 P.K. Larsen, J.F. van der Veen, A. Mazur, J. Pollman and B.H. Verbeek; Solid State Commun. 40, 459 (1981)
44 F. Cerrina, J.R. Myron and G.J. Lapeyre: Phys.Rev. B29, 1798 (1984)
45 J. Ihm, D.J. Chadi, J.D. Joannopoulous; Phys. Rev. B27, 5119 (1983)
46 R.L. Anderson: Solid State Electron. 5, 341 (1962)
47 W.A. Harrison: Electronic Structure and the Properties of Solids (Freeman, San Francisco, 1980), p. 252
48 W.R. Frensley and H. Kroemer: Phys.Rev. B16, 2642 (1977)
49 J. Pollman and A. Mazur: Thin Solid Films 104, 257 (1983)
50 A.D. Katnani and G. Margaritondo: Phys.Rev. B28, 1944 (1983)
51 E.A. Kraut, R.W. Grant, J.R. Waldrop and S.P. Kowalczyk; Phys.Rev.Lett. 44, 1620 (1980)
52 R.W. Grant, E.A. Kraut, S.P. Kowalczyk and J.R. Waldrop: J.Vac. Sci.Technol. B1, 320 (1983)
53 P. Chiaradia, A.D. Katnani, H.W. Sang and R.S. Bauer: Phys. Rev.Lett. 52, 1246 (1984)
54 R.S. Bauer, P. Zurcher and H.W. Sang: Appl.Phys.Lett. 43, 663 (1983)
55 J.R. Waldrop, E.A. Kraut, S.P.Kowalczyk and R.W. Grant: Surf.Sci. 132, 513 (1983)
56 R.S. Bauer and J.C. Mikkelsen: J.Vac.Sci.Technol. 31, 491 (1982)
57 P.K. Larsen and J.F. van der Veen: Surf.Sci. 126, 1 (1983)
58 R. Ludeke, T.-C. Chiang and D.E. Eastman: Physica 117B/118B, 819 (1983)
59 J.W. Davenport, R.E. Watson, M.L. Perlman and T.K. Sham: Solid State Commun. 40, 999 (1981)
60 T. Miller, E. Rosenwinkel and T.-C. Chiang: Solid State Commun. 47, 935 (1983)
61 J.F. van der Veen, L. Smit, P.K. Larsen and J.H. Neave: Physica 117B/118B, 822 (1983)

Metal-Semiconductor Interfaces and Schottky Barriers

G.W. Rubloff

IBM Thomas J. Watson Research Center, P.O. Box 218
Yorktown Heights, NY 10598, USA

Over the past several years, experimental techniques from surface science, materials science, and solid state physics have brought significant progress in understanding the microscopic properties and behavior of metal/semiconductor interfaces. The outstanding conclusion from this research is that these interfaces are chemically reactive and that this reactivity plays the dominant role in determining both microscopic properties of the interface (such as composition, atomic structure, and electronic structure) and macroscopic characteristics of the contact (such as electrical properties, lattice phase of the reaction product, and epitaxy). The interfacial reaction chemistry can be rather diverse, involving interdiffusion of metal and semiconductor, the formation of compound phases at the interface, as well as partial reaction and mixed phases. The case of transition-metal/Si interfaces is discussed in particular detail, because the elemental character of the semiconductor and the formation of well-defined silicide compounds has made the interpretation of results more definitive. Though necessarily more complex, metal/III-V semiconductor interfaces display corresponding chemical behavior. Attempts to understand the microscopic mechanisms which control the Schottky barrier electrical properties of metal/semiconductor interfaces have also begun to focus on the relation between interfacial chemistry and barrier height, although a clear explanation is not yet available, either for elemental or III-V semiconductors.

I. INTRODUCTION

Motivated by both fundamental scientific issues and critical technological applications, considerable research has been directed at revealing the microscopic properties and behavior of metal/semiconductor interfaces. By applying a number of surface spectroscopy and structural techniques in ways which obtain the (nearly) atomic resolution required for truly microscopic interface studies, much important insight has been obtained. In this paper we present a general but brief overview of some of the most important issues and conclusions which have arisen from this significant research effort. For further overviews and references to specific research articles, the reader is referred to a number of valuable review papers on silicide formation[1], silicide/Si interfaces[2,3,4], noble-metal/Si interfaces[5], metal/compound-semiconductor interfaces[6,7], and the Schottky barrier problem[4,8,9,10].

The two crucial *scientific* questions to be addressed are:

1. What are the essential microscopic chemical properties and behavior of metal/semiconductor interfaces?
2. What microscopic mechanisms are responsible for determining the Schottky barrier height of the metal/semiconductor interface?

Both of these questions have direct technological counterparts related to microelectronics applications[3].

Insofar as the interface chemistry is concerned, the scientific goal is to identify the composition, electronic structure, and atomic structure and microstructure of the interfacial region as well as the chemical reactivity responsible for these properties. In technological applications, the materials reaction processes inherent in the chemistry of the contact must be predictable, controlled, and sufficiently stable to insure long-term contact reliability as well as reproducibility in processing, whether the contact is used as an active Schottky device or simply as an Ohmic contact to bring electrical signals in and out of the semiconductor device.

Relative to the second question, the microscopic mechanisms which determine the band-bending and Schottky barrier height (thereby the overall electrical properties) have remained a basic, unsolved problem for decades. In spite of this, the inherent speed advantages of Schottky contacts (as majority carrier devices) for circuit applications[11] have already led to their use without fundamental understanding of the underlying mechanisms.

II. METAL/SEMICONDUCTOR INTERFACE REACTIONS

That metal/semiconductor interfaces can be chemically reactive has been known for some time. This is most clearly evident from studies of thin film silicide formation processes[1], in which the growth of silicide (metal-Si) compounds occurs at the transition-metal/Si interface by thermal reaction. These thin film studies, although lacking the essentially atomic depth resolution of recent experiments in metal/semiconductor interfaces, have been an invaluable aid in understanding the results of experiments on transition-metal/Si and corresponding silicide/Si interfaces.

Other kinds of interfacial or materials reactions have also been known to occur at metal/semiconductor interfaces, including interdiffusion and more complex reaction. The dissolution of Si into Al is perhaps the most classic example: in fact, it is the inherent instability of this interface (with respect to interdiffusion) which led to the widespread use of silicide contacts for large-scale and very large-scale integrated circuits. Because Si dissolves in Al, pits develop in the Si, become filled with Al, and eventually penetrate - as conducting links - through one Si layer to the buried p-n junction below, thereby shorting the junction. The stability of silicide contacts, a consequence of the formation of well-defined, stable compound phases at the interface, has led to their adoption widely in the industry. Atomic intermixing reactions also occur for noble-metal/Si interfaces and for metals on III-V semiconductors.

A schematic description of the kinds of interfacial reaction which occur at metal/semiconductor interfaces is given in Fig. 1, which depicts both compound formation and interdiffusion processes in the case of the metal/Si interface. The stages of interfacial compound formation are shown in Fig. 1a - 1b. First, the unreacted metal is deposited on the Si substrate (Fig. 1a). Then the sample is annealed to promote interfacial reaction. First a thin layer of silicide is formed at the metal/Si interface (Fig. 1b), representing the *initial stage* of the reaction - the formation of the first silicide layers at the *metal/Si* interface. With further annealing, more silicide is produced and a thicker interfacial silicide layer grows (Fig. 1c). This *second stage*, the *growth* of a

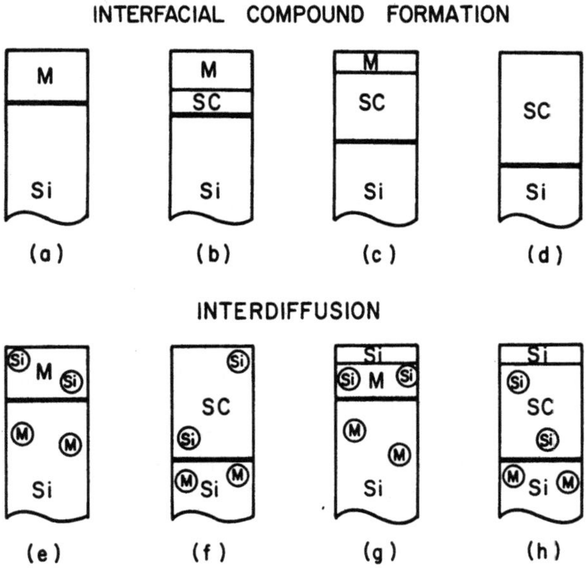

Fig. 1. Chemical reactions and atomic intermixing at metal/semiconductor interfaces, as depicted specifically for the case of transition-metal/Si.
The normal scheme of interfacial compound formation is shown in (a) - (d). Starting with the unreacted metal/Si interface (a), an initial interfacial silicide layer a few unit cells thick (~ 3 - 10 Å) is formed at the metal/Si interface. With further progress of the reaction (usually involving thermal processing), further silicide is grown (c) until the reaction is completed and all the deposited metal overlayer is fully reacted (d).
Interdiffusion processes which may also occur are depicted in (e) - (h). In (e), metal atoms diffuse into the Si (and/or Si atoms diffuse into the metal) without interfacial silicide formation. Such interdiffusion may also occur at the silicide/Si interface (f). In these situations a Si overlayer may be created at the top of the metal (g) or of the silicide (h).

thicker silicide layer, involves mass transport of metal and/or Si atoms through the silicide as well as reaction steps for these atoms at the *silicide/Si* and *metal/silicide* interfaces. Eventually the entire metal overlayer is consumed as it is converted into silicide (Fig. 1d).

It should be emphasized that in the process of silicide formation the metal/semiconductor interface changes from the *metal*/Si into the *silicide*/Si interface (nearly all silicides are metals). At the same time, this metal/semiconductor interface - which controls the electrical characteristics - typically moves below the original Si surface on which the metal was deposited. Therefore, contamination and defects normally present on the original Si surface in manufacturing do not appear at the final (reacted) metal/semiconductor interface. This makes silicide/Si contacts very reproducible and attractive for manufacturing. It also means that the highly controlled conditions of basic research are indeed relevant to actual conditions in silicide contact technology.

In the case of the transition-metal/Si interface, this interfacial compound formation process *dominates* the interfacial chemistry and properties. A variety of experiments have demonstrated that it often produces interface profiles (e.g., of stoichiometry, lattice structure, and electronic states) which are sharp, even when viewed on an atomic scale. Indeed, the most important results of research in this field over the last several years are probably the observations: (i) that the initial stages of microscopic chemical interaction at the interface are essentially those of silicide formation[12], and (ii) that the silicide/Si interface can be atomically abrupt[13].

Compound formation is also known to occur at the metal/III-V semiconductor interface, although much less documentation and details are available. In this case, the abruptness of properties of the reacted interface is much more open to question, in view of the higher complexity of the reaction (which may include preferential outdiffusion, multiple phases of cation and anion with metal, etc.).

Alternatively, the interfacial reaction might well involve primarily interdiffusion or single-atom motion across the interface. This would suggest graded interface profiles as depicted in Fig. 1e and 1f for the metal/Si and silicide/Si interfaces respectively. Indeed, there exists evidence for such processes in transition-metal/Si interface systems, but these are typically of *secondary importance* compared to the interfacial compound formation. Another manifestation of such interdiffusion is the segregation of Si to the top surface of the metal (Fig. 1g) or of the silicide (Fig. 1h).

Such interdiffusion is particularly important in the reaction at simple-metal/Si, noble-metal/Si, and metal/III-V compound semiconductor interfaces. For the compound semiconductors, it is further complicated by the existence of preferential outdiffusion, i.e., when one semiconductor component diffuses into the overlayer more rapidly than the other.

III. METAL/Si INTERFACE CHEMISTRY

A. Simple-Metal/Si and Noble-Metal/Si Interfaces

Surface spectroscopy studies of the simple-metal/Si interface were among the first to show how metal atom deposition quenched the intrinsic surface states of the clean semiconductor surface and formed new chemical bonds[14,15]. Since these experiments were confined to room temperature, the question of whether interdiffusion with the semiconductor occurs was not fully addressed. In the case of the Ag/Si interface, the growth mechanism of the overlayer has been addressed more completely, including metal coverages ranging from the submonolayer level to several thousand Å. This system follows a classic Stranski-Krastanov growth mechanism as seen by surface spectroscopy techniques[16], medium energy ion scattering measurements[17], and surface EXAFS experiments[18].

In contrast, the Au/Si[19] and Cu/Si[20] interfaces both display a notable amount of metal-Si intermixing already at room temperature. The presence of Au at the Ag/Si interface even initiates the intermixing of Ag with Si[21]. At elevated temperatures, the noble-metal/Si interfaces show additional evidence for agglomeration of metal. A more complete discussion of the noble-metal/Si interface can be found elsewhere[5].

B. Transition-Metal/Si and Rare-Earth/Si Interfaces

As indicated above, the dominant reaction path at the transition-metal/Si interface is that leading to the formation of interfacial silicide compounds, whose growth and properties have been the subject of substantial study in thin films research[1,22]. Recent microscopic results in this area have been reviewed elsewhere[2,3,4] and will be summarized below. At this point it is sufficient to note the overall chemical trends for transition-metal silicide formation by contact reaction. For the near-noble transition metals, interfacial silicide growth begins at relatively low temperatures (~ 200 - 500°C), produces mainly metal-rich phases, and gives high Schottky barrier heights on n-Si (~ 0.70 - 0.85 eV); in contrast, refractory transition metals require higher temperatures for silicide growth (over 500°C), tend to form Si-rich phases, and have relatively low barrier heights (0.50 - 0.70 eV). For the near-noble metals, the metal atom is the primary moving species, while Si dominates in the growth of refractory metal silicides. Finally, both classes display low-temperature reactivity (i.e., interfacial reaction well below normal silicide growth temperatures), but with different characteristics. For the near-noble metals, the low temperature reaction is limited to a few unit cells near the interface (typically 10 - 15Å)[2,23]. For the refractory metals, the low temperature reaction gives intermixing of metal and Si atoms over a considerable distance from the interfaces (≥ 100 Å) and produces distinctly different microstructural characteristics compared to normal silicide compound formation at higher temperatures[2,24].

In addition, recent investigations have shown that silicide compounds are also formed by contact reaction at the rare-earth-metal/Si interface[22,25,26,27,28,29]. The silicide formation temperatures are relatively low (~ 300 - 400°C), yet disilicides (MSi_2) are produced. Growth of rare earth silicide growth does not follow a simple layered mode (shown in Fig. 1a-d)[25]. The Schottky barrier heights of the rare earth silicides on Si are low, ~ 0.40 eV[26].

C. Interfacial Silicide Formation

The interfacial silicide formation process most thoroughly studied to date, using the widest variety of experimental techniques, is that of the Pd/Si interface, and we choose this example to illustrate the reaction steps and properties involved in interfacial reaction leading to silicide compound formation.

One way to identify interfacial reaction is to measure the composition of the Si surface during the initial formation of the metal overlayer. This can be done effectively in several ways, such as Auger electron spectroscopy or medium energy ion scattering. Ion scattering results for the Pd/Si(111) system[30,31] are illustrated in Fig. 2. Fig. 2a shows that the number of Si atoms displaced at room temperature to form new chemical bonds varies linearly with the number of deposited Pd atoms, and furthermore the ratio of atoms participating in this rebonding implies a stoichiometry ~ Pd_2Si. In addition, the ion scattering spectra in Fig. 2b reveal essentially no shift or splitting of the high-energy edge of the Si spectrum as up to 15 Å of Pd are deposited. Since this edge corresponds to the top-most atomic layers (with a depth sensitivity ~ 3 Å) one can conclude that the overlayer experiences fairly uniform growth and consists of a strongly mixed Pd-Si composition, namely ~ Pd_2Si.

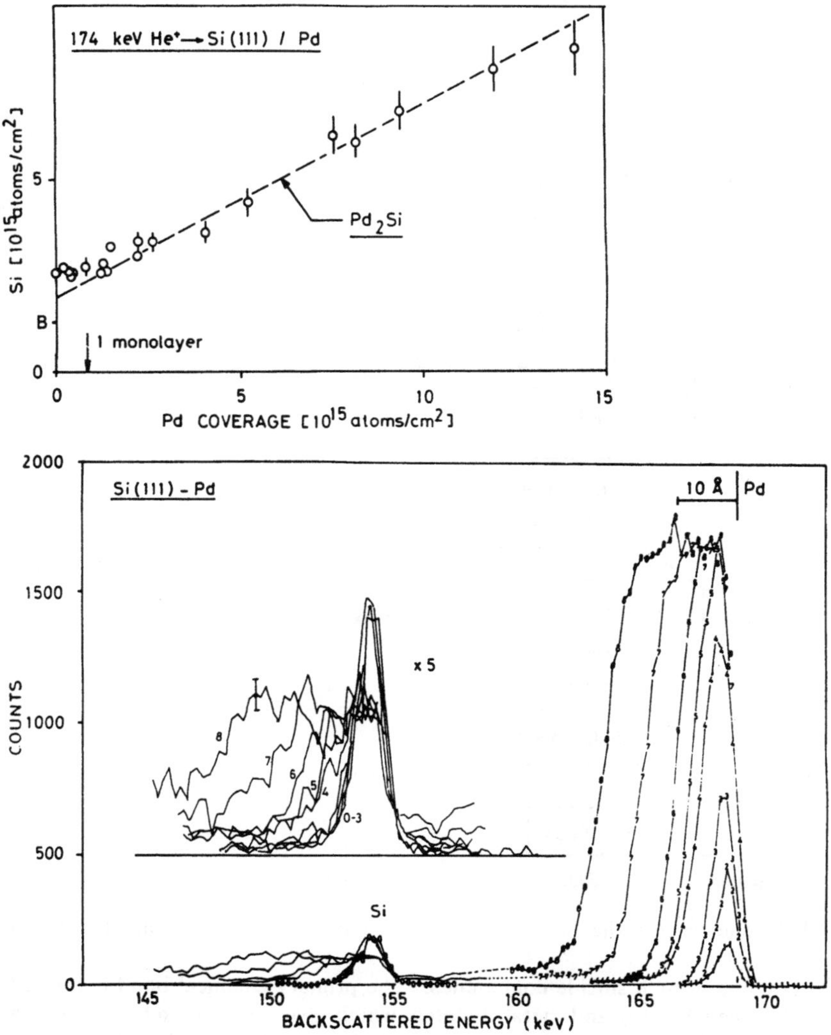

Fig. 2. (a) Composition of the Pd/Si(111) interface as determined from medium energy (174 keV) He^+ ion scattering. Upon deposition of Pd at room temperature, the relative increase in Si areal density to that of Pd (over the initial clean Si value) matches that expected for formation of a Pd_2Si compound;
b) Backscattering spectra for Pd deposition on Si. Since the leading edge of the Si peak stays at the same position independent of Pd coverage, the Si is not covered over by the Pd; this indicates strongly that substantial interface mixing has occurred. (Both from R. M. Tromp et. al.[30]).

The presence of interfacial reaction and the growth of a silicide compound can also be distinguished readily from the electronic structure of the interface, as seen from UPS (ultraviolet photoemission spectroscopy), XPS (x-ray photoemission spectroscopy), or AES (Auger electron spectroscopy), since changes in the electronic structure reflect

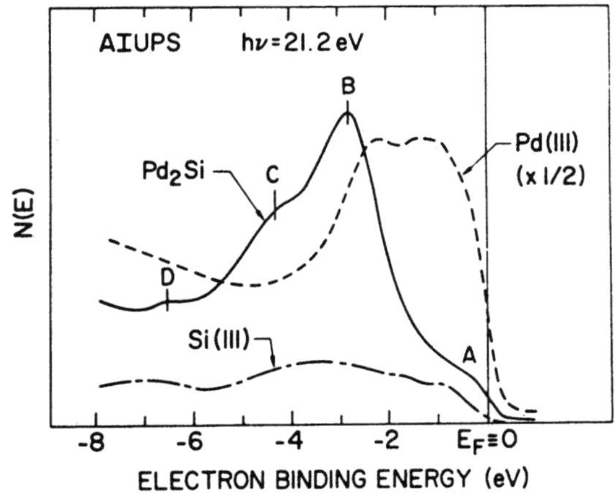

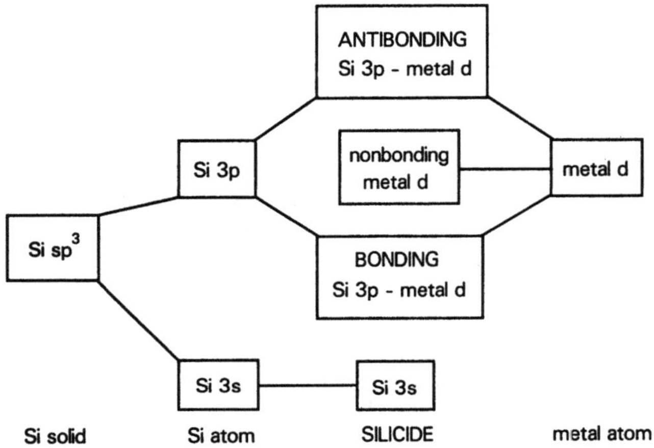

Fig. 3. (a) UPS spectra for the Si(111) surface, the Pd(111) surface, and the Pd_2Si surface.
(b) Chemical bonding in bulk silicide compounds. The primary chemical bond is formed by interaction between Si (3p) and metal d states, which give bonding and antibonding states below and above the main manifold of metal d states (nonbonding to the Si). Because of the local atomic structure of the silicide, the sp^3 hybridization in solid Si does not occur, so the Si (3s) states are essentially uninvolved in metal-Si chemical bonding.

directly differences in local chemical bonding and atomic environment. An example of this for the Pd-Si case is shown in Fig. 3a, which contrasts the valence electronic structure (seen by UPS) for the clean Si surface, the clean Pd surface, and the surface of fully reacted Pd_2Si compound. The presence of metal d-states, which causes the strong intensity just below the Fermi energy E_F in Pd metal, also gives rise to strong intensity in the silicide compound, but it is markedly shifted in energy as compared to pure Pd metal. Thus the reacted Pd-Si mixture, in this case the Pd_2Si compound, produces a UPS spectrum which is significantly different from both pure Si and pure Pd - and thereby readily distinguished. Differences between different intermixed (reacted) states are more difficult to interpret, but such electronic structure evidence as shown here is very useful for distinguishing the presence of interfacial chemical reaction.

In the course of studies of metal/Si interfaces over the past few years, a simple picture has emerged from both experiment and theory to describe the bulk electronic structure of silicide compounds[3,32,33,34,35]. This picture is illustrated schematically in Fig. 3b. The sp^3 hybridization of the Si atoms is removed in the silicide, and the primary bonding occurs between Si(3p) and metal(d) states (for transition-metal silicides), forming bonding and antibonding states which lie respectively below and above the main nonbonding d-states of the metal. In the near-noble metal silicides, where most of the states are occupied and therefore accessible to UPS, one can see that the nonbonding metal d-derived states produce the main peak in the spectrum (near -2.75 eV in Pd_2Si), while the bonding and antibonding states are weaker and lie at lower (~ -6 eV) and higher (just below E_F) energies respectively.

From the sensitivity of the electronic structure to the presence of intermixing, such spectra can then be used to evaluate interfacial reaction. An example for the Pd/Si(111) interface is given in the UPS spectra in Fig. 4. Besides quenching intrinsic surface states[36] and attenuating the bulk Si features, the Pd deposition causes strong emission several eV below E_F, forming a peak at -2.75 eV; the spectrum observed between ~ 3 and 20 Å equivalent Pd coverage is essentially that of the bulk Pd_2Si compound[23,37]. At higher coverage (see the 32 Å spectrum) the peak shifts toward E_F due to unreacted Pd in the top of the overlayer. The corresponding UPS incremental difference curves in Fig. 4b, obtained by subtracting spectra for successive depositions in

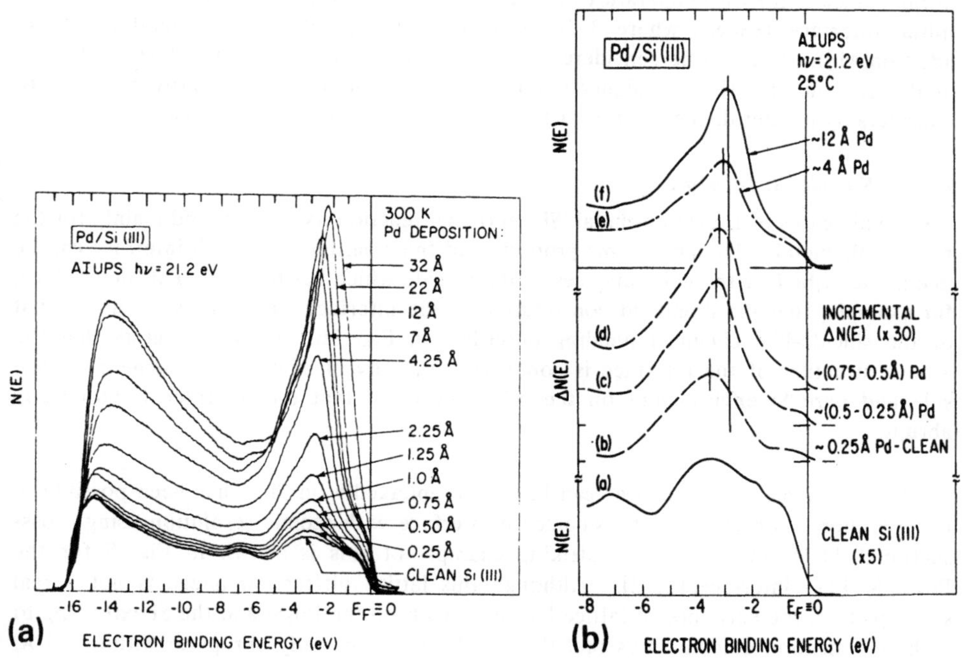

Fig. 4. (a) UPS spectra for the clean Si(111) surface and for various Pd coverages (equivalent metal thickness) deposited at room temperature.
(b) UPS incremental difference curves for the spectra in (a) for coverages 0 to 0.75 Å, together with the UPS spectra for ~ 4 and 12 Å Pd coverage.
(From G. W. Rubloff et. al.[23]).

Fig. 4b, show that the interface electronic structure (seen in the (sub)monolayer range) is essentially that of the bulk silicide, including a main peak well below E_F (near -2.5 to -3.0 eV) and a shape very similar to that for bulk Pd_2Si. This striking similarity demonstrates that the chemical bonding of atoms at the initial interface is essentially the same as in the bulk silicide, leading to the conclusion that the electronic properties of the interface are determined primarily by the silicide compound formation reaction[12].

This initial reaction of the Pd/Si interface at room temperature to form Pd_2Si has also been revealed by a variety of other techniques. Surface extended absorption fine structure (SEXAFS) measurements have determined that the initial Pd atoms on Si have the same first and second nearest-neighbor distances (to within 0.02 Å) as those for Pd atoms in bulk Pd_2Si[18]. This result can be interpreted as suggesting that the Pd atoms move into the Si lattice. Transmission electron microscopy (TEM) investigations in the flat-on mode (electron beam normal to the substrate surface) reveal diffraction features of Pd_2Si for coverages as low as ~ 5 Å[12,38,39]. For the chemically similar Ni/Si interface, high energy (MeV) ion channeling experiments, which observe the true buried interface, have shown that ~ 10 monolayers of the metal at the Ni/Si interface are reacted at room temperature[40].

For the near-noble transition-metal/Si interface, this initial reaction occurs below normal silicide growth temperatures and is limited in extent[2] to ~ 10 - 15 Å from the initial interface (except where 3-dimensional island growth of the metal occurs in addition[41]). This can be considered the initial reaction stage at the metal/Si interface as depicted in Fig. 1b. At higher temperatures ($\geq$ 200°C), silicide growth occurs to considerably greater distances from the interface, as depicted in Fig. 1c-1d.

D. The Silicide/Si Interface

For simple cases of transition-metal/Si interfacial reaction like this, found mainly for the near-noble metals, the microscopic properties of the resulting silicide/Si interface can be mainly described by the consequences of the silicide which has formed. This determines, for example, that the dominant contribution to the interface electronic structure is that of the silicide-like chemical bonding described in Fig. 3b. It also suggests that the atomic structure of the interface is composed primarily of silicide environment on one side and pure Si environment on the other, provided that the interface is atomically abrupt.

Because some silicide phases can be grown epitaxially on Si, it has been possible to observe atomic structure of the silicide/Si interface with atomic resolution using cross-sectional TEM techniques. A striking example of this is shown in Fig. 5 for the Pd_2Si/Si(111) interface[13,42]. Although this lattice diffraction image is not a real space picture, the existence of lattice fringes from both the Pd_2Si and the Si extending to a sharp transition demonstrates that the interface is atomically abrupt to within ~ 3 Å, with no amorphous layer present. Misfit dislocations and atomic steps occur over longer distances along the interface plane.

Atomic models consistent with these lattice images can be constructed[13,42], as also shown in Fig. 5. In fact, the lattice images have been replicated by detailed calculations[43,44] of the electron diffraction process, assuming particular atomic models. Thus detailed structural arrangements consistent with experiment can be specified for epitaxial silicide/Si interfaces, making them especially attractive for realistic theoretical calculations of metal/semiconductor interface properties.

As mentioned above, the electronic structure of such silicide/Si interfaces must be primarily determined by the existence of the silicide lattice extending directly to the abrupt interface. In fact, detailed AES studies of the growth of this interface for the Pd_2Si/Si(111) case have permitted a detailed study of both the composition of the interface (to search for variations in stoichiometry) as well as the presence of additional electronic states (other than silicide-like bonds and pure Si states) which might represent true interface states. In these experiments, both the intensity and the spectral shape of the Si $L_{2,3}VV$ AES transition were decomposed as a function of Pd coverage into three contributions - (1) the Si substrate, (2) a uniform, stoichiometric Pd_2Si overlayer, and (3) a very thin localized region at the interface. With increasing Pd coverage (shown in terms of Pd_2Si thickness in Fig. 6a), the intensity of the interface contribution, as well as that of the Si substrate, decrease exponentially, showing that the former is indeed localized at the interface. At higher coverage ($\geq$ 10 Å), the AES spectrum looks essentially the same as that for bulk Pd_2Si, because the measurement senses only ~ 8 Å (the electron escape depth for the Si $L_{2,3}VV$ transition) from the top surface. However, for lower coverage the interface contribution is comparable to that of the silicide and markedly alters the shape of the spectrum (see Fig. 6b); the new structure near 93 and 85 eV indicates the presence of the localized ("true") interface states. They have a relatively high density (~ 0.1 states per interface atom), which could be important in affecting some properties such as the Schottky barrier height of the contact. UPS observations[45] of the Pd/Si(111) interface at low coverage have also given evidence for true interface states. However, it should be noted that these results are consistent with the picture of basically stoichiometric Pd_2Si on Si(111) - i.e., they do not provide evidence for substantive variations in stoichiometry (composition gradients) at the interface.

It is important to note that these interface electronic states can be considered consistent with the atomic structure of the interface as described in Fig. 5. No matter what the detailed atomic structure is which gives rise to the lattice diffraction images shown, the atoms residing directly at the interface experience local chemical environments which are markedly different than those inside bulk silicide or bulk Si. Therefore, their electronic structure must deviate in some notable way from that of silicide or Si. In fact it is plausible that part of their chemical bonding to neighboring atoms is like that in silicide or Si while other parts are qualitatively different, thereby giving rise to the true interface states observed.

E. Low Temperature Interdiffusion

In contrast to the relatively straightforward silicide growth processes observed typically for the near-noble transition-metal/Si interface, the refractory transition metals exhibit low temperature reactivity which is more complex in that it extends considerably further (~ 100 Å or more) from the initial interface and produces microstructural properties quite different from that of the normal silicide growth process[2,24,46]. These charac-

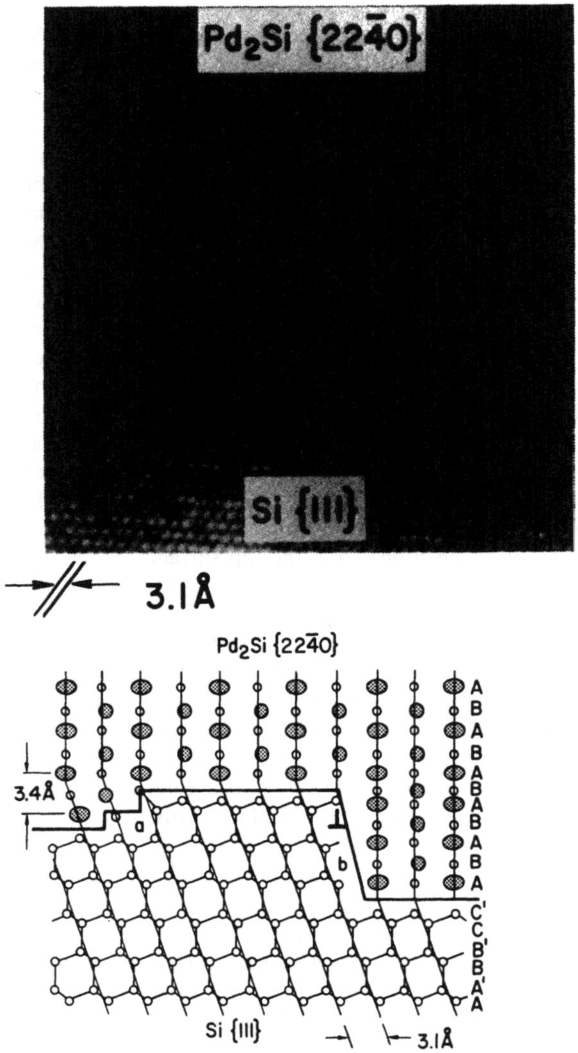

Fig. 5. Atomic structure of the epitaxial $Pd_2Si/Si(111)$ interface (from P. E. Schmid et. al.[42]).
Top: Cross-sectional TEM micrograph showing the lattice structure of the interface, which is structurally abrupt on an atomic scale (~ 3 Å).
Bottom: A schematic atomic model of the interface which could produce the observed lattice diffraction fringes. Atomic steps and misfit dislocations accompany the locally abrupt $Pd_2Si/Si(111)$ interface.

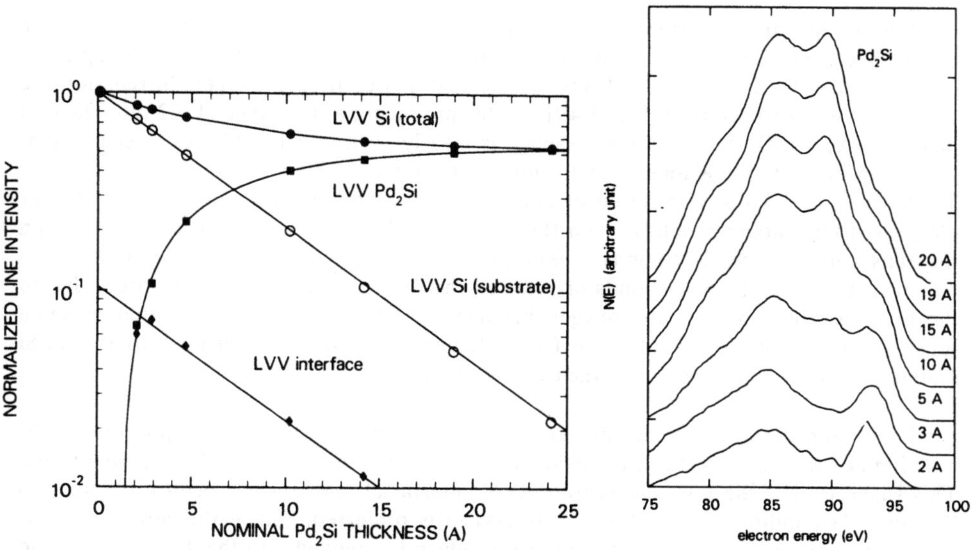

Fig. 6. (a) Decomposition of the total intensity of the Si $L_{2,3}VV$ AES line for Pd/Si(111) at room temperature into contributions from (i) the Si substrate, (ii) an ultrathin (~ 3 - 5 Å) Si-rich interfacial region, and (ii) the Pd_2Si compound formed by interfacial reaction. Note that the interface contribution (diamonds) grows briefly with the Pd_2Si and then is attenuated exponentially with further coverage beyond ~ 3 Å.
(b) Coverage dependence of the portion of the Si $L_{2,3}VV$ spectrum for Pd/Si(111) not attributable to the Si substrate. At higher coverage this is essentially just the Pd_2Si reaction product, but at very low coverage the interface portion contributes strongly and produces a very different spectral shape. (From P. E. Schmid et. al.[42]).

teristics may be considered as involving more complex reaction processes, such as those described as interdiffusion in Fig. 1e-1h, or alternatively a different materials reaction mechanism to nucleate silicide growth within a mixed metal-Si overlayer.

Finally, it should be noted that other structural evidence evidence exists for the motion of metal atoms *into* the Si substrate. Ion channeling results[47] show that a significant number of Ni atoms move into tetrahedral interstitial sites in the Si substrate from the as-deposited Ni/Si interface (which is Ni_2Si/Si). Since these sites are inconsistent with any Ni atom sites in Ni silicides, these results indicate diffusion of the metal atoms from the silicide into the Si substrate as in Fig. 1f. Composition analysis of the V/Si interface reacted at low temperatures (compared to normal silicide formation temperatures) indicates that a considerable portion of the V atoms move well into the Si substrate[48], as depicted in Fig. 1e.

F. Influence of Interface Contamination

The capability for silicide growth to produce a new metal/Si interface (in fact the silicide/Si interface) below the original Si surface is very important to practical application in technology, since typical starting surfaces of Si contain chemical contaminants and/or defects. In this way, as depicted in Fig. 1, a clean buried interface intrinsic to the silicide/Si system is obtained. Silicide growth is known to occur in spite of the presence of native oxide on the initial Si surface.

Microscopic investigations can be carried out to determine the influence of such chemical contaminants at the interface, as demonstrated in the following example for oxygen at the Ti/Si interface. The essential features of the Ti/oxygen/Si reaction are shown in the UPS spectra in Fig. 7[49]. After producing a thin layer (1 - 2 monolayers) of oxide-like (Si-O) bonds, their presence is clearly indicated by the peak occurring at -11 eV in curve b. It is important to note that no peaks occur in this energy range for Ti-O bonds[49]. Upon deposition of only 5 Å of Ti metal at room temperature, the -11 eV peak disappears completely, even though such a thin layer is insufficient to cover up the emission from the Si-O bonds as seen in UPS. Therefore, it is concluded that the Ti has in fact broken the Si-O bonds or reduced the Si surface at room temperature due to the high reactivity of Ti; the oxygen becomes dissolved into the Ti, so that in some sense the unreacted Ti/Si interface (room temperature) is made intrinsic to the Ti-Si system even in the presence of oxygen contamination.

Upon annealing to sufficient temperatures (~ 250 - 300°C) to promote Ti-Si reaction at the clean interface, the evidence for Si-O bonds (the -11 eV peak) reappears. This suggests that the low temperature Ti-Si interfacial reaction has been promoted, so that now Ti bonding to Si instead of to oxygen is preferred. Since the supply of Ti is limited, the oxygen then rebonds to excess Si which has moved into the Ti overlayer, and the system again contains Si-O bonds, primarily nearer the top surface of the overlayer. In this situation, the interface is again intrinsic to the Ti-Si system, with reacted material at the Si interface.

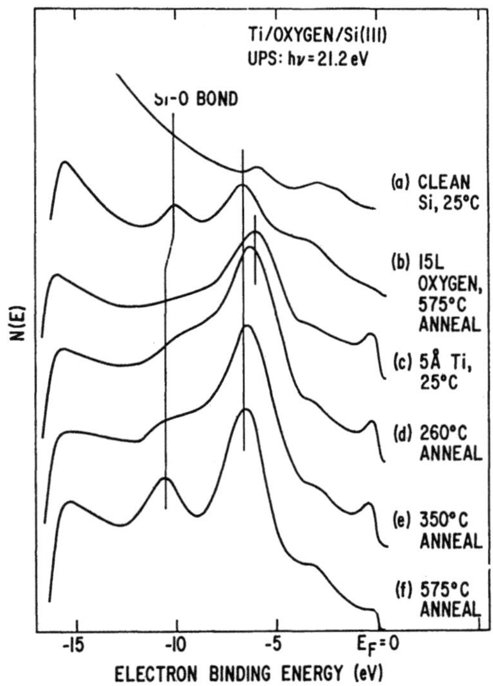

Fig. 7. UPS spectra for: (a) clean Si(111); (b) Si surface covered by 1 -2 monolayers of Si-O bonds, which produce a characteristic peak at -11 eV; (c) 5 Å Ti deposited at room temperature; (d - f) annealing at progressively higher temperatures which promote Ti-Si reaction. (From R. Butz et. al.[49]).

To understand the fundamental chemical interactions at a real metal/semiconductor interface, it is important to carry out basic studies like these of the role of controlled chemical contaminants on the metal/semiconductor interactions, as well as investigating the chemistry of the atomically clean interface.

IV. METAL/III-V INTERFACE CHEMISTRY

The chemistry of the metal/III-V interface is complicated considerably by additional reaction possibilities, most of which have been observed. These extra complexities include: (i) interdiffusion involving two semiconductor atomic constituents; (ii) preferential out-diffusion of one semiconductor constituent relative to the other; (iii) the formation of binary compound phases between metal and one semiconductor constituent; (iv) the formation of ternary phases involving metal and both semiconductor constituents; (v) exchange (displacement) reactions in which the metal atom replaces the cation of the compound semiconductor (for group III metals); and (vi) more complex microstructure and growth modes due to simultaneous growth of several reaction phases.

Nevertheless, the metal/III-V semiconductor interface shows many chemical similarities to metal/Si interfaces. First, there is a case - namely - in which the interface appears to show essentially no chemical reactivity[50,51]; Instead, the growth is three-dimensional (nucleated). This case provides a valuable counterpoint to the second chemical similarity between metal/III-V and metal/Si interfaces, which is that chemical reaction at the interface is the rule rather than the exception. The third similarity, discussed below, is that compound phases are often formed by the interfacial reaction.

The interfaces of simple metals (especially group III) with III-V compound semiconductor surfaces have been given considerable study[6]. The interpretation of these results has been the subject of some debate because of the difficulty in verifying assumptions about the structure of the interface in order to interpret spectroscopic results[52]. However, the existence of preferential outdiffusion of semiconductor species has certainly been documented[6]. The case of Al/GaAs is especially intriguing in that the exchange reaction to form Ga/AlAs at the interface is energetically favorable, but the presence or absence of this reaction at the interface appears to depend on detailed kinetics, which in turn depend on specific structural characteristics of the interface. When the kinetics is too slow to permit the exchange reaction, the interface growth may instead be interpreted as simple Stranski-Krastanov growth without atomic intermixing[52].

Noble-metal interfaces with III-V semiconductors exhibit strong in-diffusion of metal atoms as well as preferential outdiffusion of semiconductor constituents, especially with annealing[6]. The degree of chemical reaction can be altered markedly by the presence of thin interlayers of other metals. However, developing a consistent picture of these reactions, which can depend on structural and compositional details of the semiconductor surface as well as on metal overlayer thickness and annealing history, is no easy task. As with metal/Si interfaces, the absence of metal-semiconductor compound phases as reaction products may make the job more difficult.

Recently a number of studies have been carried out on transition-metal/III-V semiconductor interfaces[53,54,55,56,57]. Compound phases involving transition-metal atoms with the semiconductor constituents have been identified, so that a better understanding of the chemical reactions and products may be obtained for such interfaces. Again, the existence of well-defined reaction products, such as compounds, improves the chance of understanding what is intrinsically a complex chemical situation.

An example which is especially useful for comparison to the transition-metal/Si case is that of the Pd/GaAs interface. Recent studies[55,57] have shown that the interfacial reaction produces only the binary PdGa phase following relatively high temperature annealing (≥ 350°C). At lower temperatures additional products are formed, including a ternary phase and some Pd_2Ga and $PdAs_2$, with an overall composition ~Pd_2GaAs.

The chemical bonding and electronic structure of the resulting interfaces are indicated by the UPS spectra in Fig. 8. In Fig. 8a, the spectrum evolves from that of pure Pd metal (from 150 Å Pd on GaAs(100) without annealing) to a markedly different spectrum upon annealing to ~ 200°C; note that the latter spectrum is dominated by a peak well below the Fermi energy, as is the case for Pd_2Si formation at the Pd/Si interface. With annealing to higher temperatures, the spectrum changes somewhat as the reaction product phase distribution changes (as verified by TEM), but similar character-

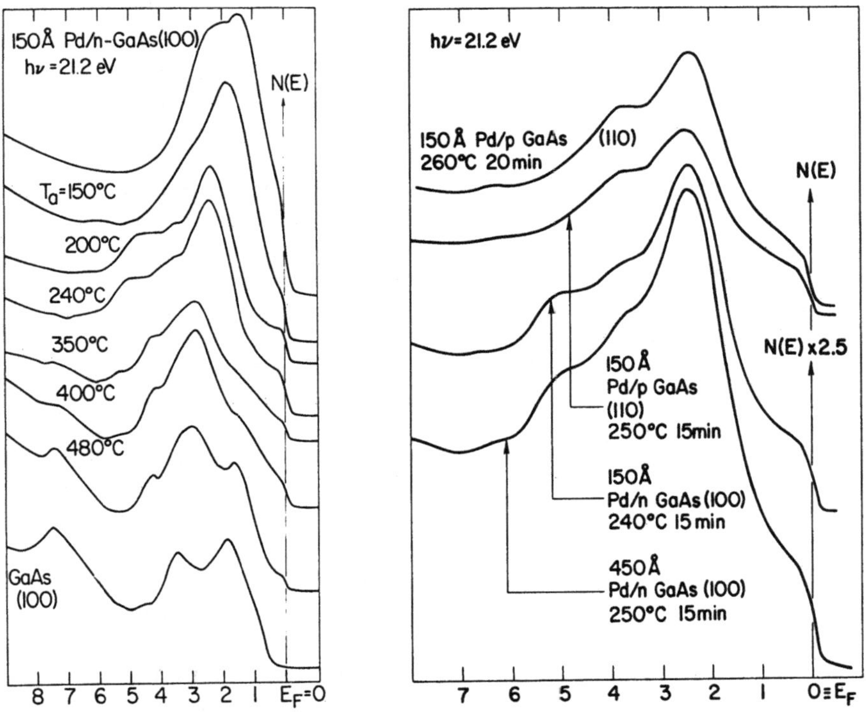

Fig. 8. (a) UPS spectra for a relatively thick (150 Å) Pd overlayer on GaAs(100) surface as deposited at room temperature and then annealed at different temperatures, and compared with the spectrum for the clean GaAs(100) surface;.
(b) UPS spectra for 150 Å Pd reacted near 250°C on n- and p-type GaAs (110) and (100) surfaces. (From P. Oelhafen et. al.[55]).

istics of transition-metal compounds with semiconductor constituents remain. At sufficiently high temperatures (near 500°C) the spectrum resembles that of the clean GaAs surface, suggesting redistribution of the reaction products and surface enrichment by the semiconductor, again similar to the behavior of the Pd/Si system.

The results in Fig. 8b show that the chemical bonding at the interface following reaction at ~ 250°C is essentially independent of whether the substrate was (110) or (100), or n- or p-type. This demonstrates that the interfacial reaction dominates the properties of the resulting interface for such systems, regardless of the details of initial surface structure, etc.. Again, the transition-metal/III-V semiconductor interface shows a distinct parallel to the behavior of the transition-metal/Si interface.

Although in principle an Ohmic contact to a semiconductor might be achieved simply by choosing a metallurgy with a sufficiently low Schottky barrier height, this cannot always be done reliably. For example, Ohmic contacts to p-Si are readily obtained by choosing one of many metals which give high barriers to n-Si. On the other hand, intrinsically low barriers to n-Si are not easy to find (except perhaps for rare earth metals, which are not yet used in technology). The standard way to achieve effectively Ohmic contacts to n-Si is therefore to simply produce a heavily doped region in the Si, so that the depletion width shrinks greatly and carriers can tunnel easily through it, thereby making the barrier invisible.

Unfortunately, GaAs (the dominant III-V material for microelectronics) does not incorporate sufficiently large concentrations of electrically active dopants to permit this trick for making ohmic contacts, while on the other hand its interface Fermi level is pinned well within the band gap. Consequently, Ohmic contacts to GaAs present a very formidable challenge required for technological application.

The primary metallurgy to achieve Ohmic contacts to GaAs has been a specific combination and treatment of Ni-Au-Ge, which involves very complex alloying reactions with the GaAs. To understand the microscopic chemistry which occurs and develop more reliable, reproducible contacts of this kind, the distribution of phases has been studied for three allowing processes which span a relatively small range of annealing times (90 - 200 sec) and temperatures (390 - 450°C) in the range of interest[56]. Cross-sectional TEM micrographs of these reacted contacts show numerous phases, including Au, NiGe, NiAs, and Ni_2GeAs, with the phase distribution depending on fairly small changes in the annealing parameters. Furthermore, these phases contain differing amounts of Ga, As, and Ge impurities. An example of such an alloy contact is given in Fig. 9[56], which illustrates clearly the complexity of both the interfacial reaction and the microscopic properties (structure, composition, chemical bonding, etc.) that must be understood and controlled to effectively use such contacts.

V. THE SCHOTTKY BARRIER PROBLEM

We turn now to the essence of the Schottky barrier problem and the resulting electrical characteristics of the interface. The change in semiconductor band-bending with deposition of a metal contact is depicted schematically in Fig. 10a for the specific case of the metal/Si interface. Before the metal is deposited, the surface states and charge distribution of the clean Si surface determine the band-bending in the semiconductor space charge region (of order 1000 Å thick). This gives rise to the Schottky barrier

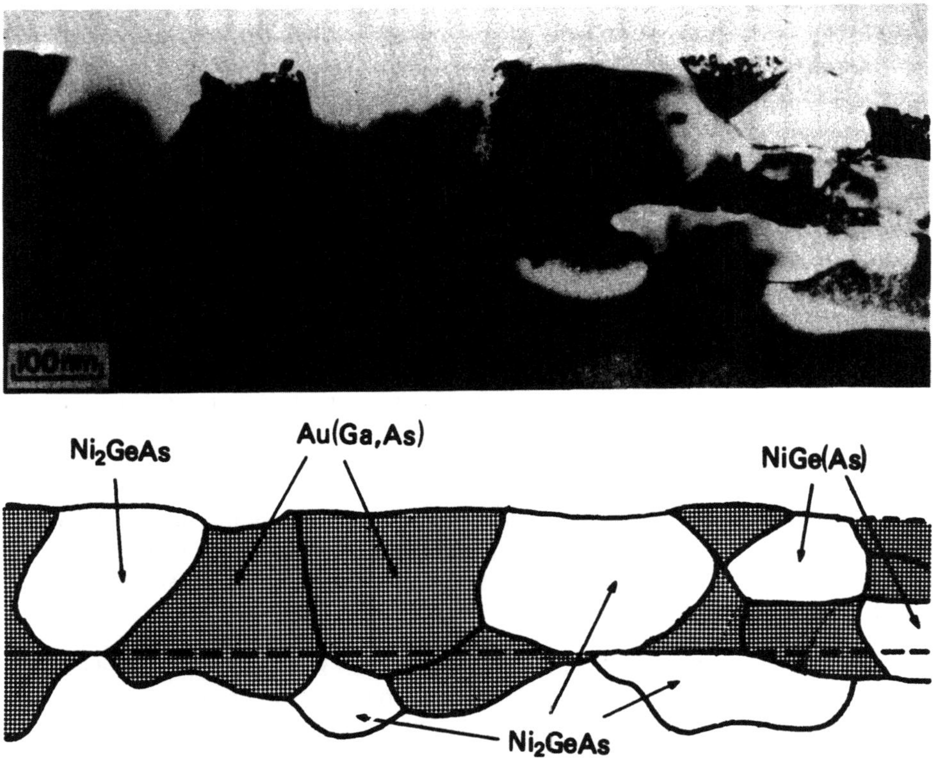

Fig. 9. Cross-sectional TEM bright-field image of an alloyed Au/Ni/Au-Ge Ohmic contact to GaAs(100), annealed 200 sec. at 450°C. The distribution of phases and their chemical compositions are indicated in the schematic diagram, together with the impurities (in parentheses) present in each phase. (From T.S. Kuan [56]).

height ϕ_{bn}, i.e. the energy separation at the surface between the conduction band minimum E_C and the Fermi energy E_F. Upon deposition of the metal contact, the interface charge is redistributed, so that the band-bending and Schottky barrier height are altered. This new barrier height ϕ_{bn}' acts as the energy barrier to electrical transport across the interface and hence determines the primary electrical characteristics of the contact. The Schottky barrier problem is then to understand the mechanisms which determine the barrier height ϕ_{bn} for the metal/semiconductor contact. This has posed a challenging and unsolved problem for decades[8,58].

Numerous models have been presented to explain the chemical trends in Schottky barrier heights with different metals and semiconductors, starting from Schottky's original explanation[58] in terms of metal-semiconductor work functions and Bardeen's involving semiconductor surface states[59]. Since the experimental data generally fall between the predictions of these two models, other theories have been presented, and comprehensive reviews have been given[8,9]. Many approaches have been thermochemical in nature, i.e. directed at recognizing the role which the interfacial chemistry must play in determining the charge redistribution and barrier height at the contact.

For example, several such theories have been advanced to explain the Schottky barrier height of the silicide/Si interface. Andrews and Phillips[60] related these barrier

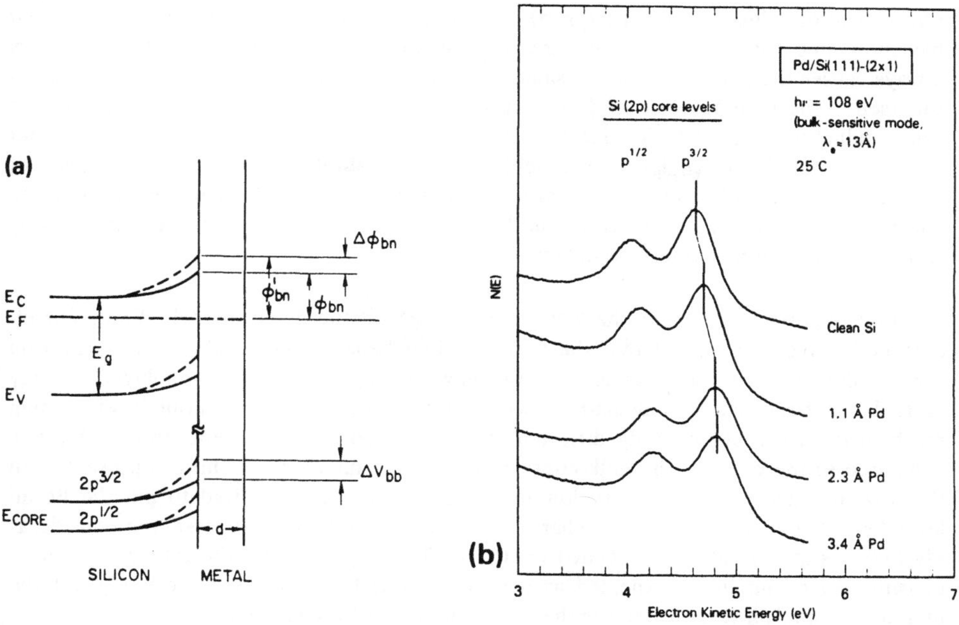

Fig. 10. (a) Schematic energy level scheme for the clean Si surface (solid lines, Schottky barrier height ϕ_{bn}) and for a thin metal overlayer (dashed lines, barrier height ϕ_{bn}'). The barrier height change $\Delta\phi_{bn}$ is given by the change in band-bending ΔV_{bb} as measured by the shift in the Si (2p) core lines.
(b) Schottky barrier height dependence on Pd coverage for the Si(111)-(2×1) and Si(111)-(7×7) surfaces as determined from shifts of the Si (2p) core levels observed by synchrotron radiation photoemission. The saturation value of the barrier height is reached at a few Å metal coverage. (From R. Purtell et. al.[48]).

heights to the heat of formation of the silicide reaction product. Freeouf[61] reinterpreted Schottky's work function model on the basis of Si-rich silicides present at the intimate metal/Si interface. Ottaviani, Tu, and Mayer[62] related the barrier height to the eutectic temperature of the metal-Si system. All three approaches give a reasonable explanation of the observed barrier heights, so no preferred model emerges, and microscopic insight has not yet been reached.

More recently, attempts to explain Schottky barrier heights have turned to more specific microscopic aspects of the reactive metal/semiconductor interface. These efforts have really concentrated on three directions: (i) the role of interface defects in controlling Schottky barrier properties; (ii) the influence of strong chemical interactions associated with interface reactivity; and (iii) the consequences of microscopic inhomogeneity of the reaction product on observed electrical characteristics.

A. Defect Model

The behavior of III-V Schottky barrier heights has been the motivation for the defect model of Schottky barriers[63,64], which has been discussed widely[10,65]. First, some evidence suggests that the Fermi level pinning position on III-V's, especially GaAs, does not depend on which metal (or oxygen) is in contact with the semiconductor. This result has motivated the idea that upon creating the interface the semiconductor chemistry always causes the generation of the same defects, which depend only on the semiconductor itself. The pinning position then varies only with the doping type (n- vs. p-) of the semiconductor. Second, for III-V semiconductors the final band-bending (barrier height) of the interface is already achieved with notably less than a monolayer of coverage, so that relatively low densities of states are needed to pin the Fermi level. Chemical processes like preferential outdiffusion might reasonably represent the driving force for generation of interface defects.

Two important arguments have been raised against the defect model. First, Zur et. al.[66,67] have pointed out that the barrier height behavior noted above as motivation for the defect model pertains only to the low coverage regime (~ monolayer). They carried out electrostatics calculations which demonstrate that the Fermi level pinning mechanisms are fundamentally different in this low coverage ("chemisorption") range compared to the case for the bulk contact (where true metallic screening is present). In the low-coverage case, only very low densities of states are required to pin the Fermi level (~ 10^{12} cm^{-2}), whereas at higher coverages involving metallic screening considerably higher state densities are needed (~ 10^{14}). This suggests that the rather low defect densities which might be expected as a result of interfacial reaction are likely not the pinning mechanism responsible for the barrier height at bulk contacts.

Second, if defect generation were intrinsic to the reactivity of III-V interfaces, it is difficult to understand how epitaxial growth (e.g., MBE) could ever result in device-quality material (which can be grown). Indeed, a recent study[68] of an epitaxial "metal"/GaAs interface (in which the "metal" was n^+-GaInAs), showed that the Schottky barrier height of the contact varied smoothly with misfit dislocation density. At negligible dislocation density, the interface Fermi level was unpinned, demonstrating that thermal (evaporant) beams of atoms impinging on the GaAs surface do not generate intrinsic (metallurgical) defects in sufficient quantity to affect the Schottky barrier height. Instead, with sufficiently high defect density generated by a specific structural defect (the misfit dislocation, extrinsic to the interface and its inherent chemistry), the barrier height could be changed.

The case of metal/Si Schottky barriers presents an interesting contrast to the case of Fermi level pinning at low metal coverage on III-V semiconductor surfaces. On Si the band-bending changes require ~ 0.5 to a few monolayers of metal for saturation. The change in Si substrate band-bending has been measured using high-resolution synchrotron radiation photoemission for the Si(2p) core level, examples of which are shown in Fig.10b [2,3,48]. As can be seen from the band-bending/barrier height changes for Ni, Pd, and Pt on Si(111) in Fig. 11, the barrier is determined over a coverage range of 0.5 - 5 Å metal[2,3,69].. Therefore a considerable density of interface states is required to determine the Fermi level position at the transition-metal/Si interface. Furthermore, the barrier height does depend notably on the identity of the metal atom itself, so that a simple defect structure related only to the Si substrate cannot explain the Schottky barrier. From this point of view, a defect model for Si Schottky contacts, as has been suggested[70], must involve defects whose characteristics (energy in the gap, etc.) depend on the metal atom identity as well.

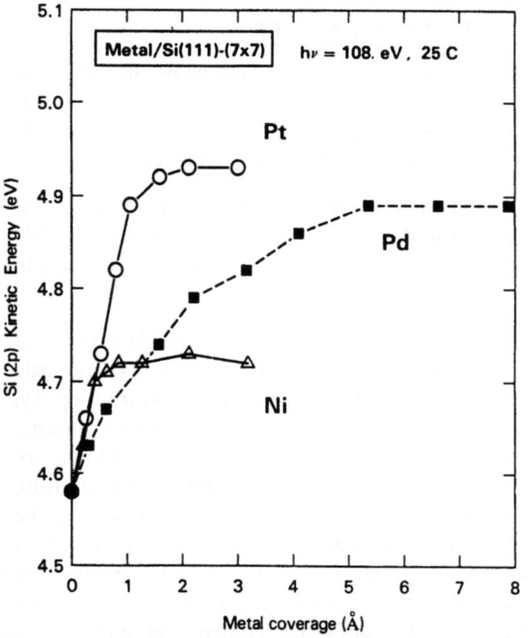

Fig. 11. Si (2p) photoelectron kinetic energy versus equivalent metal coverage for Pt, Pd, and Ni overlayers deposited on Si(111)-(7×7) at room temperature, as obtained from synchrotron radiation photoemission experiments in bulk sensitive mode, hν = 108 eV. Changes in Si(2p) kinetic energy are equivalent to changes in the Schottky barrier height at the interface. (From R. Purtell et. al.[69]).

B. Chemical Bonding and Reaction

At this point, however, the explanation for interface states responsible for determining the Schottky barrier height becomes closely related to the detailed chemical bonding and reaction at the interface. In fact, for both metal/Si[2,3,4] and metal/III-V[6,71,72] Schottky barriers, the reactive chemistry of the interface has been proposed to play the decisive role in determining barrier heights.

C. Inhomogeneity of Reaction Products

Finally, with the recognition of high chemical reactivity at metal/Si and metal/III-V interfaces and its likely importance in determining Schottky barrier properties, the role of interface inhomogeneity on those properties has been considered. First, the combination of non-interacting parallel diode regions along the interface was noted, especially as it relates to reacting alloy contacts[73,74]. Then it was recognized that lateral inhomogeneity along the interface would require a more microscopic treatment of electrostatic potential energy contours and transport in the depletion region when the lateral inhomogeneity scale became comparable to or smaller than the depletion length in the semiconductor[75,76]. These effects are especially important in the more complex cases of interfacial reaction, such as when the interface structure contains facetting, atomic steps and misfit dislocations, and mixed reaction phases, for which the microstructure in Fig. 9 is a striking example.

VI. MICROSCOPIC ASPECTS OF THE SCHOTTKY BARRIER

At this stage it is useful to summarize what factors have been shown to be important or unimportant in determining the Schottky barrier electrical properties of metal/semiconductor contacts. Many of these results have come directly from correlating microscopic interface chemistry and reaction with Schottky barrier heights.

First, it may well be that the Fermi level pinning mechanisms which dominate at metal/Si interfaces are different from those which control the barrier at metal/III-V interfaces, as seen from the relative sensitivity and insensitivity to metal atom identity in the two cases respectively.

Second, the Schottky barrier height can be determined either by intrinsic aspects of the interface chemistry or by extrinsic factors. For metal/III-V contacts, intrinsic pinning is suggested by the absence of barrier height changes with metal atom identity, while the example of misfit dislocation pinning demonstrates a case of extrinsic pinning. For metal/Si contacts, the growth of silicides produces barrier heights which are independent[77,78,79] of many details (such as initial Si surface structure, orientation, and cleanliness, reacted silicide phase and microstructure, etc.). This shows that the reacted interface barrier is determined by chemical aspects intrinsic to the metal-Si system, although extrinsic factors in sufficient quantity may be able to change the barrier height[2,4]. The transition from an unreacted to a reacted state at the interface plays an important part in changing the Fermi level pinning mechanism from extrinsic to intrinsic, e.g. as in the case of interfacial compound formation.

Third, the Schottky barrier height appears to be strongly affected or determined within a few monolayers of metal coverage, both for metal/Si and metal/III-V contacts. This fact and the independence of barrier height on silicide phase both suggest that the Schottky barrier height is a true interface property[4]. Furthermore, the barrier height seems to be unrelated to the presence of a true metal on the semiconductor surface[69,80].

These factors suggest a picture for microscopic Schottky barrier mechanisms in which specific chemical bonds intrinsic to the interface play a dominant role. For the case of metal/Si contacts, these may be either silicide-like bonds or true interface states. The determination of how such electronic states at the interface can affect the Schottky barrier height may be aided considerably by studies of highly controlled epitaxial systems like that of n^+-GaInAs/GaAs described above[68] or $NiSi_2$/Si(111)[81]. In the latter case, epitaxial silicide/Si interfaces have been prepared with two different local atomic orientations (i.e., all A-type or all B-type interface arrangements), and the Schottky barrier height was found to differ by 140 meV (a considerable amount on Si) in the two cases. A possible explanation for this difference is that the two arrangements provide different sets of true interface electronic states to pin the Fermi level, but in any case experiments like these should give much opportunity for progress in the understanding of Schottky barrier heights.

Acknowledgement

I am very pleased to thank Paul Ho for his valuable contributions to our joint work on silicide interfaces. I am also grateful for close collaboration with R. M. Matz, Y. Yokota, J. Tsang, M. Liehr, R. Butz, J. G. Clabes, R. J. Purtell, P. E. Schmid, V. L. Moruzzi, A. R. Williams, F. J. Himpsel, G. Hollinger, and T. Y. Tan. The interest and support of the Office of Naval Research is appreciated.

REFERENCES

*Supported in part by the U.S. Office of Naval Research.

1. K. N. Tu and J. W. Mayer, in Thin Films - Interdiffusion and Reactions, edited by J. M. Poate, K. N. Tu, and J. W. Mayer (Wiley, New York, 1978), p. 359.
2. G. W. Rubloff, Festkorperprobleme (Advances in Solid State Physics, Volume XXIII, p. 179, P. Grosse (ed.) (Vieweg, Braunschweig, F. R. Germany, 1983).
3. G. W. Rubloff, *Surface Science* ***132***, 268 (1983).
4. P. S. Ho, *J. Vac. Sci. Technol. A* ***1***, 745 (1983).
5. G. Le Lay, *Surface Science* ***132***, 169 (1983).
6. L. J. Brillson, *Surface Science Reports* ***2***, 2 (1982).
7. I. P. Batra and F. Herman, *J. Vac. Sci. Technol. A* ***1***, 1080 (1983).
8. M. Schlüter, *Thin Solid Films* ***93***, 3 (1982).
9. R. H. Williams, *Contemp. Phys.* ***23***, 329 (1982).
10. R. H. Williams, *Surface Science* ***132***, 122 (1983).
11. S. M. Sze, Physics of Semiconductor Devices, 2nd edition (Wiley, N.Y., 1981).
12. J. L. Freeouf, G. W. Rubloff, P. S. Ho, and T. S. Kuan, *Phys. Rev. Letters* ***43***, 1836 (1979).
13. P. S. Ho, P. E. Schmid, and H. Föll, *Phys. Rev. Letters* ***46***, 782 (1981).
14. J. E. Rowe, S. B. Christman, and G. Margaritondo, *Phys. Rev. Letters* ***35***, 1471 (1975).
15. G. Margaritondo, J. E. Rowe, and S. B. Christman, *Phys. Rev. B15* ***14***, 5396 (1976).
16. A. McKinley, R. H. Williams, and A. W. Parke, *J. Phys. C* ***12***, 2447 (1979).
17. E. J. van Loenen, M. Iwami, R. M. Tromp, and J. F. van der Veen, *Surface Science* (in press).
18. J. Stöhr and R. Jaeger, *J. Vac. Sci. Technol.* ***21***, 619 (1982).
19. L. Braicovich, C. M. Garner, P. R. Skeath, C. Y. Su, P. W. Chye, I. Lindau, and W. E. Spicer, *Phys. Rev. B15* ***20***, 5153 (1979).
20. F. Ringeisen, J. Derrien, E. Daugy, J. M. Layet, P. Mathiez, and F. Salvan, *J. Vac. Sci. Technol. B* ***1***, 546 (1983).
21. A. Cros, F. Salvan, and J. Derrien, *J. Appl. Phys.* ***52***, 4757 (1981).
22. R. D. Thompson and K. N. Tu, *Thin Solid Films* ***93***, 265 (1982).
23. G. W. Rubloff, P. S. Ho, J. L. Freeouf, and J. E. Lewis, *Phys. Rev. B15* ***23***, 4183 (1981).
24. J. G. Clabes, G. W. Rubloff, and T. Y. Tan, *Phys. Rev. B15* ***29***, 1540 (1984).
25. R. D. Thompson, B. Y. Tsaur, and K. N. Tu, *Appl. Phys. Letters* ***38***, 535 (1981).
26. K. N. Tu, R. D. Thompson, and B. Y. Tsaur, *Appl. Phys. Letters* ***38***, 626 (1981).
27. G. Rossi, J. Nogami, I. Lindau, L. Braicovich, I. Abbati, U. del Pennino, and S. Nannarone, *J. Vac. Sci. Technol. A* ***1***, 781 (1983).
28. G. Rossi, J. Nogami, J. J. Yeh, and I. Lindau, *J. Vac. Sci. Technol. B* ***1***, 530 (1983).
29. A. Franciosi, J. H. Weaver, P. Perfetti, A. D. Katnani, and G. Margaritondo, *Solid State Commun.* ***47***, 427 (1983).
30. R. Tromp, E. J. van Loenen, M. Iwami, R. Smeenk, and F. W. Saris, *Thin Solid Films* ***93***, 151 (1982).
31. R. M. Tromp, E. J. van Loenen, M. Iwami, R. G. Smeenk, F. W. Saris, F. Nava, and G. Ottaviani, *Surface Science* ***124***, 1 (1983).
32. G. Rossi, I. Abbati, L. Braicovich, I. Lindau, and W. E. Spicer, *Phys. Rev. B15* ***25***, 3627 (1982).

33. P. J. Grunthaner, F. J. Grunthaner, and A. Madhukar, *J. Vac. Sci. Technol.* ***20***, 680 (1982).
34. J. H. Weaver, V. L. Moruzzi and F. A. Schmidt, *Phys. Rev. B15* ***23***, 2916 (1981).
35. J. H. Weaver, A. Franciosi, and V. L. Moruzzi, (submitted).
36. D. E. Eastman, F. J. Himpsel, J. A. Knapp, and K. C. Pandey, Proceedings of the 14th International Semiconductor Conference, Edinburgh, 1978, edited by B. L. H. Wilson (Institute of Physics and Physical Society, London, 1979).
37. P. S. Ho, G. W. Rubloff, J. E. Lewis, V. L. Moruzzi, and A. R. Williams, Thin Film Interfaces and Interactions, ed. by J. E. E. Baglin and J. M. Poate (Vol. 80-2, The Electrochemical Society, 1980), p. 85.
38. P. S. Ho, T. Y. Tan, J. E. Lewis, and G. W. Rubloff, *J. Vac. Sci. Technol.* ***16***, 1120 (1979).
39. J. L. Freeouf, G. W. Rubloff, P. S. Ho, and T. S. Kuan, *J. Vac. Sci. Technol.* ***17***, 916 (1980).
40. N. W. Cheung, R. J. Culbertson, L. C. Feldman, P. J. Silverman, K. W. West, and J. W. Mayer, *Phys. Rev. Letters* ***45***, 120 (1980).
41. R. Matz, R. J. Purtell, Y. Yokota, G. W. Rubloff, and P. S. Ho, *J. Vac. Sci. Technol. A* ***2***, 253 (1984).
42. P. E. Schmid, P. S. Ho, H. Föll, and G. W. Rubloff, *J. Vac. Sci. Technol.* ***18***, 937 (1981).
43. D. Cherns, D. A. Smith, W. Krakow, and P. E. Batson, *Phil. Mag.* ***45***, 107 (1982).
44. W. Krakow, *Thin Solid Films* ***93***, 109 (1982).
45. L. Braicovich, *Surface Science* ***132***, 315 (1983).
46. R. Butz, G. W. Rubloff, T. Y. Tan, and P. S. Ho, *Phys. Rev. B15* (in press).
47. N. W. Cheung and J. W. Mayer, *Phys. Rev. Letters* ***46***, 671 (1981).
48. R. J. Purtell, P. S. Ho, G. W. Rubloff, and P. E. Schmid, *Physica* ***118B***, 834 (1983) (Proceedings of the 16th International Conference on the Physics of Semiconductors, Montpellier, France, Sept. 6-10, 1982).
49. R. Butz, G. W. Rubloff, and P. S. Ho, *J. Vac. Sci. Technol. A* ***1***, 771 (1983).
50. R. Ludeke, T.-C. Chiang, and D. E. Eastman, *J. Vac. Sci. Technol.* ***21***, 599 (1982).
51. R. Ludeke, T.-C. Chiang, and T. Miller, *J. Vac. Sci. Technol.* (in press).
52. R. Ludeke, *Surface Science* ***132***, 143 (1983).
53. J. R. Waldrop, S. P. Kowalczyk, and R. W. Grant, *J. Vac. Sci. Technol.* ***21***, 607 (1982).
54. C. Fontaine, T. Okumura, and K. N. Tu, *J. Appl. Phys.* ***54***, 1404 (1983).
55. P. Oelhafen, J. L. Freeouf, T. S. Kuan, T. N. Jackson, and P. E. Batson, *J. Vac. Sci. Technol. B* ***1***, 588 (1983).
56. T. S. Kuan, P. E. Batson, T. N. Jackson, H. Rupprecht, and E. L. Wilkie, *J. Appl. Phys.* ***54***, 6952 (1983).
57. T. S. Kuan, Proc. of MRS Symposium on Electron Microscopy, 1983, , ed. by W. Krakow and D. A. Smith, p.
58. W. Schottky, *Z. Physik* ***113***, 367 (1939).
59. J. Bardeen, *Phys. Rev.* ***71***, 717 (1947).
60. J. M. Andrews and J. C. Phillips, *Phys. Rev. Letters* ***35***, 56 (1975).
61. J. L. Freeouf, *Solid State Commun.* ***33***, 1059 (1979).
62. G. Ottaviani, K. N. Tu, and J. W. Mayer, *Phys. Rev. Letters* ***44***, 284 (1980).
63. W. E. Spicer, I. Lindau, P. R. Skeath, C. Y. Su, and P. W. Chye, *Phys. Rev. Letters* ***44***, 420 (1980).
64. W. E. Spicer, P. Skeath, C. Y. Su, and I. Lindau, *J. Phys. Soc. Japan* Suppl. A, ***49***, 1079 (1980).
65. R.H. Williams, *J. Vac. Sci. Technol.* ***18***, 929 (1981).

66. A. Zur, T. C. McGill, and D. L. Smith *Phys. Rev. B15* ***28***, 2060 (1983).
67. A. Zur, T. C. McGill, and D. L. Smith, *Surface Science* ***132***, 456 (1983).
68. J. M. Woodall, G. D. Pettit, T. N. Jackson, C. Lanza, K. L. Kavanagh, and J. W. Mayer, *Phys. Rev. Letters* ***51***, 1783 (1983).
69. R. J. Purtell, G. Hollinger, G. W. Rubloff, and P. S. Ho, *J. Vac. Sci. Technol. A* ***1***, 566 (1983).
70. O. F. Sankey, R. E. Allen, and J. D. Dow, *Solid State Commun.* (in press).
71. L. J. Brillson, C. F. Brucker, A. D. Katnani, N. G. Stoffel, R. Daniels, and G. Margaritondo, *Surface Science* ***132***, 212 (1983).
72. P. J. Grunthaner, F. J. Grunthaner, and A. Madhukar, *Physica* ***118B***, 831 (1983) (Proceedings of the 16th International Conference on the Physics of Semiconductors, Montpellier, France, Sept. 6-10, 1982).
73. K. N. Tu, *J. Vac. Sci. Technol.* ***19***, 766 (1981).
74. T. Okumura and K. N. Tu, *J. Appl. Physics* ***54***, 922 (1983).
75. J. L. Freeouf, T. N. Jackson, S. E. Laux, and J. M. Woodall, *J. Vac. Sci. Technol.* ***21***, 570 (1982).
76. J. L. Freeouf, T. N. Jackson, S. E. Laux, and J. M. Woodall, *Appl. Phys. Letters* ***40***, 634 (1982).
77. P. S. Ho and G. W. Rubloff, *Thin Solid Films* ***89***, 433 (1982).
78. P. E. Schmid, P. S. Ho, and T. Y. Tan, *J. Vac. Sci. Technol.* ***20***, 688 (1982).
79. G. Ottaviani, K. N. Tu, and J. W. Mayer, *Phys. Rev. B15* ***24***, 3354 (1981).
80. R. Purtell, J. G. Clabes, G. W. Rubloff, P. S. Ho, B. Reihl, and F. J. Himpsel, *J. Vac. Sci. Technol.* ***21***, 615 (1982).
81. R. T. Tung, J. M. Gibson, and J. M. Poate, *Phys. Rev. Letters* ***50***, 429 (1983).

Contribution

Catalytic Effect of Near-Noble Metals on Si Oxidation

P. Lomellini and S. Valeri

Dipartimento di Fisica, Università di Modena, GNSM del CNR, Via Campi 213/A
I-41100 Modena, Italy

It is well known that Si oxidation is enhanced by the presence of metal atoms, even at RT (room temperature). Up to now only few results are available on the oxygen exposure and metal concentration dependence of this effect, and on the catalytic ability of different metals [1,2] .
We studied by AES and XPS the RT oxidation of Ni silicides of different Ni content ($NiSi_2$, NiSi, Ni_2Si and Ni_3Si) versus oxygen exposure in the range 0.2 - 10^4 Langmuir (L). Near-noble metals (Ni,Pd,Pt), silicides with the same metal content (M_2Si), were also comparatively investigated, after an oxygen exposure of 10^4L at RT. Finally we also studied the oxidation of pure Si and near-noble metal surfaces.

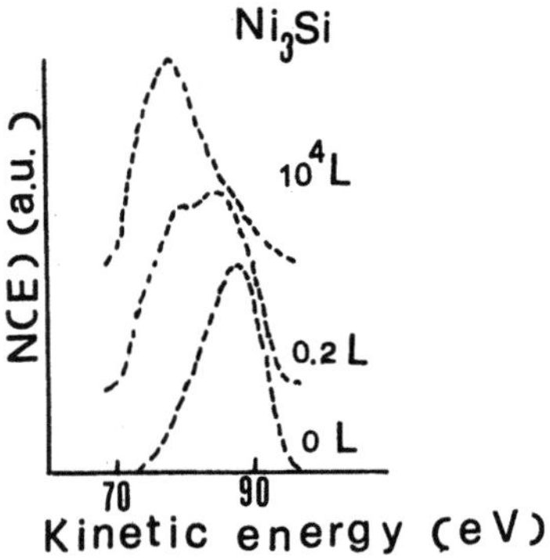

Fig. 1: Normalized Si(LVV) Auger in Ni_3Si in clean sample and at two different exposures

The changes in the Si(LVV) Auger lineshape under oxygen exposure (fig. 1) indicate an involvement of the Si valence band in the oxygen interaction, which is in silicides than in pure Si and increases in Ni silicides with higher metal content. In general, no changes in shape and position occur in the core-valence-valence metal Auger lines, indicating that the metal valence band is not involved. However, in the Ni rich compounds there is evidence in the Ni(MVV) line of an onset of metal oxidation at exposures >100 L.
Ni,Pd and Pt silicides of comparable metal content (M_2Si) exhibit a very different oxidation behaviour, reflected in the Si oxidation state, in the number of Si surface atoms involved in oxygen interaction and in the oxygen uptake intensity. The metal catalytic effect is stronger in Ni_2Si than in Pd_2Si, and is very weak in Pt_2Si: the oxidation in Pt silicides is similar

Table 1: Comparative effects of 10^4 L oxygen exposure on M_2Si near-noble silicides and pure Si. Pox is the percentage of Si(LVV) signal coming from oxidized Si atoms in the probed layer of the surface. The shift refers to the distance between the main structure of the spectrum of the clean sample and the main one of the difference spectrum between oxidized and clean sample

SAMPLE	H_{pp} O(KLL) [a.u.]	Si(LVV) SHIFT [eV]	P_{ox}
Ni_2Si	100	10.5	77
Pd_2Si	60	10.5	63
Pt_2Si	31	7.0	40
Si	28	7.0	32

to that of pure Si (table 1). XPS results are consistent with the Auger ones. In particular the Si 2p core line shows a shoulder on the high binding energy (BE) side (fig. 2) which evolves both in intensity and BE shift at increasing Ni content and oxygen exposure. We can refer to the BE of Si suboxides peaks (Si_2O,SiO,Si_2O_3) and of SiO_2 peak quoted in the literature [3,4] to identify this shoulder as a convolution of four different contributions, whose relative weight changes with metal content and oxygen exposure. Our results allow to establish the following points: (i) the Si oxidation in Ni silicides is enhanced over that of pure SI at all metal concentrations; (ii) this enhancement depends on Ni content, being higher in Ni rich silicides both for the number of Si atoms involved and for the oxidation state; (iii) while in general the oxygen bonds with Si leaving the metal unoxidized, evidence for the occurrence of Ni-oxygen bonds at high exposures is found for Ni rich compounds; (iv) the Si oxidation enhancement is higher in Ni silicides than in Pd and Pt silicides. Several explanations were proposed for the metal catalyzed enhancement of Si oxidation [1] . The driving force was suggested to be proportional to the difference between the heats of formation of SiO_2 and of silicides.

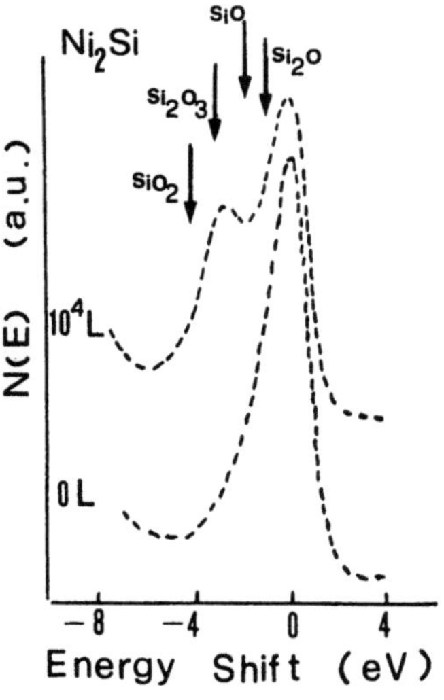

Fig. 2: Si 2p core line of clean and oxidized Ni_2Si; the zero of the energy scale has been taken to coincide with the position of the Si 2p core before the oxygen exposures

This thermodynamic approach accounts for the Si preferred oxidation with respect to the metal one in silicides, being the heat of formation of SiO_2 higher than that of near-noble metal oxides. On the other hand it doesn't account for the abserved trend versus metal concentration,and for the different behaviour of silicides of different near-noble metals.

On the basis of our results,we are inclined to refer the catalytic effect of metal atoms to their ability to break the O_2 molecule, so that atomic, more reactive oxygen is available at the silicide surface to react with Si atoms. At RT, O_2 is dissociatively chemisorbed on Pt,Pd and Ni with a sticking coefficient of about 0.1, 0.4 and 1.0 respectively [5,6] , in reasonable agreement with our results concerning the oxidation behaviour of the different near-noble metal silicides. This effect, at a first approximation, is also proportional to the number of dissociation sites (i.e. metal atoms sites) and so accounts for our results about Ni silicides of different metal content, too.

REFERENCES

1) S.Valeri,U. delPennino,P.Lomellini and P.Sassaroli; to be published on Surface Sci., and references quoted therein
2) S.Valeri,U. delPennino,P.Sassaroli; Surface Sci. 134, L537 (1983)
3) F.J.Grunthaner,P.J.Grunthaner,R.P.Vasquez,B.F.Lewis, J.Mserjian and A.Madhukar; J.Vacuum Sci. Technol. 16(5), 1443 (1979)
4) G.Hollinger and F.J.Himpsel; Phys.Rev. B28(6), 3651 (1983)
5) P.W.Davies and R.M.Lambert : Surface Sci. 110, 227 (1981)
6) B.E.Nieuvenhuys: Surface Sci. 126, 307 (1983)

Part V

Optical Properties

Raman and Infrared Spectroscopy of Layered Structures

J.D. Swalen
IBM Research Laboratory, San Jose, CA 95193, USA

Recent advances in Raman and infrared spectroscopy of thin films have led to new and better ways to probe layered structures of dielectrics, that is, organic and inorganic layers, perhaps even including metals. The characterization of specific groups present, their packing and orientation, and interfacial interactions, can many times be done from the observed spectra. Thin layers, however, present a problem of sensitivity because of the small number of absorbing species present and hence, a low signal to noise ratio. Various optical techniques have been developed to enhance the optical fields at a specific interface or in a specific thin film layer. Some of these techniques and results to be discussed are infrared absorption at glancing angles, guided optical waves using integrated optical structures, and surface electromagnetic waves, such as surface plasmons.

First, some mathematical details important for thin films and interfaces in order to derive dispersion curves and mode structures including the optical field configurations will be covered. Then the preparation of thin films of polymers and of long chain fatty acids and their salts, called Langmuir-Blodgett or simply L-B films, will be briefly explained. Recent experiments with glancing angle infrared spectroscopy of thin films of L-B monolayers and polymers has shown that films as thin as 25 A can be observed. Raman spectroscopy of thin films and surfaces as part of waveguide structures has been used to probe monolayers and thin layers of polymers on optical guides. Different optical modes sample different regions of laminate structures consisting of two layers, one in contact with another. Surfaces supporting surface electromagnetic waves are very sensitive to small changes in the dielectric function or changes in structure at a surface. Techniques for measurement and analysis of some recent SEW experimental results will be given.

1. Introduction

Vibrations of molecular or ionic species give in their infrared or Raman spectra signatures of the species present and their orientation, determined by the polarization of the radiation and analyzing optics. The difference between the two methods is that infrared spectroscopy measures the absorption of radiation, while Raman measures inelastic scattering of photons by a molecule, the difference being the energy difference between two vibrational energy levels. Selection rules are different for these two types of spectroscopy and as such, the combination gives more information. All this is well known, and these spectroscopic techniques have been used for a number of years to measure the bulk optical properties of materials in the phonon region.

Recently it has become apparent that these techniques could be used to probe surfaces. This is especially true with the development of Fourier transform infrared spectroscopy (FTIR) and laser Raman scattering. The discovery and much heralded prospects for surface enhanced Raman spectroscopy certainly sparked enthusiasm for its application. At the present time the understanding does not, however, appear to be sufficient, in our opinion, for its systematic use and application. Maybe in the near future we will see this approach refined and routinely used. On the other hand, recent advances in optical and electronic techniques to be described here make surface spectroscopy a reality.

Basically, molecular and ionic vibrations can be sensitive to changes in local environment and can be used to observe surfaces, that is, changes in frequency, intensity, or polarization giving indications of intermolecular interactions. Their interpretation is many times neither easy nor straightforward. Nevertheless, they offer the hope that perturbations from a surface near a vibrating species can be detected. Interest in new applications of thin films has added a strong driving force for their investigation. Some of these are insulating layers in integrated circuits, photochemical and electron beam resists, integrated optics, photoactive layers, piezoelectric sensors, electroluminescent films, antireflection coatings, second harmonic generating layers, and adhesive or lubricating layers.

Unfortunately, when one is presented with a film a few nanometers thick,it can become quite difficult to measure and to understand what is taking place at the interface. The number of species absorbing or scattering at the surface is small and many high vacuum techniques, though very sensitive, may not be applicable. We hope to show that both infrared and Raman methods can successfully address some surface problems. To accomplish this,we took advantage of glancing angle reflection and the techniques developed for integrated optics and surface electromagnetic waves. Our purpose has been to concentrate the radiation at the interfacial region, the region of interest for any surface interaction. Without these, or similar methods, the observations are either too weak to observe or else masked by effects due to bulk species not at the interface.

Before we approach the actual experimental arrangements and the new results on thin films, some preliminary mathematical details will be outlined, some methods of film preparation illustrated and the procedures for optically measuring films explained. This latter case involves the determination of the optical constants, i.e., the index of refraction which can be complex, and the film thickness. These are all needed for both spectroscopic evaluations and for any linear or non-linear effects. Frequently films and/or substrates are rough,causing scattering losses or absorption, attenuating the light propagation. Detailed analysis can then be quite difficult,and some of the methods used to resolve these situations will be illustrated.

2. Mathematical Considerations

For the spectroscopic observation of a specific region of a thin film or surface,we must know the optical field configuration for a given polarization. This requires some understanding and manipulation of Maxwell's equations. Fortunately, much of the details are contained in many references; see, for example, references 1 - 3. There are several points, however, which need some clarification and explanation for their use in the analysis of thin films. From matching the E and H fields for a traveling wave $e^{i(\omega t - k_z z - k_x x)}$ at some angle θ at an interfacial boundary, we can derive the well known Fresnel reflection and transmission coefficients. Normally these are given for real constants, i.e., nonabsorbing media. For our problems dealing with metals and absorbing films,it is desirable to recast them into a form convenient for complex parameters. We do this with the complex propagating vectors, k_i's, which include absorption as well as propagation:

$$r_{12} = \frac{1-\beta_{12}}{1+\beta_{12}} \quad \text{and} \quad t_{12} = \frac{2}{1+\beta_{12}} \,. \tag{1}$$

Here β has different forms for the two polarizations: labeled s ("senkrechte") or TE (transverse electric) or $\perp$ and labeled p (parallel) or TM (transverse magnetic) or $\parallel$. These are the two different polarizations with respect to the plane of incidence.

$$\beta^s_{12} = \frac{k_{z2}}{k_{z1}} \quad \text{and} \quad \beta^p_{12} = \frac{k_{z2}/e_{x2}}{k_{z1}/e_{x1}} \,. \tag{2}$$

The orientation of the axes are such that z is normal to the surface, x is in the forward propagating direction, and y is perpendicular to the plane of incidence. The propagation vector in the x direction is conserved and set by an angle in a prism or on a grating. The vectors in the z direction across the film and surface structure are

$$k_z^2 = \varepsilon_y k_o^2 - k_x^2 \qquad \text{(for s polarization) and} \tag{3}$$

$$k_z^2 = \varepsilon_x k_0 - \varepsilon_x k_x^2/\varepsilon_z \qquad \text{(for p polarization).} \tag{4}$$

Note that the vector for TM waves involve two dielectric functions,and they can be different in anisotropic materials. Frequently films by their very nature of formation become anisotropic, especially between the in-plane direction and the direction normal to the surface. The importance of these k_z vectors is that,with the distance in the z direction,they give the exponential factors for the electric and magnetic field across a thin film structure. For example, for a three layer system with layers 1, 2 and 3 the electric field will vary as

$$\frac{(e^{ik_{z1}z} + r_{12}e^{-ik_{z1}z})(1 + r_{23}e^{-2i\delta_2})}{1 + r_{12}r_{23}e^{-2i\delta_2}} \qquad \text{(layer 1),} \tag{5}$$

$$\frac{t_{12}(1 + r_{23}e^{-2i\delta_2})}{1 + r_{12}r_{23}e^{-2i\delta_2}}\, e^{-ik_{z2}z} \qquad \text{(layer 2) and,} \tag{6}$$

$$\frac{t_{12}t_{23}e^{-i\delta_2}}{1 + r_{12}r_{23}e^{-2i\delta_2}}\, e^{-ik_{z2}(z-d_2)} \qquad \text{(layer 3),} \tag{7}$$

where t_{12} is the transmission from the first layer to the second t_{23} is the transmission from the second layer to the third and r_{12} and r_{23} are the reflectivity coefficients, respectively. Here δ $(= k_{z2}d_2)$ is the phase shift from the second layer and $z = 0$ is defined at the interface between layer 1, where $z < 0$, and layer 2, where $z > 0$. Remember that the k_z's, r's, and t's can each be complex,so that this expression must be evaluated with complex arithmetic. The extension to multilayer systems is straightforward. For each layer the evaluations of the total r and t are done for all the combined layers above the specific layer,and the combined total r is evaluated for the all layers below the layer in question. The field for each layer is subsequently calculated by a generalization of (6) to generate the whole field configuration.

Another point worth stating is that the optical constants for each layer can be cast in a matrix form,from which total reflectivity and transmission coefficients can be generated by matrix multiplication. This also allows the group of layers above and the group below to be conveniently lumped and considered as a single layer. The general form of this matrix was written by Heavens [2] in a convenient form,but other authors, for example, Born and Wolf [1] have written similar expressions:

$$\frac{1}{t_{ij}} \begin{pmatrix} e^{i\delta_j} & r_{jk}e^{i\delta_j} \\ r_{jk}e^{-i\delta_j} & e^{-i\delta_j} \end{pmatrix} . \tag{8}$$

Here the first row of the matrix is for the forward propagating wave,and the second row is for the backward propagating wave. The matrix product of the layers multiplied on the right by the transmitted wave gives the incident and reflected wave. Normally the transmitted backward propagating wave is zero. The composite reflectivity coefficient follows from the complex division of the 2,1 element by the 1,1 element of the product matrix and the composite transmission coefficient follows from the complex product of each individual transmission coefficient divided by the 1,1 element. This procedure we have found to be particularly adaptable to computer programming in a variety of different high level languages. Hence,we use it extensively for evaluating the optical properties of a layered structure,or by a non-linear least squares technique we search the parameter space for the "best fit" to some optical observed values, such as optical waveguides or surface plasmons. These will be treated more completely later in this paper.

3. Organic Thin Film Preparation

Thin films of inorganic, semiconductor or metallic materials are usually prepared by thermal evaporation, electron beam evaporation, or sputtering. Chemical vapor deposition and electrochemistry are also used. These techniques and methods are also applicable to organic films, but in addition other methods, such as spraying, extruding, casting, coating, or spinning can be used. For example, most of the polymeric films we have studied were made by either spinning or "doctor blading". In the first method drops of a solution containing a polymer in low concentration are slowly dropped onto a substrate, spinning at some 2000 to 5000 rpm. The solution spreads by centrifugal force and the solvent evaporates, leaving a thin polymer film. The second method uses a knife edge to spread a solution of a polymer across a substrate. Here both viscosity of the solution and the separation of the knife edge from the substrate, called the "wet gap" are important factors leading to the final film thickness. In contrast, viscosity is the main determining factor in the case of spinning. Now spinning generally produces films in the range of several nanometers while "doctor blading" produces films usually in the micron range of thickness. With regard to solvent, one should not select one which evaporates too rapidly,because this leads to an "orange peel" effect caused by the surface drying with solvent trapped in the film. This solvent then diffuses from the film randomly and causes a rough surface. Sometimes drying films in the presence of a high concentration of solvent vapor can help to eliminate this "orange peel" effect. A better way is to use a less volatile solvent. Following these simple principles we have made good optical quality films of poly (methyl methacrylate) (PMMA) and poly (styrene) from chlorobenzene solutions and poly (vinyl alcohol) from water solutions.

Another class of thin, i.e., very thin, organic films are those of monolayer assemblies of long chain fatty acids, their salts, and related type compounds; see references 4 and 5 for more details. These are called Langmuir-Blodgett films, or simply L-B films, named after the two well known early investigators of these materials. Deposition is by transfer from a water surface or possibly another liquid surface, though water is preferable. Up to this time they have not seen much practical use,but many are considering various technological applications [6]. They have been, however, used as accurate spacers,and as such used to probe the spatial dependence of energy-transfer and charge-transfer processes [4]. Their preparation is based on their unique molecular properties. One end of the molecule is a long chain hydrocarbon and soluble in organic solvents but not water; this effect is referred to as hydrophobic. The other end of the molecule has a functional group more soluble in water,and exhibits a hydrophillic effect. When these molecules are placed on a water surface by carefully adding drops of it in a solvent which subsequently evaporates, the hydrophillic end contacts the water and the hydrophobic end floats above the water. Compression of this surface film by a float or weir leads to a close packed and aligned, probably randomly oriented, structure. Lowering and raising a clean hydrophobic substrate through the interface (or raising and lowering a clean hydrophillic substrate through the interface), keeping a close to constant surface pressure, leads to a transfer of two monolayers, one for each direction of travel. These layers are about 3 nm or slightly less in thickness, the exact value depending on the specific molecule. Temperature, pH, and cleanliness are crucial factors for good film preparation,and are really the key to good L-B film growth. Much of their properties have been determined by the spectroscopic techniques described below,as well as those described in reference 4.

4. Infrared Spectroscopy of Thin Films

It was pointed out by Greenler [7] in 1966 that the infrared spectra of thin films could be observed by measuring the reflectivity of a thin dielectric film on a highly reflecting metal. The earlier works were somewhat limited with dispersion spectrometers. Fourier transform infrared (FTIR) instrumentation has given many improvements in signal collecting abilities,and has led to some of the advances reported here. Greenler further showed that by a consideration of the phase-shifts at a metallic surface,the optical electric field for s polarized light would be significantly reduced,because the reflected wave would almost cancel the incident wave. For p polarized waves the same is true at grazing angles, but some few degrees away from grazing the phase difference between the incident and reflected beam could be such that only a partial cancellation would occur and, in fact, absorption signals could be detected with reasonable signal to noise ratios. Greenler did a numerical calculation for the Fresnel coefficients of reflection for a thin organic layer on a highly reflecting metal, like silver. The reflectivity ratio $(R_0-R)/R_0$, where R_0 is the reflectivity of bare metal and R is the reflectivity of the metal overcoated with a thin film,was calculated and found that this ratio has a maximum at about 88°, very close to grazing. We have verified this fact for reasonable optical parameters for thin films in the infrared region of the spectrum. Two other factors seemed relevant [8], especially when experimentally attempting to record spectra at an angle of 88°. Calculations of the maximum optical field showed that a maximum occurred at about 80°, not 88°. Further with an FTIR, the beam is approximately circular in cross-section. Filling factors must be considered, i.e., at shallow angles the beam fills $w/(\ell \cos \theta)$, where w is the width of the beam, ℓ is the length of the sample and θ is the angle of incidence. At steep angles the beam misses part of the sample and the factor is the inverse. Clearly,when this factor equals unity the maximum situation exists. This also occurs at about 80° for the standard conditions we considered. The third factor is the sampling time. As the angle increases the signal levels drop and one must collect scans for much longer times. In fact,in one experiment to test these ideas,a point at 84° (we never reached 88°) required some 6 hours of signal accumulation. Greater grazing angles would seem to be impractical. Hence,we have concluded that measurements at about 80° are about optimal considering the optical field, the filling factor such that the whole sample is irradiated, and the collection time for signal averaging is not unreasonable. Another point which we noted is that different metals have slightly different optimal angles. Simply, the smaller the imaginary part of the dielectric constant the closer the optimal angle is to grazing incidence. For example, silver is optimal at about 88° while chromium, a slightly poorer reflector, is optimal at 84°. One might be drawn to the conclusion that using a poor reflecting metal would be better because its angle is more optimal. This is not the case. The better reflecting metals are better substrates for this grazing angle infrared spectroscopy.

We have measured the infrared spectra of several samples of PMMA on various metal substrates, for example silver, aluminum, copper and chromium [8]. Only in the last case did we see any wavelength shifts, indicative of a chemical/physical surface reaction. Other evidence has been presented to substantiate this situation. Currently our record [9] for the thinnest film is 25A of PMMA on silver; the infrared spectrum is shown in the figure below.

Infrared reflection and transmission measurements [10] on monolayers of cadmium arachidate, the cadmium salt of arachidic acid (a long chain fatty acid of 20 carbon atoms) have shown from the spectral assignment that the molecules are oriented to within 5° of the normal. This is in contrast with that reported by Chollet [11],where the acid was found to be inclined at an angle of 25° ± 5°. Two molecules of cadmium arachidate per unit cell was confirmed by a splitting in the $C-H_2$ wagging and rocking modes. This could mean that either we have a orthorhombic or monoclinic structure for a microcell of this material. The small angle, however, is consistent only with the orthorhombic structure. Other molecules, such as tetradecyl benzoic acid, do not exhibit this splitting of the modes,indicating that the symmetry is lower.

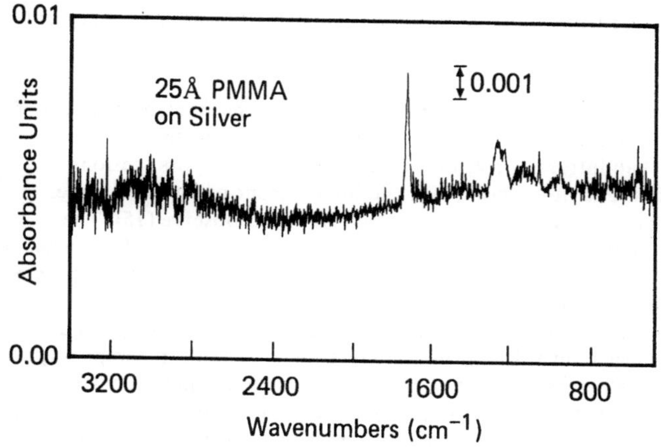

Raising the temperature while recording the spectrum,we interestingly observed [12] that melting occurs in two sequential steps: First at about 60°C the chains start to move as indicated by changes in the C-H stretching region. The groups almost parallel to the surface begin to exhibit a sizable component perpendicular to the surface, i.e., the chains are becoming mobile and the orientations of the chains are modified. On the other hand, in the carbonyl region of the spectrum,we see that the heads are still fixed to the substrate or the next adjoining layer, i.e., the lattice is still intact. Only the chains are exhibiting motion. Finally, the head groups melt at a temperature over 100°C., the known melting point of the bulk compound. This type of behavior is, no doubt, important for diffusion through membranes in biological systems.

5. Raman Spectroscopy with Optical Guided Waves

Integrated optics techniques, developed for communications application, can be adapted to study thin films. Initial investigations showed that this method was convenient and accurate for determining film thickness and refractive index, at a specific wavelength [3,12,13]. A corollary to this is that by making measurements at a series of wavelengths, one, in effect, measures the dispersion of the real part, $n(\lambda)$, and the absorption spectrum, $k(\lambda)$, from the imaginary part of the index of refraction [14]. Experimentally,light is coupled into a optical waveguide conveniently by a prism or a grating. A change in angle changes the propagation vector so as to match those for the propagating modes at a given polarization, i.e., TE or TM. These modes for non-absorbing films are defined as solutions to the eigenvalue equation:

$$\tan^{-1}\beta_{21} + \tan^{-1}\beta_{23} + m\pi = k_{z2}d_2 \ . \qquad (9)$$

The three terms of this equation represent half of the phase-shift at each of the two interfaces and the phase of the light in transversing the film (layer 2); the superstrate is labeled 1 and the substrate 3. When these combine to be a multiple of π, a mode can propagate, the number being how many multiples of π can be supported. There is a minimum thickness below which no modes are supported,and this minimum thickness is roughly about a third of the wavelength of light used for propagation, i.e., a fraction of a micron for visible light. The solution of the above eigenvalue equation can be obtained in a number of ways,but we have found that one of the best ways is to assume a value for the index of refraction of the film slightly larger than expected. The film thicknesses calculated for the modes will, undoubtedly, be all different. Slowly reducing the magnitude of the index of refraction, we reach a minimum in the sum of squares of the deviation from the average thickness. We call this thickness at minimum deviation the "best fit", and for good films it can be quite small - a small fraction of a percent of the thickness.

Knowing the optical properties of a film, we are now ready to measure the Raman spectrum of the whole film, of a layer on the surface or from a layered region within the film. Raman cross-sections are much smaller than infrared or visible absorption spectral cross-sections because it is a three photon process and quite weak,with the involvement of a virtual state. Hence,our first goal was to determine how thin a polymer film we could observe [15]. The situation is not quite as bad as it may seem,because the waveguide geometry does offer some distinct advantages. The optical fields are higher, because we have coupled the laser light from a fraction of a millimeter in one direction to a few microns. A large number of sites are irradiated for Raman scattering - those along the few centimeter path length of the guided wave. This scattering streak is imaged onto the slit of a monochromator, and can completely fill it. In addition, the quality of detectors is continually improving,and computer control of the spectrometer allows the averaging of many scans, improving the signal to noise ratio. As a consequence,we were able to measure the ring modes of a 80 nm film of poly (styrene) on a glass (Corning 7059) waveguide with an argon ion laser operating at 488nm. The signal to noise ratio was about 4:1. The limit appears to be about a 20 nm film.

Next,we measured the Raman spectrum of a L-B monolayer of a cyanine dye molecule on a glass waveguide [16,17]. Besides the enhancement from the waveguide technique,we took advantage of the resonance Raman effect from dye itself. With this added feature of being close to an absorption band but not absorbing too much radiation,we bring a real electronic state close to the virtual electronic state to increase the cross-section. Not only could we observe the ring modes of the cyanine molecule,but we observed a wavelength shift of several wavenumber units. This, we believe, is an indication of some interaction with the glass surface, e.g., hydrogen bonding. A subsequent experiment with poly (vinyl alcohol) instead of glass exhibited a similar shift,and here we know there are many OH groups. Although not proven, there is a strong supposition that hydrogen bonding between the cyanine ring containing a nitrogen atom and the OH groups is taking place. Finally, observations made with each polarization (TE and TM) showed the that electronic transition dipole moment of the interacting group is oriented parallel to the plane of the film. Raman scattering from the film with TM polarization was extremely weak compared to that from TE polarized radiation. Hence,these two related experiments show that both interfacial interactions exist,and that the orientation of functional groups can be determined.

Not only can one study the air surface of a film,but also it is possible to observe the interfacial region between two layers, both of which are optically guiding,or when one is guiding and the other has evanescent waves [18]. By forming a two-layered film of poly (styrene) and poly (vinyl alcohol), the light guided ($m = 0$ mode) first in the poly (styrene) layer and we observed its spectrum. The second mode ($m = 1$) switched to the other layer,and guided predominately in the poly (vinyl alcohol) and the spectrum was, therefore, that of poly (vinyl alcohol), with only a little trace of the spectrum from poly (styrene). The higher order modes exhibited spectra from both layers in superposition. By calculating the optical field contours,we could see the reasons for our observations. It still is surprising that the light switches from one layer to the another with little radiation in the opposite layer. It had been our first thought that the optical field would spread over the two layers as if it were a single guiding layer with variation of index. This was clearly not the case. Mathematical deconvolution of our signals and subtraction techniques showed that some scattering came from interfacial species. This method should be useful for their investigation and related to the physical and chemical bonding between two layers.

6. Raman Scattering at an Surface by Surface Plasmons

Although optical waveguides have many advantages for spectroscopy, plasmon surface polaritons also have advantages, some of which are distinct and unique. As such,we have used both approaches depending on the specific situation. Matching E and H fields for radiation at an interface,we see that surface electromagnetic waves require both a dielectric function which is strongly negative in some spectral region and a p (TM) polarized wave.

For s polarization

$$k_{z1} = - k_{z2} , \tag{10}$$

so that the only solution is $k_z = 0$, i.e., there is no propagation.

While for p polarization

$$\frac{k_{z1}}{\varepsilon_{x1}} = - \frac{k_{z2}}{\varepsilon_{x2}} . \tag{11}$$

Of course, highly reflecting metals, such as silver, gold, aluminum, or copper satisfy this relation over most of the visible spectral region. Other metals, such as the alkali or alkaline earths, also are appropriate, except they require more care as to environment. Assuming isotropic materials,and substituting in the k_z values given above, we obtain the dispersion relation for surface plasmons.

$$k_{sp}^2 = k_0^2 \left(\frac{\varepsilon_1 \varepsilon_2}{\varepsilon_1 + \varepsilon_2} \right) . \tag{12}$$

From the Drude expression for the dielectric function of a metal

$$\varepsilon = 1 - \frac{\omega_p^2}{\omega^2} , \tag{13}$$

we find the asymptotic limit for the surface plasmon energy at a metal-air interface as

$$k_{sp} \rightarrow \infty \text{ when } \omega_{sp} \rightarrow \frac{\omega_p^2}{\sqrt{2}} . \tag{14}$$

Surface electromagnetic waves (surface plasmons, surface polaritons, and surface excitons) are, as such, very sensitive to the optical properties adjacent to the surface, e.g., overcoatings a few nanometers result in an angular shift of the order of a degree and are easily observable [19,20] Most measurements are done in reflection with a prism or grating,and close to total attenuation of the reflected radiation is possible at the angle matching the propagation vector (k) for SEW's. Absorption in a film leads to an angular shift, a broadening of the reflectivity versus angle curve, and a reduction in the depth of the minimum, i.e., assuming close to optimal conditions exist [21]. If the metallic film is too thin,then adding another absorbing material could increase the depth of the reflectivity minimum - a condition which we have not seen in all our experiments because we attempt to make metallic film close to optimal.

Raman scattering from a metallic surface capable of propagating surface electromagnetic waves has been proposed as a mechanism for surface enhanced Raman scattering (SERS) [22,23]. In order to explain all the data,some other close range mechanisms, in addition to this long range mechanism of SEW's, are required. No doubt, they contain some specific molecular-metal interactions. Possibly even more than two mechanisms or effects are operative. Nevertheless the high optical fields at the surface of a metal from surface electromagnetic waves can lead to significant Raman scattering. For example, a 10 nm film of poly (styrene) is clearly observable at the surface plasmon angle,and shows a strong angular dependency following the surface plasmon [24]. This is about 30 times greater then comparable waveguide results.

L-B monolayers have also been investigated by surface plasmons,and they show a strong amplification from coupling energy from the surface mode to the Raman mode. In fact,as one changes angle the surface mode moves through the various Raman bands, adding increased intensity as it goes [25]. Intensity increases with film thickness,and follows the optical field increases. Also certain regions can be shown to be forbidden. Coupling to molecular fluorescence can also occur [26]. Here,as the surface wave energy passes through fluorescence bands, each are increased in intensity. Energy is exchanged between the surface mode, the molecular levels, and those of the metal. Clearly these transfers, as with the quenching of fluorescence and Foerster type dipole-dipole energy-transfer, can be used both for informative and practical uses.

7. Conclusion

By the examples shown, albeit organic films, I think one can easily see that surface vibrational spectroscopy can be an effective tool to unravel some of the complexities of surface interactions. Not only can we solve interesting scientific problems relating to molecular interaction at an interface,but these techniques hold considerable promise for solving some key technical problems involving surfaces.

8. Acknowledgements

I wish to thank the many colleagues who have collaborated on much of the work: J. F. Rabolt, N. E. Schlotter, M. R. Philpott, W. Knoll, and M. Jurich.

9. References

1. M. Born and E. Wolf, Principles of Optics, (Pergamon Press, Oxford, 1975).

2. O. S. Heavens, Optical Properties of Thin Solid Films, (Dover Publications, Inc., New York, 1965).

3. J. D. Swalen, J. Phys. Chem. 83, 1438 (1979).

4. H. Kuhn, D. Mobius, and H. Bucher, "Spectroscopy of Monolayer Assemblies", in Physical Methods of Chemistry, A. Weissberger and B. W. Rossitier, Ed., Part IIIB, Wiley-Interscience, New York, 1972, Chapter VII.

5. G. L. Gaines, Insoluble Monolayers as Liquid-Gas Interfaces, Interscience, New York, 1966.

6. P. S. Vincett and G. G. Roberts, Thin Solid Films 68, 135 (1980)

7. R. G. Greenler, J. Chem. Phys. 44, 310 (1966).

8. J. D. Swalen and J. F. Rabolt, "Order in Thin Films by Fourier Transform Infrared Spectroscopy", in Fourier Tranforms Infrared Spectroscopy: Applications to Chemical Systems, J. R. Ferraro and L. J. Basile, Ed., Academic Press, New York, 1985.

9. J. F. Rabolt M. Jurich and J. D. Swalen (to be published).

10. J. F. Rabolt, F. C. Burns, N. E. Schlotter, and J. D. Swalen, J. Chem. Phys. 78, 946 (1983).

11. P. Chollet, Thin Solid Films 52, 343 (1978).

12. J. F. Rabolt and J. D. Swalen,to be published

13. J. D. Swalen, M. Tacke, R. Santo, and J. Fischer, Opt. Commun. 18, (1976).

14. J. D. Swalen, R. Santo, M. Tacke, J. Fischer, IBM J. Res. Dev. 21, 168 (1977).

15. J. F. Rabolt, R. Santo, and J. D. Swalen, Appl. Spectrosc. 34, 517 (1980).

16. J. F. Rabolt, R. Santo, N. E. Schlotter, and J. D. Swalen, IBM J. Res. & Dev. 26, 209 (1982).

17. J. F. Rabolt, N. E. Schlotter, J. D. Swalen, and R. Santo, J. Polymer Science - Polymer Phys. Ed. 21, 1 (1983).

18. J. F. Rabolt, N. E. Schlotter, and J. D. Swalen, J. Phys. Chem. 85, 4141 (1981).

19. I. Pockrand, J. D. Swalen, J. G. Gordon II and M. R. Philpott, Surf. Sci. 74, 2324 (1978).

20. J. D. Swalen, J. G. Gordon, II, M. R. Philpott, A. Brillante, I. Pockrand, and R. Santo, Am. J. Phys. 48, 669 (1980).

21. I. Pockrand, J. D. Swalen, R. Santo, A. Brillante, and M. R. Philpott, J. Chem. Phy. 69, 4001 (1978).

22. M. R. Philpott, J. Chem. Phys. 62,1812 (1975).

23. J. C. Tsang, J. R. Kirtley, and T. N. Theis, Solid State Commun. 35, 667 (1980).

24. A. Girlando, M. R. Philpott, D. Holtmann, J. D. Swalen, and R. Santo, J. Chem. Phys. 72, 5187 (1980).

25. W. Knoll, M. R. Philpott, J. D. Swalen, and A. Girlando J. Chem. Phys. 77, 2254 (1982).

26. W. Knoll, M. R. Philpott, and J. D. Swalen, J. Chem. Phys. 75, 4795 (1981).

Brillouin Scattering in GaAs-$Ga_{1-x}Al_xAs$ Superlattices

J. Sapriel

Laboratoire de Bagneux, C.N.E.T., 196 rue de Paris, F-92220 Bagneux, France

This is a general presentation of light interaction with acoustic waves in GaAs-$Ga_{1-x}Al_xAs$ superlattices. Two kinds of vibrations are considered : the Rayleigh surface waves propagating parallel to the free surface, and the bulk acoustic waves with wavevectors perpendicular to the layers of the superlattice. The folding of the longitudinal acoustic branch allows the probing of bulk acoustic waves up to 1000 GHz. Two theoretical approaches are used for the study of the vibrational modes in superlattices : the theory of elasticity and a linear diatomic chain model, only the former being developed here. This lecture contains a description of the high contrast and high resolution Brillouin scattering set up used for the investigation of acoustic waves in these semiconductor structures. A brief insight on light scattering by modes of the optical branches is also given.

I - INTRODUCTION

Superlattices structures are prepared by alternately depositing predetermined thickness of (at least) two different compounds, thus creating an artificial period in the direction perpendicular to the layers. We are concerned here with the vibrational modes of these systems, mainly the acoustical phonons and their interaction with light.

The "superlattice effect" actually occurs when there is an interaction between the atoms belonging to the different layers. It may consist of modification of the elastic constants, shift of the phonon frequencies and occurrence of additional modes due to the introduction of a new period in the direction of growth z.

The investigations are limited to the available GaAs-$Ga_{1-x}Al_xAs$ superlattices though many results of the study can be generalized to other systems. They were made of layers of several tens Angstroms thickness of GaAs and $Ga_{1-x}Al_xAs$ and x ranged between 0.25 and 0.4.
Two kinds of vibrations have been particularly investigated : the bulk phonons belonging to the longitudinal acoustic branch along z and the vibrations corresponding to acoustic surface waves of the Rayleigh type propagating parallel to the free surface of the semi-infinite superlattice.

II - BULK VIBRATIONS IN GaAs - $Ga_{1-x}Al_xAs$ SUPERLATTICES

A - General features of the phonon dispersion curves

In the GaAs-AlAs superlattices, one can distinguish two ranges of phonon frequencies in the folded Brillouin-zone in the z direction (see the diagram of Fig.1). The first one is composed of dispersion branches, whose frequencies fall in the range of the acoustic branch of GaAs and AlAs. These dispersion curves are called the "folded acoustic branches". The second one contains the branches originating from the optical phonon branches, which are clearly separated in GaAs and AlAs. If one considers a superlattice of N_1 monolayers of GaAs and N_2 mono-

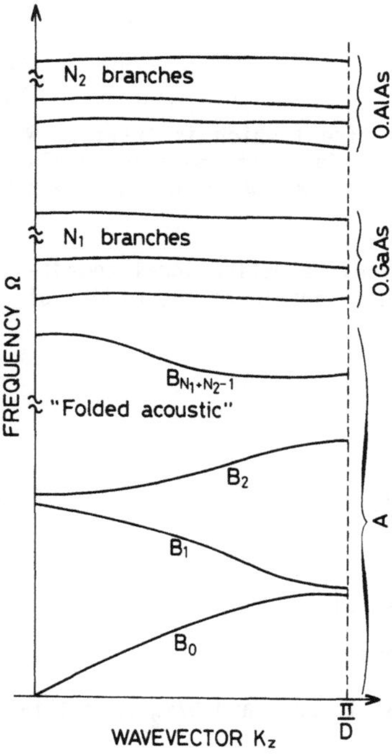

Fig.1: Schematic representation of the bulk phonon dispersion curves in the folded Brillouin-zone along the direction of growth of a superlattice which contains N_1 monolayers of GaAs and N_2 monolayers of AlAs in the period D. The results are obtained from a linear chain model [1] assuming the same elastic constant for GaAs and AlAs. One can distinguish 3 ranges of frequency. The folded acoustic branches B_1, B_2,.. $B_{N_1+N_2-1}$ fall in the frequency range A of the acoustic branch of GaAs and AlAs. Bo is the acoustic branch of the superlattice ($\Omega \to 0$ for $K \to 0$). O-GaAs and O-AlAs are the frequency range of the optical branch in GaAs and AlAs respectively. In A the branches are dispersive and separated one from the other by small gaps at K=0 and K=π/D, though the branches belonging to O-GaAs and O-AlAs are very flat. The elasticity theory gives the same results as the chain model for the low-frequency folded acoustic branches, provided that N_1 and $N_2 > 5$.

layers of AlAs in the period there are N_1+N_2-1 folded acoustic branches, N_1 modes of the GaAs type and N_2 modes of the AlAs type. The lowest frequency branch ($\Omega \to 0$ for $K \to 0$) is the acoustic branch of the superlattice.

The features displayed on Fig.1 are obtained with the help of linear diatomic chain [1] model along the z axis. Such a model is suitable since in the zinc-Blende cubic structure to which GaAs and AlAs belong, all the atomic planes perpendicular to z are equidistant by a/2, 2a being the edge of the cube (a=2.83Å for GaAs and AlAs). Moreover, each of them is occupied by the same element (Ga, As or Al). A comparative study of two methods of determination of the dispersion curves of bulk waves have been performed, one based on the elasticity theory and the other derived from the linear-chain model. It was established [1] that for a number of monolayers of GaAs and AlAs larger than 5 per period, the two models give the same values for the dispersion curves as long as the lowest energy "folded acoustic branches" are considered. As the present study is mainly focused on long period superlattices, we shall develop the elastic model which is easier to carry out.

For alternating (001) layers of GaAs and $Ga_{1-x}Al_xAs$ of symmetry $\bar{4}3m$, the point group of the superlattice is tetragonal $\bar{4}2m$. Group theory shows that for the wavevector parallel to [001], there is a complete separation of transverse and longitudinal vibrations belonging to different representations.

B - <u>The theory of elasticity and the folding of the longitudinal branch</u> [1-3]

Let us consider a superlattice (Fig.2) made of a medium 1 of thickness d_1, density ρ_1 and a medium 2 (d_2,ρ_2). For the longitudinal propagation along z the velocities are v_1 and v_2 respectively. The displacement W and the stress T are expressed in the form of Bloch functions

$$W = g(z) \exp i\,(Kz-\Omega t) \qquad (1)$$

$$T = f(z) \exp i\,(Kz-\Omega t) \qquad (2)$$

where $g(z+D) = g(z)$ and $f(z+D) = f(z)$ (3)

D being the period d_1+d_2

In the following we will omit the dependence in the time t which is trivial. W and T must fulfill the continuity equations at all interfaces of the superlattice ; but owing to the periodicity, the problem can be limited to two adjacent interfaces.

Consider now the Nth cell of the infinite superlattice. We define local coordinates $x_{N,1}$, $y_{N,1}$, $z_{N,1}$ and $x_{N,2}$, $y_{N,2}$, $z_{N,2}$ with their origin at the middle of the layers(see Fig.2).

We thus have :

$$W_{N,1}(z_{N,1}) = A_1 \exp i K_1 z_{N,1} + B_1 \exp -i K_1 z_{N,1}$$

$$T_{N,1}(z_{N,1}) = i \Omega Z_1 [A_1 \exp i K_1 z_{N,1} - B_1 \exp -i K_1 z_{N,1}]$$

$$W_{N,2}(z_{N,2}) = A_2 \exp i K_2 z_{N,2} + B_2 \exp -i K_2 z_{N,2}$$

$$T_{N,2}(z_{N,2}) = i \Omega Z_2 [A_2 \exp i K_2 z_{N,2} - B_2 \exp -i K_2 z_{N,2}]$$

where $Z_1 = \rho_1 v_1$, $Z_2 = \rho_2 v_2$ and A_1, B_1, A_2, B_2 are constants.

K_1 and K_2 which are the wavector in 1 and 2 are equal to Ω/v_1 and Ω/v_2 respectively.

The continuity relations are given by :

$W_{N,1}(-d_1/2) = W_{N,2}(+d_2/2)$ (4)

$T_{N,1}(-d_1/2) = T_{N,2}(+d_2/2)$ (5) for interface Q

and :

$W_{N,1}(+d_1/2) = W_{N+1,2}(-d_2/2) = W_{N,2}(-d_2/2) \exp iKD$ (6)

$T_{N,1}(d_1/2) = T_{N+1,2}(-d_2/2) = T_{N,2}(-d_2/2) \exp iKD$ (7) for interface R

Obviously the problem of the propagation of the longitudinal acoustic wave in this layered structure is completely solved by the set of Equ. (4), (5), (6), (7) where only the layered 1 and 2 of the unit cell are involved.

This is a linear and homogeneous system of 4 equations with 4 unknown variables A_1, B_1, A_2, B_2. The solution is obtained by solving the secular equation. One finds finally :

$$\cos K D = \cos\frac{\Omega d_1}{v_1} \cos\frac{\Omega d_2}{v_2} - \frac{1}{2}(Z+1/Z)\sin\frac{\Omega d_1}{v_1} \sin\frac{\Omega d_2}{v_2} \quad (8)$$

where $Z = Z_1/Z_2$ is the ratio of the acoustic impedances.

One must keep in mind that the vector K is defined in the Brillouin zone, whose extension is now $(-\Pi/D, +\Pi/D)$ in the z direction. Equation (8) gives the dispersion of the acoustical branch B_0 (for which Ω tends to zero for vanishing values of K) as well as the variation of Ω versus K for the "folded acoustic branches" in the frequency range A of Fig.1. Small gaps also appear between the folded acoustic branches at K=0 and $K=\frac{\Pi}{D}$. The velocity of the acoustic branch is given by :

$V = (d\Omega/dK)_{K=0}$ Equ.(8) thus gives :

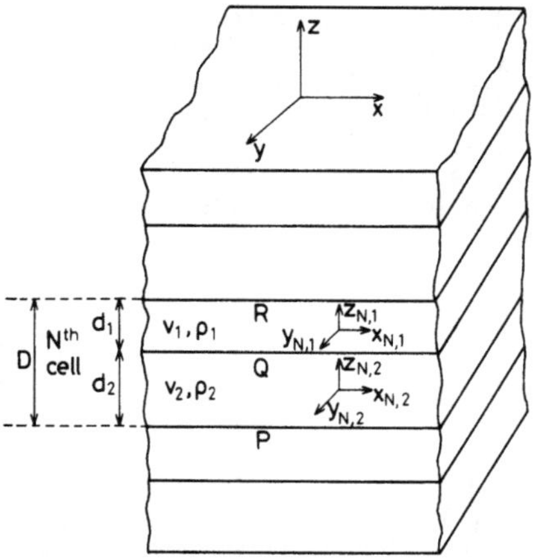

Fig.2 : Superlattice made of two media 1 and 2;x,y,z are the general coordinates and $x_{N,1}, y_{N,1}$ $z_{N,1}; x_{N,2}, y_{N,2}, z_{N,2}$ the local coordinates. The continuity equation for the displacement and the stress are written for interface. Q and R.

$$V = D\left[\frac{d_1^2}{v_1^2} + \frac{d_2^2}{v_2^2} + \frac{d_1 d_2}{v_1 v_2}(Z+1/Z)\right]^{-\frac{1}{2}} \qquad (9)$$

In the investigated superlattices x ≃ 0.3 and the gaps at K=0 and $K=\frac{\pm\Pi}{D}$ become so small that they can be neglected. The expression of the dispersion branches of Equ(8) is given then by

$$\Omega_n = V\left[(-1)^n K + \frac{2\Pi}{D} E\left(\frac{n+1}{2}\right)\right] \qquad (10)$$

where E(n+1) is the integral part of n+1
n=1 designates the first folded branch, n=2 the second folded branch etc...

C - Light scattering by the modes of the folded acoustic branch [4,5]

In ordinary Brillouin scattering, one observes the interaction between light and the acoustic branch. The energy of the probed phonons is at the utmost of the order of a few cm^{-1} ($1cm^{-1} \simeq 30GHz$). These phonon peaks are observed with a Fabry Perot interferometer, and the width of the brillouin line in the case of transparent media can be related to the acoustic attenuation. The situation is quite different for semiconductors, where the experiments are limited to backscattering because of the light absorption. This geometry enhances the stray light which comes from the elastic scattering (Rayleigh line). Only a high contrast experimental set up can then be used for the Brillouin line observation. Another bothering problem consists of the severe broadening of the Brillouin line due to the optical absorption, which limits the penetration depth to a value $\Delta x \simeq 1/\alpha$. The uncertainty principle indicates that the wavevector K of the phonon can be determined only to an accuracy of $\Delta K \sim \alpha$ (i.e. phonons within the range α contribute to the scattering). The Brillouin peak occurring at $\Omega = vK$ will than have a width $\Delta\Omega \simeq v\alpha$. For $\alpha > 10^5 m^{-1}$ this width (which is actually asymmetric) prevents accurate measurements.

In GaAs-$Ga_{1-x}Al_xAs$ superlattices the modes which belong to the folded longitudinal acoustic branches have the same properties as the well-known acoustic modes at K≃0. This is because the longitudinal acoustic branch in III-V compounds is linear over a long range of the Brillouin zone. All the atoms undergo nearly the same displacement for such modes. Such modes are therefore of symmetry A_1. As GaAs and AlAs are strongly absorbing semiconductors, only backscattering experiments are possible. The Brillouin scattering selection rules [6,7] then show that the light scattered by longitudinal acoustic waves is polarized in the same

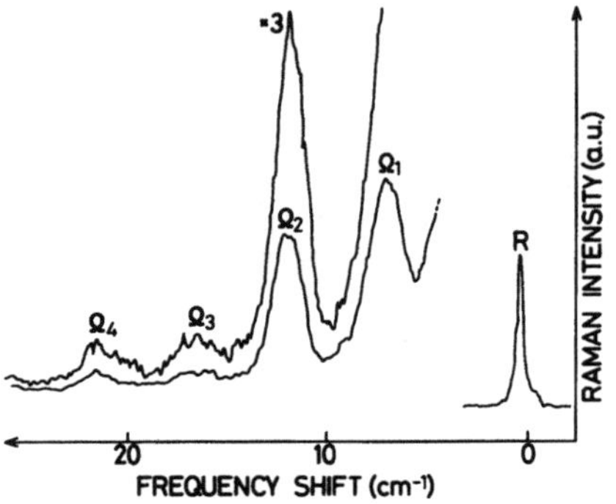

Fig.3: Raman scattering in a GaAs-$Ga_{1-x}Al_xAs$ superlattice. The modes $\Omega_1,\Omega_2,\Omega_3,\Omega_4$ belong to the 1st, 2nd 3rd and 4th folded longitudinal branch (laser at Brewster incidence λ=5145Å, resolution $1cm^{-1}$) R is the reduced Rayleigh line at the laser frequency.

direction as the incident light, and the scattered intensity is independent of the polarization (Pockels photoelastic constant P_{13}). In the case of transverse acoustic wave, the interaction with light is forbidden for backscattering along [001]. Such is exactly the selection rules we observed for the coupling of light with the folded acoustic branches in GaAs-$Ga_{1-x}Al_xAs$ superlattices [8]. Besides, there are certain features characteristic of the interaction that we develop here after.

The wavevector K of the phonon involved in a backscattering experiment with bulk vibrations is twice the wavevector of the light in the crystal (i.e. $K=4\pi n/\lambda$) and its frequency is obtained by replacing K by its value in Equ.(10). The frequency of the first phonons probed are $\Omega_1 = \frac{-4\pi n}{\lambda} V + VT^*$,

$\Omega_2 = \frac{4\pi n}{\lambda} V + VT^*$, $\Omega_3 = \frac{-4\pi n}{\lambda} V + 2VT^*$, $\Omega_4 = \frac{4\pi n}{\lambda} V + 2VT^*$, where $T^* = \frac{2\pi}{D}$ is the

primitive vector of the reciprocal lattice in the z direction. A spectrum of the scattered intensity is represented in Fig.3 ; it is clear that Ω_1 and Ω_2 which correspond to T* have very comparable intensities ; but that Ω_3 and Ω_4 which involve 2T* have an intensity-reduced by a factor~15. The peaks corresponding to nT* with $n > 2$ are generally too small to be observed in the investigated superlattices. The additional modes due to the folding are displaced with respect to the acoustic mode by ~ VT*. Their energies thus fall in the range, which can be easily probed by a double or triple Raman monochromator ($10cm^{-1}$) and this is still now the best way to observe them. The width due to the optical absorption is less than the experimental resolution ($1cm^{-1}$) and the relative uncertainly $(\frac{\Delta\Omega}{\Omega})$ is good enough to perform accurate measurements. Actually, rather sharp peaks are observed in GaAs-$Ga_{1-x}Al_xAs$ superlattices.

One must keep in mind that the magnitude $(\frac{4\pi n}{\lambda})$ of the K vector of the phonon involved in the scattering process cannot always be neglected with respect to the dimensions $(\frac{\pi}{D})$ of the folded Brillouin-zone. For a period D=150Å, taking λ=5000Å and n=4, one realizes that the extremity of K falls in the middle of the Brillouin zone. Besides, for a given superlattice, one can scan a large part of the Brillouin zone only by varying the laser wavelength in the scattering experiments [9].

III - BRILLOUIN SCATTERING BY RAYLEIGH SURFACE WAVES :

A - Rayleigh surface wave propagation

Rayleigh surface waves (RSW) are modes of propagation of elastic energy along the free boundary of an infinite half space. The amplitude of their displacement which is polarized parallel to the sagittal plane (containing the k vector and the normal to the surface) undergoes an exponential decay with depth below the free surface and vanishes within a distance of the order of one wavelength. In a cubic crystal for a (001) cut, the velocities v_R of the RSW propagating along

[100] and [110] are given respectively by :

$$C_{11}[v_R^2-C_{44}/\rho]\,[v_R^2-C_{11}/\rho+C_{12}^2/C_{11}\rho]^2 = C_{44}\,v_S^4 \tag{11}$$

$$C_{11}[v_R^2-C_{44}/\rho]\,[v_R^2-(2C_{44}+C_{12}+C_{11})/\rho+C_{12}^2/C_{11}\rho]^2 =$$

$$C_{44}\,v_R^4\,[v_R^2-(2C_{44}+C_{12}+C_{11})/\rho] \tag{12}$$

Equations (11) and (12) which are of third degree in v_R^2 contain the 3 independent constants of the crystal.

Consider now a superlattice made of two alternating cubic media. For long acoustic wavelength (λ_a D) the system behaves as a tetragonal homogeneous crystal. Explicit expressions for the complete set of 6 independent elastic constants have not yet been found in terms of the thickness, the density and the elastic constants of the two constituents. Only the two constants C_{33}^{eff} and C_{44}^{eff} relative to the propagation along the [001] axis can be obtained easily. The mean homogeneous crystal equivalent to the superlattice has a longitudinal bulk velocity V = C_{33}^{eff}/ρ along [001] and a density $\rho=d_1\rho_1/D+d_2\rho_2/D$. Replacing V and ρ by their values in (9) one obtains

$$1/C_{33}^{eff}= (d_1/D)/C_{33,1} + (d_2/D)/C_{33,2}$$

where $C_{33,1}$ and $C_{33,2}$ are the elastic constants in the crystals 1 and 2 respectively.

For the transverse bulk acoustic velocity propagating along [001] one similarly obtains :

$$1/C_{44}^{eff}= (d_1/D)/C_{44,1} + (d_2/D)/C_{44,2}$$

Let us consider now the crystal $Ga_{1-x}Al_xAs$. It is generally assumed that the elastic constants C_{11}, C_{12} and C_{44} are the same as for GaAs because the lattice parameters are very similar in the whole range of x. The superlattice made of alternating layers of GaAs and $Ga_{1-x}Al_xAs$ is then equivalent to a crystal of density $\rho=d_1\rho_1/D + d_2\rho_2/D$ and of elastic constants C_{11}, C_{12} and C_{44}. Thus the superlattice is equivalent to a $Ga_{1-\bar{x}}Al_{\bar{x}}As$ homogeneous crystal as long as the acoustic properties are considered. Here $\bar{x}=d_2x/(d_1+d_2)$ refers to a crystal containing the same number of Ga, As and Al atoms as the superlattice.
Once the equivalent crystal is defined and its elastic constants are calculated, the study of Rayleigh surface wave propagation in superlattices is reduced to the well-known problem of homogeneous media [10].

B - Experimental results

Our Brillouin scattering spectrometer [5] uses a triple-pass plane and a confocal Fabry-Perot interferometers in tandem. The frequency resolution is 10^8 and the contrast is of the order of 10^{10}. The characteristics of this spectrometer are suitable for the study of light scattering from surface acoustic waves in semiconductors. The triple-pass plane Fabry-Perot is used like a filter whose maximum transmission is adjusted and maintained on the frequency of the Brillouin line under investigation, owing to a reference line generated with the help of an electro-optic modulator. Fast repetitive scanning of the confocal interferometer is performed to eliminate the broadening of the Brillouin lines due to slow frequency drifts originating from thermal fluctuations.

The wave-vector conservation requirement gives for the wave-number $K=\Omega/v_R$ of the surface wave [11],

$$K=k_i\,\sin\Theta_i + k_s\,\sin\Theta_s \tag{13}$$

k_i and k_s are the wave numbers of incident and scattered light in the air ; Θ_i and Θ_s are the incident and scattering angles respectively.

Mishra and Bray [12] have shown that for scattering from semiconductors, the dominant mechanism was the interaction with the ripples at the surface. The theory of the scattering of light by the ripples which occur from acoustic propagation is discussed in several papers [13,14] but is out of the scope of this lecture.

Brillouin scattering measurements of Rayleigh wave velocity as a function of x have been performed in $Ga_{1-x}Al_xAs$ epitaxial layers of several microns thickness deposited on a (001) GaAs substrate. Two directions of propagation have been investigated [100] and [110]. Contrary to the case of bulk acoustic waves, the velocity v_R is independent of the refractive index of the material as can be seen in Equ(13).

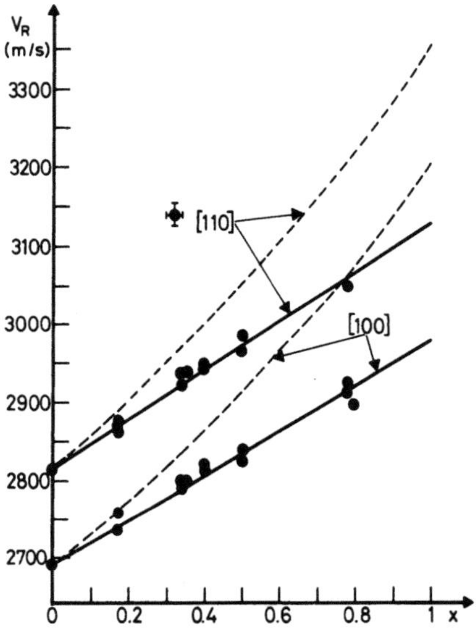

Fig.4 : Rayleigh surface wave velocity v_R as a function of x in $Ga_{1-x}Al_xAs$ epitaxial layers of several micrometers thickness. The dashed lines are calculated variations assuming the same set of elastic constants for all x values (only the density varies) ; the full lines join the experimental points obtained with a high-contrast Brillouin experimental set-up.

Two kinds of curves are plotted In Fig.4. The full lines join the experimental points of the Rayleigh wave velocity, and the broken curves are calculated by keeping the same elastic constants as in GaAs in the whole concentration range x. The broken curves are fitted for GaAs and the density variations versus x are taken into account for the velocity calculation. The variations of the measured velocities can be approximated by the following laws

v_R = 2690 + 290x (m/s) along [100]

v_R = 2817 + 315x (m/s) along [110]

Comparisons between the full lines and broken lines of Fig.4 indicate a small softening of the surface wave velocities with increasing values of x. But it is not yet possible to separate the contributions of the different elastic constants C_{11}, C_{12} and C_{44}. However, it is clear from Fig.4 that for the Gallium rich samples (x<0.4), taking the same set of constants C_{ij} in $Ga_{1-x}Al_xAs$ as in GaAs is a good approximation.

The investigated superlattices were made of alternating GaAs and $Ga_{1-x}Al_xAs$ layers with x ranging between 0.25 and 0.40. The values of the Rayleigh acoustic wave velocities v_R measured with the Brillouin spectrometer in the superlattices were compared with the values corresponding to the equivalent $Ga_{1-\bar{x}}Al_{\bar{x}}As$ crystal. A slight increase of v_R in the superlattice with respect to this mean homogeneous crystal has been observed. The value of v_R in $Ga_{1-\bar{x}}Al_{\bar{x}}As$ is actually obtained with the help of Fig.4. Possible stresses in the superlattice due to lattice mismatch between adjacent GaAs and $Ga_{1-x}Al_xAs$ layer of several tens Angstroms could be at the origin of this small increase in the velocities in superlattices. In epitaxial layers of several micrometers thickness, the stresses, which are localized at the close vicinity of the interface between the substrate and the layer, can relax at the free boundary of GaAlAs, giving the stress-free results of Fig.4.

Conclusion : This is a general study of light scattering from acoustic modes in GaAs - $Ga_{1-x}Al_xAs$ superlattices. Two experimental set-up are actually used. A Raman triple monochromator for the study of the bulk modes propagating perpendicular to the layers and belonging to the folded acoustic branches. In this case their frequencies ranges between 240 GHz and 1500 GHz. A sophisticated Brillouin spectrometer is needed for the observation of RSW (typical frequency 10 GHz) propagating parallel to the free surface of a semi-infinite superlattice. The measurement of the velocity v_R in several direction allows an acoustic characterization of the superlattice.

These light scattering investigations are useful for the determination of the crystalline quality, the period of the superlattice and the parameters of the layers. Information on the low-frequency dispersion branches are also obtainable by varying either the period of the superlattice or the wavelength of the laser radiation. The advantage of the light scattering technique is to provide to the experimentalist a non-destructive and versatile tool for characterization of microstructures.

Another field of interest is the study of the modification of the vibrational properties created by the stresses occurring between mismatched materials. The present study is focused on GaAs - $Ga_{1-x}Al_xAs$ superlattices where the mismatch is very small since the lattice parameters differ by only about a part in a thousand. Very recently G. Osbourn et al [15]. have shown the possibility of growing good superlattices with much more severely mismatched crystals. For thin layers (up to 300 Å) alternating layers with lattice constants differing by a few percent can acommodate themselves elastically to compromise lattice constants. It would thus be interesting to extend the present study to such promising systems for optical and electrical applications.

[1] J. Sapriel, B. Djafari-Rouhani and L. Dobrzynski, Surface Science 126, 197 (1982)
[2] M. Rytov, Akust. Zh. 2, 71 (1956) [Sov. Phys. Acoust. 2 68 (1956)]
[3] T.J. Delph, G Herrmann and R.K. Kaul, J. Appl. Mech. 45, 343 (1978)
[4] C. Colvard, R. Merlin and M.V. Klein, Phys. Rev. Lett. 45, 298 (1980)
[5] J. Sapriel, J.C. Michel, J.C. Tolédano, R. Vacher, J. Kervarec and A. Regreny, Phys. Rev. B28, 2007 (1983)
[6] R. Vacher and L. Boyer, Phys Rev. B6, 639 (1972)
[7] R. Loudon, Adv. Phys. 13, 423 (1964)
[8] J. Sapriel, J.C. Michel, J.C. Tolédano and R. Vacher, J. de Physique (France) Colloque C5-139 (1984)
[9] B. Jusserand, D. Paquet, A. Regreny and J. Kervarec, Solid State Comm. 48, 499 (1983)
[10] G.W. Farnell, Physical Acoustics (Academic, New York, 1973) vol.6 p.109
[11] J.R. Sandercock, Solid State Comm. 26, 547 (1978)
[12] S. Mishra, R. Bray, Phys. Rev. Lett. 39, 222 (1977)
[13] R. Loudon and J.R. Sandercock, J. Phys. C13, 2609 (1980)
[14] A.M. Marvin, V. Bortolani and F. Nizzoli, J. Phys. C13, 299 (1980) and J. Phys. C13, 1607 (1980)
[15] G. Osborn, R. Biefeld and P. Gourley, Appl. Phys. Lett. 41, 172 (1982).

Nonlinear Surface Polaritons

George I. Stegeman and Colin T. Seaton

Optical Sciences Center and Arizona Research Laboratories, University of Arizona, Tucson, AZ 85721, USA

1. Introduction

Surface polaritons are electromagnetic waves guided by a single interface between two semi-infinite media, or by single or multiple films bounded by two semi-infinite media. In all cases the electromagnetic fields decay with distance from the boundary into the semi-infinite media in an exponential-like fashion, resulting in fields localized near or at the surface(s). The 1/e penetration distances are typically less than one optical wavelength. For all surface guided waves, the propagation wavevector parallel to the surface is always larger than that for plane wave propagation in the bounding media. There are two polarization states possible in optically isotropic media: s-polarized (p-polarized) for which the electric (magnetic) field lies in the plane of the interface and is orthogonal to the propagation wavevector. Such surface guided waves have been investigated both theoretically and experimentally for many years now. In fact, waves guided by thin dielectric films have evolved into the field of integrated optics, which has many applications in electro-optics and all-optical devices.

Traditionally the optical properties, for example dielectric constant or refractive index, of all of the media into which the optical fields penetrate have been assumed to be independent of the local optical field. In this case only p-polarized surface polaritons can be supported by a single interface, and only if the dielectric constant is negative. Dielectric films bounded on both sides by media with lower refractive indices can support both s- and p-polarized guided waves, but only for propagation wavevectors less than that of plane waves in the thin film medium.

The discovery of optical bistability [1,2] with its multiplicity of applications to all-optical devices has peaked interest in nonlinear optical phenomena based on a refractive index (or dielectric constant) that varies quadratically with the local optical field. For plane waves, the optical wavevector varies linearly with the optical intensity. This is also true for surface guided waves, provided that the optically-induced change in refractive index Δn_0 is much less than the refractive index difference Δn across any of the interfaces. If $\Delta n_0 \simeq \Delta n$, it becomes necessary to solve the wave equation with a refractive index that varies quadratically with the local field $|\mathbf{E}|$, subject to the continuity of tangential electric and magnetic fields across all of the interfaces. This approach leads to many new features in the surface wave spectrum, including new waves, anomalous cut-off phenomena, and multiple surface wave branches. These will be the subject of this lecture.

2. Nonlinear Single Interface Waves

We start our discussion with one of the simplest cases, namely that of s-polarized waves guided by a single interface between two nonlinear media [3-5]. In the interests of simplicity, only s-polarized waves will be treated in these notes: for the analogous p-polarized cases (including surface plasmon polaritons) the reader is referred to references [5-6].

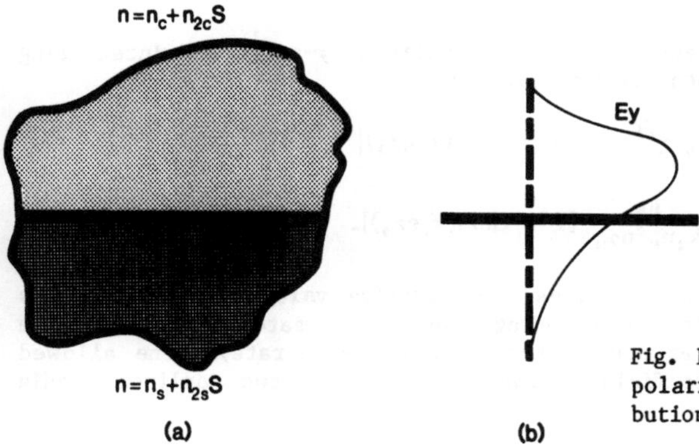

Fig. 1. (a) Single interface surface polariton geometry. (b) Field distribution

The geometry under study is shown in Fig. 1a. The relative dielectric constant is written as $\epsilon_i + \alpha_i E_y^2$ where the subscript "i" refers to the substrate ($0 \geq z$) or the cladding ($z \geq 0$).

At this point we shall assume that $\alpha_c > 0$ and $\alpha_s > 0$. The nonlinear wave equation to be solved in each medium is

$$\nabla^2 E_{iy} + k_0^2[\epsilon_i + \alpha_i E_{iy}^2]E_{iy} = 0 \tag{1}$$

where $k_0 = \omega/c$. The solutions are by now well-known [3-7], namely

$$E_{cy}(x,z) = \frac{1}{2}\sqrt{\frac{2}{\alpha_c}}\,\frac{q}{\cosh[qk_0(z_1-z)]}\,e^{i(k_0\beta x - \omega t)} + \text{c.c.} \tag{2}$$

$$E_{sy}(x,z) = \frac{1}{2}\sqrt{\frac{2}{\alpha_s}}\,\frac{s}{\cosh[sk_0(z_3+z)]}\,e^{i(k_0\beta x - \omega t)} + \text{c.c.} \tag{3}$$

where β is the effective index (βk_0 is the guided wave wavevector), $s^2 = \beta^2 - n_s^2$, $q^2 = \beta^2 - n_c^2$, and z_1 and z_3 are determined by the guided wave power. The magnetic field is obtained from $\nabla\times\mathbf{E} = i\omega\mathbf{H}$ and the guided wave Poynting vector S_0 along the interface can be used to define the parameter z_1, i.e.

$$z_1 = \pm\frac{1}{k_0 q}\,\text{arc cosh}\sqrt{\frac{q^2\beta}{n_c^2 n_{2c} S_0}}\,, \tag{4}$$

where the refractive index has been written as $n_i + n_{2i}S$ and $\alpha_i = n_i^2 c\epsilon_0 n_{2i}$. Note that $S_0 \propto \cosh^{-2}(z_1 qk_0)$ and hence both signs for z_1 are possible. Continuity of the tangential electric field across the $z = 0$ interface can be used to define z_3 as

$$z_3 = \pm\frac{1}{k_0 s}\,\text{arc cosh}\left[E_{cy}(0,0)\sqrt{\frac{\alpha_s}{2s^2}}\right], \tag{5}$$

where again both sign choices for z_3 are possible. Ensuring now the continuity of the tangential magnetic field yields the dispersion relation

$$q\tanh(qk_0 z_1) = -s\tanh(sk_0 z_3) \tag{6}$$

from which it is clear that solutions are in principle possible for $z_1 > 0$ and $z_3 < 0$, and $z_3 > 0$ and $z_1 < 0$. The useful power parameter for a surface-guided wave is the power per unit distance along the wavefront (y-axis), and integrating the Poynting vector over the depth (z) dimension gives

$$P_c = \frac{\beta q}{k_0 n_c{}^2 n_{2c}} [1 - \tanh(k_0 q z_1)] \tag{7a}$$

$$P_s = \frac{\beta s}{k_0 n_s{}^2 n_{2s}} [1 - \tanh(k_0 s z_3)]. \tag{7b}$$

What is the nature of these solutions? A negative value of z_1 (z_3) means that the field has a maximum in the cladding (substrate) material. A positive value of z_1 (z_3) signifies no maximum in the cladding (substrate). The allowed solutions therefore correspond to field maxima in one of the two nonlinear media as shown in Fig 1b.

These formulas are sufficiently general that the one nonlinear medium and all linear media cases can be easily deduced from them. As we let, for example, $\alpha_s \to 0$, then from (5) $ez_3 \to \pm\infty$. Because the fields must decay exponentially with distance from the boundary for large $|z|$, ony $z_3 \to +\infty$ is physically reasonable. In this limit

$$E_{sy}(x,z) = \frac{1}{2}\sqrt{\frac{2}{\alpha_c}} \frac{q}{\cosh(qk_0 z_1)} e^{i\beta k_0 z - i\omega t - sk_0 z} + \text{c.c.} \tag{8}$$

$$P_s = \frac{S_0}{2sk_0} \qquad q \tanh(qk_0 z_1) = -s. \tag{9}$$

Clearly, the only allowed solution is for $z_1 < 0$, which results in a field that exhibits a maximum in the cladding.

In the limit that $\alpha_c \to 0$, $z_1 \to \infty$ so all of the media are linear and the dispersion relation becomes $q = -s$, which cannot be satisfied under any conditions. Therefore s-polarized surface polaritons <u>cannot</u> be guided by the interface between two linear media. However, if one of the two media does have a positive nonlinearity, i.e. $n_{2i} > 0$, s-polarized waves can be guided by such an interface!

What about an interface separating two media with negative nonlinearities? In this case the solutions to the nonlinear wave equation are

$$E_{cy}(x,z) = \frac{1}{2}\sqrt{\frac{2}{|\alpha_c|}} \frac{q}{\sinh[qk_0(z_1 - z)]} e^{i\beta k_0 x - i\omega t} + \text{c.c.} \tag{10a}$$

$$E_{sy}(x,z) = \frac{1}{2}\sqrt{\frac{2}{|\alpha_s|}} \frac{s}{\sinh[sk_0(z_3 + z)]} e^{i\beta k_0 x - i\omega t} + \text{c.c.} \tag{10b}$$

and, following the procedure outlined above, the dispersion relation can easily be shown to be

$$q\,\text{cotanh}(qk_0 z_1) = -s\,\text{cotanh}(sk_0 z_3). \tag{11}$$

From Eq. (11) a solution is possible only if one of z_1 or z_3 is negative. However, a negative z_1 (z_3) implies that the cladding (substrate) field diverges at $z = z_1$ ($z = -z_3$), which is physically unreasonable. Therefore there are no solutions when both media have a negative nonlinearity. Furthermore, the same argument can be used to show that there are no surface guided waves in the limit that only one medium is nonlinear and it has a negative nonlinearity.

The remaining possibility is that one medium has a positive nonlinearity, the second a negative one. Assuming that $\alpha_c > 0$ and $\alpha_s < 0$, the fields are given by Eqs. (2) and (10b) respectively. Following the procedures outlined above,

$$z_3 = \frac{1}{sk_0} \text{ arc sinh}\left[E_{cy}(0,0)\sqrt{\frac{|\alpha_s|}{2s^2}}\right] \quad (12)$$

$$q \tanh(qk_0z_1) = -s \text{ cotanh}(sk_0z_3) \quad (13)$$

$$P_s = \frac{\beta s}{k_0 n_s{}^2 |n_{2s}|} [\text{cotanh}(k_0 s z_3) - 1]. \quad (14)$$

Since $z_3 > 0$ for non-diverging fields, then Eq. (14) can be satisfied only for $z_1 < 0$, which means that the field must exhibit a maximum in the positively nonlinear medium, cladding in this case.

What have we learned from this analysis? S-polarized surface polaritons cannot be guided by the interface between two linear dielectric media. They can, however, be guided if one of the media is characterized by a positive Kerr-type nonlinearity. If such a surface polariton does exist, it exists at a specific power only. All of this is quite reasonable from the point of view of the phenomena of self-focusing ($\alpha_i > 0$) and self-defocusing ($\alpha_i < 0$). Negative nonlinearities tend to defocus a field and do not lead to the field maxima required for the existence of surface polaritons. However, self-focusing nonlinearities tend to focus a field and therefore lead to field maxima. Counteracting this self-focusing effect is the interface across which there is an index discontinuity that forces continuity of the field and its first spatial derivative at the boundary. These two effects counterbalance at only one value of power, and lead to a steady-state field distribution independent of x.

3. A Film Bounded by Nonlinear Media

We now consider placing a film ($d \geq z \geq 0$) of thickness "d" and refractive index n_f between the substrate (now $z \geq d$) and cladding ($0 \geq z$). There are a number of cases: the bounding media and the film can all be nonlinear, with the nonlinearities being either positive or negative. Here we will just look at the nonlinear bounding media cases [5,8-12]. Some of the nonlinear film cases are given in [12,13] and will not be discussed here.

In the limit of all-linear media, a thin film of sufficient thickness will guide waves whose fields decay exponentially away from the film boundaries into both semi-infinite bounding media and whose fields oscillate sinusoidally within the film. Therefore in this case there are surface modes in the limit of low powers and no nonlinearities.

a. Bounding Media With Positive Nonlinearities

Based on the single interface results, we would expect that the richest set of new solutions would emerge for geometries in which one or more of the bounding media is characterized by $n_{2i} > 0$. Certainly for thick enough films and both bounding media positively nonlinear, two high power modes that degenerate into the single interface surface polaritons associated with each film boundary are anticipated.

The ensuing analysis is for the case $n_{2c} > 0$ and $n_{2s} > 0$; the $n_{2s} = 0$, $n_{2c} > 0$ and the $n_{2c} = 0$, $n_{2s} > 0$ cases are obtained in the appropriate limits of the more general results. The solution fields in the bounding media are given by Eqs. (2) and (3) with z_3+z-d replacing the argument z_3+z in the cosh function in Eq. (3). The parameter z_1 can still be defined by Eq. (4) in terms of the Poynting vector at the cladding-film interface. For the linear film, the solutions to the linear wave equation are written as

$$E_{fy}(0,z) = E_{cy}(0,0)\,[\cos(\kappa k_0 z) + B \sin(\kappa k_0 z)] \qquad \beta < n_f \quad (15a)$$

$$E_{fy}(0,z) = E_{cy}(0,0)\,[\cosh(\kappa k_0 z) + B' \sinh(\kappa k_0 z)] \qquad \beta > n_f \quad (15b)$$

where $E_{cy}(0,0)$ is given by Eq. (2) and $\kappa^2 = |\beta^2 - n_f^2|$. Equations (15) are already formulated to ensure continuity of electric field at $z = 0$ and continuity of tangential magnetic field gives $B = B' = q \tanh(qk_0 z_1)/\kappa$. Continuity of tangential fields at $z = d$ yields Eq. (3) with $E_{cy}(0,0)$ replaced by $E_{fy}(0,d)$, and the dispersion relations

$$\tan(\kappa k_0 d) = \frac{\kappa[q \tanh(qk_0 z_1) + s \tanh(sk_0 z_3)]}{\kappa^2 - sq \tanh(qk_0 z_1) \tanh(sk_0 z_3)} \qquad \beta<n_f \tag{16a}$$

$$\tanh(\kappa k_0 d) = \frac{\kappa[q \tanh(qk_0 z_1) + s \tanh(sk_0 z_3)]}{-\kappa^2 - sq \tanh(qk_0 z_1) \tanh(sk_0 z_3)} \qquad \beta>n_f. \tag{16b}$$

The total guided wave power per meter of wavefront is given by $P_c + P_f + P_s$ where P_c and P_s are given by Eq. (7) and

$$P_f = \frac{S_0}{2} [d(1+\frac{q^2}{\kappa^2} \tanh^2(k_0 q z_1)) + \frac{\sin(2k_0\kappa d)}{2k_0\kappa} (1-\frac{q^2}{\kappa^2} \tanh^2(k_0 q z_1)) + \frac{q}{\kappa^2 k_0} \tanh(k_0 q z_1)(1-\cos(2k_0\kappa d))] \qquad n_f>\beta \tag{17a}$$

$$P_f = \frac{S_0}{2} [d(1-\frac{q^2}{\kappa^2} \tanh^2(k_0 q z_1)) + \frac{\sinh(2k_0\kappa d)}{2k_0\kappa} (1+\frac{q^2}{\kappa^2} \tanh^2(k_0 q z_1)) + \frac{q}{\kappa^2 k_0} \tanh(k_0 q z_1)(\cosh(2k_0\kappa d)-1)] \qquad \beta>n_f , \tag{17b}$$

where S_0 is the guided wave Poynting vector at the film-cladding interface.

Consider first the case of one nonlinear medium, for example the cladding. The parameter $z_3 \to \infty$ in Eq. (16) and the guided wave power per meter in the substrate are given by (see Eq. (9))

$$P_s = \frac{S_0}{2sk_0} \left|\frac{E_{fy}(0,d)}{E_{cy}(0,0)}\right|^2 . \tag{18}$$

From our previous discussion for the single interface case, z_1 can be either positive or negative (with a field maximum in the cladding). Also based on that analysis, we expect that for thick enough films a single interface surface polariton wave should emerge at high powers.

Sample calculations for the effective index versus guided wave power are shown in Fig. 2 for TE_0 and TE_1 waves for a realistic system of a liquid crystal medium on top of a thin film waveguide. (The "m" in TE_m denotes the number of field zeros that occur inside the film.) For the TE_0 case, with increasing β the optical fields exhibit a maximum that moves from inside the film right into the cladding, i.e., z_1 starts positive, decreases with increasing β, vanishes when the field maximum reaches the cladding-film boundary, and becomes progressively more

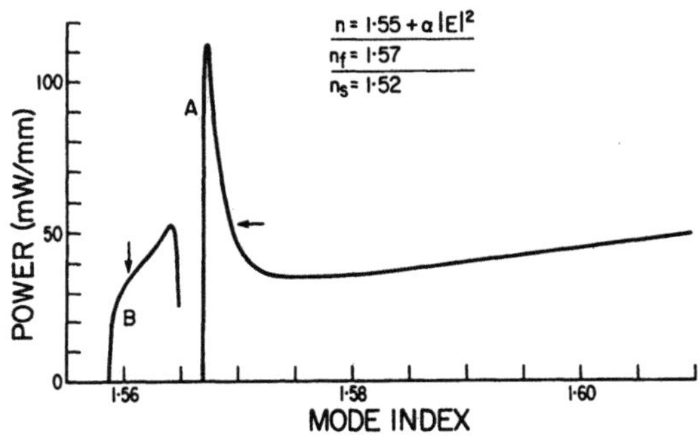

Fig. 2. The guided wave power versus effective index for TE_0 and TE_1 waves guided by a 2.0-μm thick film. The material parameters are $n_f = 1.57$, $n_c = n_s = 1.55$ and $n_{2c} = 10^{-9}$ m²/W

negative as the peak moves further into the cladding. The corresponding fields are shown in Fig. 3a. Note that solutions exist for $\beta > n_f$, that there is a maximum in the amount of power which can be transmitted via the TE_0 wave, and that at high powers this solution evolves into a single interface surface polariton. None of these features can exist for strictly all linear media. In fact, the linear case is represented by a single vertical straight line that starts at the zero power value of β.

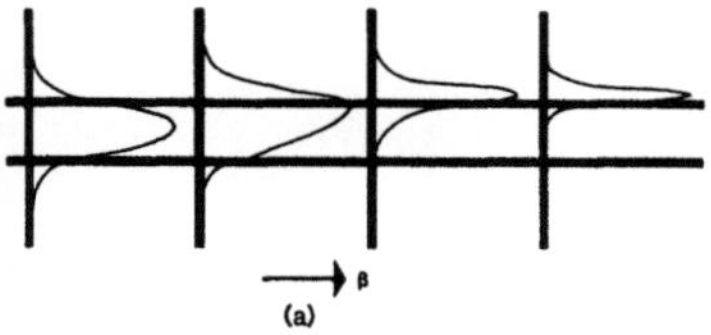

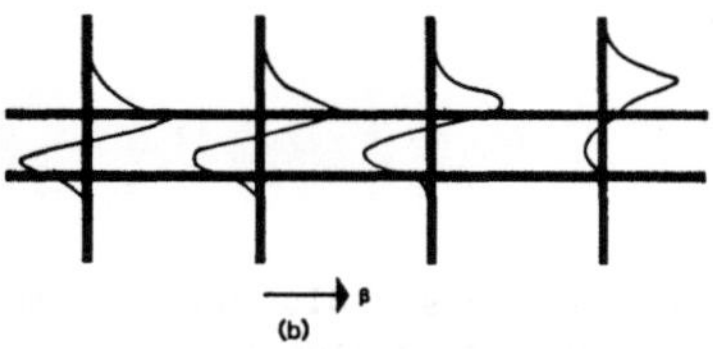

Fig. 3. The field distributions associated with the nonlinear TE_0 (a) and TE_1 (b) waves in Fig 2

The TE_1 solutions also exhibit anomalous behavior when compared to the linear waveguide case. Again z_1 starts positive, decreases with increasing β, becomes negative, and reaches a limiting negative value at the termination point of the TE_1 branch. The field distributions are shown in Fig. 3b. Note the maximum allowed value of the guided wave power. Higher order modes $m \geq 2$ exhibit essentially the same behavior as TE_1.

The richest set of new guided wave features occurs for the case in which both bounding media are nonlinear with positive coefficients. This should not be surprising! From previous discussions, the parameters z_1 and z_3 can take on both positive and negative values giving four possible combinations. In fact all four lead to physically sensible solutions.

Sample calculations and the accompanying field distributions are shown in Figs. 4 and 5 for $n_c = n_s$. Consider first the TE_0 solution waves labelled A and B. As β increases for branch A, the single field maximum that occurs for low powers in the film splits into two maxima, both of which move out of the film into the bounding nonlinear media, i.e. $z_1 > 0$ and $z_3 > 0$ evolve into $z_1 < 0$ and $z_3 < 0$. Branch B has a power threshold and is characterized by asymmetric field distributions. For the case shown, $n_c = n_s$ and this branch actually consists of two degenerate waves whose field distributions are similar, but peaked on

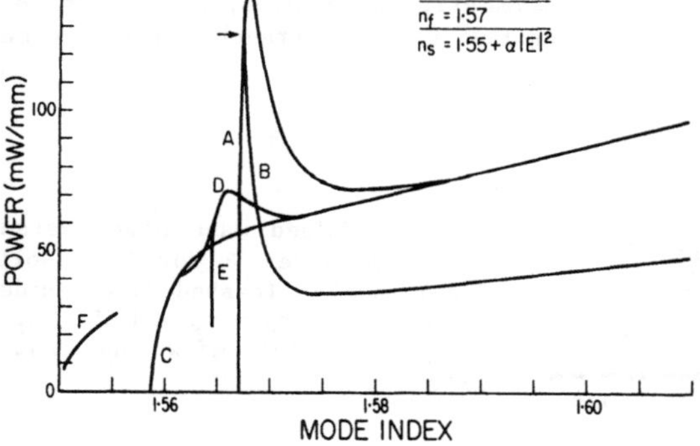

Fig. 4. The guided wave power versus effective index for TE_0 (A,B), TE_1 (C,D,E) and TE_2 (F) waves guided by a 2.0-μm film bounded by nonlinear media. The material parameters are $n_f = 1.57$, $n_c = n_s = 1.55$ and $n_{2c} = n_{2s} = 10^{-9}\ m^2/W$

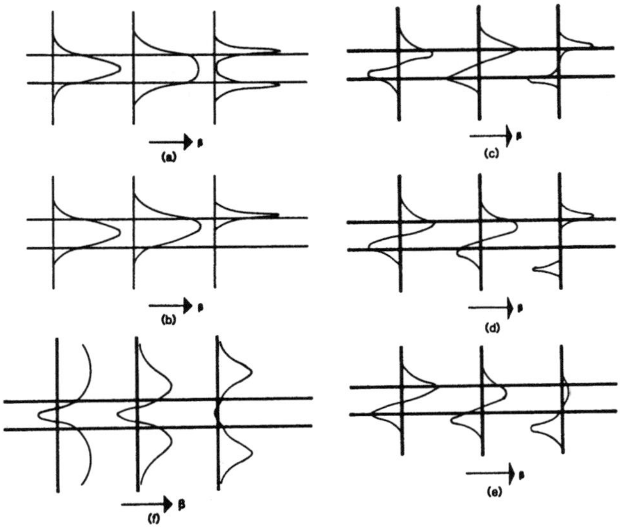

Fig. 5. The field distributions associated with the nonlinear waves in Fig. 4

different sides of the center of the film. As β increases, these solutions become identical to those obtained for TE_0 waves with only one medium nonlinear.

The TE_1 wave solutions exhibit even more bizarre behavior. The usual low power mode (branch C), which is characteristic of all-linear media evolves with increasing β into a wave in which both field maxima occur in the bounding nonlinear media. Branch D contains two degenerate solutions with asymmetric field distributions. For large values of β these field distributions remain asymmetric with both field maxima outside the film, but as shown in Fig. 5 for each solution the distances at which the two peaks occur from the film boundaries are different. Branch E corresponds to $\kappa k_0 d = \pi$ and is characterized by two degenerate solutions with $z_1 > 0$, $z_3 = -z_1$ and $z_1 < 0$, $z_3 = -z_1$. A field maximum occurs in one of the nonlinear media for each solution. Note the power thresholds which occur for branches D and E (and for B).

The TE_2 mode for a waveguide made up of all-linear media is normally cut off at this film thickness, i.e. cannot exist. When both media are nonlinear, this wave can now propagate over a limited range of powers. The fields are characterized by field maxima in both nonlinear media.

One of the very interesting features that the nonlinearities have introduced is to break the symmetry of the waveguide. Even for identical nonlinear media on both sides of the film, asymmetric field distributions become possible. The reason, of course, is that an asymmetric field distribution breaks the symmetry of the refractive indices.

The situation changes dramatically for asymmetric waveguides, i.e. $n_c \neq n_s$ or $n_{2s} \neq n_{2c}$. Sample calculations are shown in Fig. 6. The branches now become

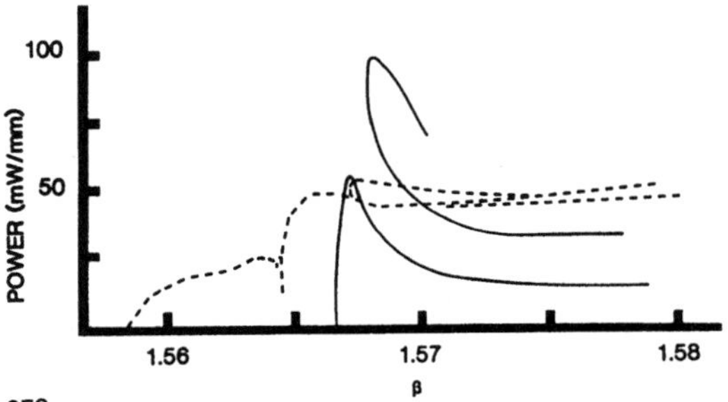

Fig. 6. Guided wave power versus effective index β for TE_0 (solid line) and TE_1 (dashed line) guided waves for $n_c = n_s = 1.55$, $n_f = 1.57$, $n_{2s} = 10^{-9}$ m^2/W and $n_{2c} = 2\times10^{-9}$ m^2/W

separate. The field distributions associated with these waves no longer exhibit any symmetry with respect to the center of the film.

b. Bounding Media With Negative Nonlinearities

Negative nonlinearities lead to self-defocusing, and the effective index decreases monotonically with increasing power. Recall from our previous discussion that non-divergent fields require that both z_1 and z_3 must be positive.

The fields in the bounding media are given by Eq. (10) with the argument of the sinh function replaced by z_3+z-d for the substrate field. The film field is given by Eq. (15) with $B = B' = q\ \mathrm{cotanh}(qk_0z_1)/\kappa$. Solving for the dispersion relations as outlined above gives

$$\tan(\kappa k_0 d) = \frac{\kappa[q\ \mathrm{cotanh}(qk_0z_1) + s\ \mathrm{cotanh}(sk_0z_3)]}{\kappa^2 - qs\ \mathrm{cotanh}(qk_0z_1)\ \mathrm{cotanh}(sk_0z_3)} \qquad \beta<n_f \qquad (19a)$$

$$\tanh(\kappa k_0 d) = \frac{\kappa[q\ \mathrm{cotanh}(qk_0z_1) + s\ \mathrm{cotanh}(sk_0z_3)]}{-\kappa^2 - qs\ \mathrm{cotanh}(qk_0z_1)\mathrm{cotanh}(sk_0z_3)} \qquad \beta>n_f. \qquad (19b)$$

The power carried by the wave is given by Eq. (14) for P_s, by Eq. (14) with the subscript "s" replaced by "c" for P_c, and by Eq. (17) for P_f with $\tanh(qk_0z_1)$ replaced by $\mathrm{cotanh}(qk_0z_1)$.

Because z_1 and z_3 must both be positive, no new modes with $\beta > n_f$ can occur. The new interesting behavior occurs as β decreases towards cutoff (case I: $\beta \to n_c$ if $n_c > n_s$; case II: $\beta \to n_s$ if $n_s > n_c$). For linear modes, as $\beta \to$ cutoff the fields penetrate progressively deeper into one of the bounding media (n_c for case I, n_s for case II) and the guided wave power required to sustain a field diverges. For the nonlinear media cases, cutoff occurs as $q \to 0$ (case I) or as $s \to 0$ (case II). It is relatively straightforward to show from Eq. (14) and the dispersion relations that $z_1 > 0$ (and $z_3 > 0$) at cutoff and that $P_c \to (k_0^2 z_1 n_c^2 n_{2c})^{-1}$ for case I and $P_s \to (k_0^2 z_3 n_s^2 n_{2s})^{-1}$ for case II. That is, cutoff is characterized by finite cutoff powers.

If only one of the media is nonlinear, say $n_{2c} < 0$, then the cutoff behavior depends on whether $n_c > n_s$ or $n_s > n_c$. For $n_c > n_s$, cutoff occurs in the cladding (nonlinear) medium and therefore at finite powers. However, if $n_s > n_c$, the cutoff occurs in the substrate (linear) medium and is characterized by $s \to 0$, which leads to a diverging power since P_s, given by Eq. (18), is proportional to s^{-1}. If $n_c = n_s$, again divergence occurs via the substrate cutoff.

c. Bounding Media With Positive and Negative Nonlinearities

These cases closely resemble the single nonlinear medium one with a positive nonlinearity. If the absolute magnitude of the negative nonlinearity is much larger than the magnitude of the positive nonlinearity, β may initially decrease with increasing power. Because self-defocusing reduces the fields in the negatively nonlinear medium, eventually the self-focusing action of the positively nonlinear medium dominates, and the guided wave power versus effective index curves resemble those for TE_0 and TE_1 in Fig. 2.

4. Experimental Verification

One experiment [14] has been reported that lends credence to the nonlinear wave solutions discussed above. The waveguide consisted of a drop of liquid crystal MBBA placed on top of a thin film glass waveguide, Fig. 7. The guided waves were first launched in the linear part of the waveguide. They are transmitted through the air-MBBA boundary into the nonlinear waveguide region as nonlinear surface guided waves, presumably of the TE_1 type. They are then converted back to linear

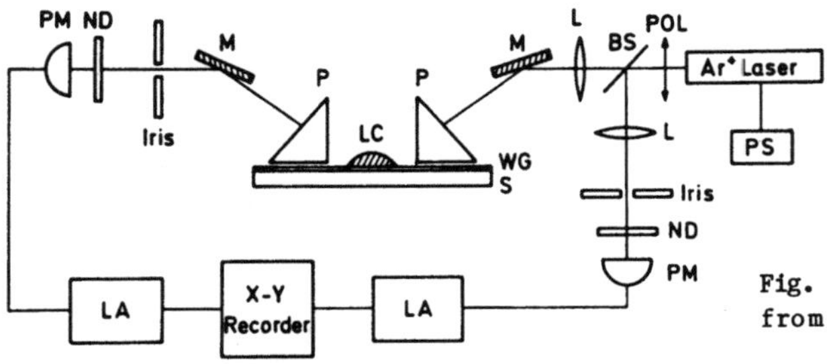

Fig. 7. Experimental geometry from [14]

waves at the second liquid crystal-air boundary. The nonlinear effect in MBBA is thermal in nature at $\lambda \approx 0.51$ µm (argon ion laser) with $n_{2c} \approx 10^{-9}$ m²/W. Since MBBA has an absorption coefficient of ≈100 cm⁻¹, as more of the guided wave energy is pulled into the cladding by the material self-focusing effect, the guided wave transmission decreases.

The experimental results are shown in Fig. 8, and the relevant theoretical curve for comparison is TE_1 in Fig. 2. As the incident power is increased from zero, the curve in Fig. 2 is traced out and progressively, more guided wave power is localized in the liquid crystal, which results in a reduced guided wave transmission coefficient. When the peak in the TE_1 curve is reached, further increase in incident power does not increase the transmitted power. As the power is now decreased, waves on both sides of the curve are excited. The waves on the high β side have progressively more of their field in the liquid crystal, and hence have a lower transmission than the waves on the low β side. This leads to the optical hysteresis observed. As the incident power is decreased further, progressively less of the high β branch waves are excited due to field mismatch at the liquid crystal interfaces. When the high β branch terminates, the transmission characteristics return to those of the low β branch.

Therefore the predictions of the theoretical calculations have been verified, and one can expect further experiments to investigate other features of the theory in the near future.

5. Discussion

Clearly the use of media with nonlinear refractive indices enriches considerably the phenomenon of guided wave propagation. Some of these phenomena will find application in all-optical devices.

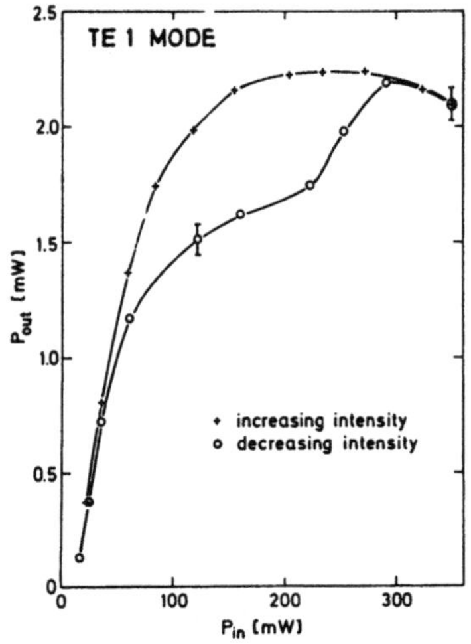

Fig. 8. Transmitted versus incident power for nonlinear TE_1 surface guided waves.

There are numerous examples of regions in which small changes in power produce large changes in the effective index and hence propagation wavevector. This occurs in the vicinity of every peak in the power versus effective index characteristics. Such regions become excellent candidates for optical bistability where it is desirable to produce large changes in optical phase inside a cavity with small changes in total power. The existence of power maxima versus effective index raises the question of the stability of the solutions in such regions.

There is the fascinating question of the propagation properties, when two or more waves coexist in a nonlinear waveguide under conditions such that one of the waves is sufficiently intense to radically affect the refractive index distribution in the waveguide. Clearly, the intense beam will affect the guided mode properties of all of the other waves present, because it has altered the refractive index distribution in the waveguide. This raises the possibility of modulating one wave by another, and utilizing power limiters at wavelengths away from material resonances.

This research was sponsored by NSF (ECS-8117483 and ECS-8304749) and the Joint Optics Program of the Army Research Office and the Air Force Office of Scientific Research.

References

1. H.M. Gibbs, S.L. McCall, T.N.C. Venkatesan, A.C. Gossard, and W. Wiegmann, Appl. Phys. Lett. **35**, 451 (1979)
2. D.A.B. Miller, S.D. Smith, and A. Johnston, Appl. Phys. Lett. **35**, 658 (1979)
3. W.J. Tomlinson, Opt. Lett. **5**, 323 (1980)
4. A.A. Maradudin, Zeit. Phys. B. **41**, 341 (1981)
5. A.A. Maradudin, "Nonlinear Surface Electromagnetic Waves," in Proceedings of the Second International School on Condensed Matter Physics, Varna, Bulgaria (World Scientific Publishers, Singapore, 1984) in press
6. V.M. Agronovich, V.S. Babichenko and V. Ya. Chernyak, Sov. Phys. JETP Lett. **32**, 512 (1981)
7. A.E. Kaplan, Sov. Phys. JETP **45**, 896 (1977)
8. N.N. Akhmediev, Sov. Phys. JETP **56**, 299 (1982)
9. A.D. Boardman and P. Egan, J. Physique Colloq., in press (1984)
10. D.J. Robbins, Opt. Comm. **47**, 309 (1983)
11. G.I. Stegeman, C.T. Seaton, J. Chilwell and S.D. Smith, Appl. Phys. Lett. **44**, 830 (1984)
12. A. Boardman and P. Egan, private communication, 1984
13. U. Langbein, F. Lederer and H.E. Ponath, Opt. Comm. **46**, 167 (1983)
14. H. Vach, G.I. Stegeman, C.T. Seaton, and I.C. Khoo, Opt. Lett. **9**, 238 (1984)

Surface Plasmon Emission as a Probe of Hot-Electron Dynamics

J.R. Kirtley, T.N. Theis, D.J. DiMaria, J.C. Tsang, and M.V. Fischetti

IBM Thomas J. Watson Research Center, P.O. Box 218
Yorktown Heights, NY 10598, USA

S.D. Brorson

Massachusetts Institute of Technology, Cambridge, MA 02139, USA

Surface plasmon mediated light emission from rough metal films can be driven in a number of different ways. They all show characteristic features independent of the driving mechanism. The unifying feature is that each mechanism injects hot electrons into the metal. We describe surface plasmon mediated emission from tunnel junctions in terms of a hot-electron picture. We show how this emission can be used as a spectroscopic probe of hot electron dynamics in tunnel junctions. We then describe how we used this tool to probe hot electron processes in electron injector structures, leading to important new insights into electron heating in SiO_2.

I. Introduction

It has been known for a number of years that rough metal surfaces emit light when excited by a beam of energetic electrons[1,2,3]. Similar radiation from randomly rough surfaces occurs when the films are excited optically[4,5,6]. Data also exists in the literature for emission from metal films on gratings excited by electron beams[7,8], and by lasers[9,10]. In this paper we will add two more systems to this list: light emission from rough films excited by tunneling electrons, and light emission from rough films excited by current across a conduction band discontinuity. One interesting thing about this collection of data is that remarkably similar emission intensities are reported. A few examples will make this clear. HEITMAN[8] reported emission from Ag films on gratings when the films were excited by 80 keV electrons. He observed sharp emission peaks using a 0.16 ° aperture, 26 Å spectrometer resolution, a grating amplitude of 500 Å and a beam current of 0.3 μA. His observed peak intensity of 125 counts/sec at 4500 Å is directly comparable to the signal we observe from light emitting tunnel junctions on gratings (which we will discuss below) if we make the appropriate corrections for differences in angular aperture, spectrometer resolution, and grating amplitude, and correct for the power pumped into the films: an electron beam at 80 keV and 0.3 μA has 24 mW of power, a tunnel junction at 2.6 V and 86 mA dissipates 224 mW. Similar comparisons can be made for tunnel and optical pumping of Ag films on gratings[11]. These similarities result because the physics of the emission process is the same for all of these systems, once the excitation process heats up the electrons.

In this paper we will discuss our results for light emission from rough tunnel junctions and charge injection structures in detail, showing how our data can be analyzed to infer hot-electron dynamics from the emission spectra.

II. Light Emission from Tunnel Junctions

LAMBE and MCCARTHY[12,13] showed in 1976 that metal-insulator-metal tunnel junctions emit light when biased at sufficiently high voltages[14]. The light is broad-band, with a characteristic linear spectral cutoff as the emission energy $\hbar\omega$ approaches the bias voltage energy eV. The emission becomes more intense if noble metals are used for the second metal electrodes, and if the films are intentionally roughened. The emission becomes brighter and (because of the increase in the spectral cutoff energy) bluer as the junction bias increases. External quantum efficiencies for the devices are, unfortunately, quite low, at best of order 10^{-4} emitted photons per electron tunneling through the junction[15,16,17].

Lambe and McCarthy proposed that these junctions emit light because electrons tunnel inelastically, losing energy to surface plasmons in the barrier region, and that in turn the surface plasmons radiate light through the mediation of surface roughness. This explains why junctions with noble metal counter electrodes are more efficient, since dielectric losses at optical frequencies are small in these metals. It also explains why rough junctions are more efficient: Surface plasmons are electromagnetic resonances of the junction structures that travel parallel to the interface at speeds less than that of light. They can only phase match to external radiation in the presence of surface roughness. Lambe and McCarthy coined the acronym LEIT, Light Emission by Inelastic electron Tunneling, for this process. A number of authors have treated this model theoretically[18,19, 20,21,22].

We will argue that a more fruitful view of the light emission process involves not the two steps of Lambe and McCarthy; 1) inelastic excitation of surface plasmons, and 2) coupling of those plasmons to external radiation; but rather three (see Fig. 1). In this view the electrons: 1) tunnel elastically into the (in this case) silver electrode. 2) The electrons then relax in energy by emitting phonons, by scattering with other electrons, and by exciting surface plasmons. 3) The plasmons then radiate light. In this (hot electron) view[11], the reason that light emitting tunnel junctions are inefficient is that relaxation to phonons and less energetic electrons is a faster process than relaxation to surface plasmons. The final step, radiation of light by the surface plasmons, can be quite efficient[23].

A few words should be said at this point about surface plasmons. They are traveling wave solutions of Maxwell's equations that have their energy densities localized at an interface. In a multiple interface structure like a metal-insulator-metal tunnel junction, there are multiple solutions[24]. In Fig. 2 we plot the dispersion curves (wavevector parallel to the interface vs energy), for the three lowest energy modes of an aluminum-aluminum oxide-silver tunnel junction. Mode 1 has its energy density localized primarily in the tunneling barrier region. Its phase velocity (the derivative of the frequency with respect to wavevector) is much slower than the speed of light in vacuum. For the sake of brevity we will call mode 1 the "slow" mode. Mode 2 is localized primarily at the Ag-vacuum interface. Its speed of propagation is only about 3 percent slower than that of light at 2 eV. The dispersion curve for this mode bends over as the energy approaches the Ag interband transition region at about 3.7 eV. We will call mode 2 the "fast" mode. Mode 3 is localized between the aluminum and the substrate, in this case taken to have the dielectric properties of photoresist. It is almost as fast as mode 2 in the optical region, only bending away from the light line when the mode energy approaches 10 eV. Some authors refer to modes far from the light line as "surface

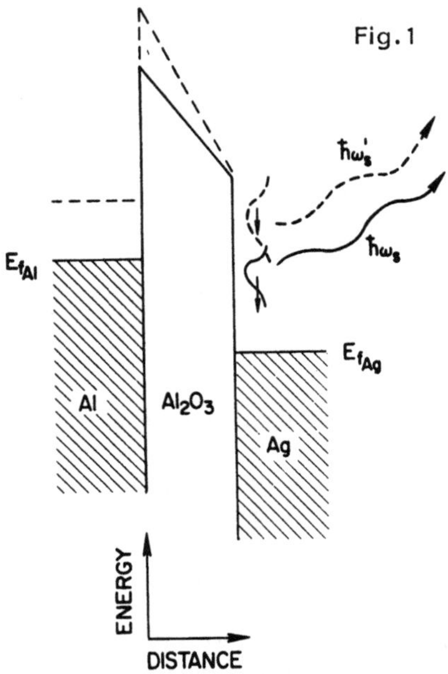

Fig. 1

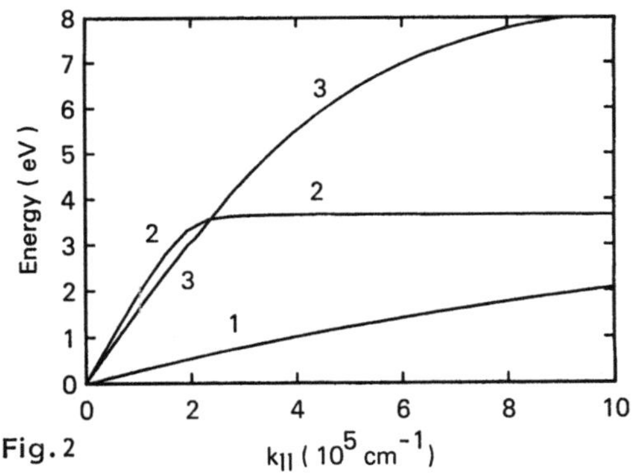

Fig. 2

Fig. 1. Schematic diagram of the 3 step (hot-electron) picture of light emission from tunnel junctions. When a voltage bias V is applied across the junction electrons tunnel across the oxide. The tunneling process is most likely for electrons near the Fermi energy of the cathode. The electrons relax in energy, partially through the emission of surface plasmons. The surface plasmons then emit radiation. As the bias voltage across the junction is increased (dashed line), electrons are injected into the Ag film at higher energies relative to the Ag Fermi level, and more energetic photons can be emitted.

Fig. 2. Dispersion curve for the 3 lowest energy modes of an Al-Al_2O_3-Ag tunnel junction. Mode 1, the "slow" mode, is localized in the barrier region. Mode 2, a "fast" mode, is localized at the Ag- vacuum interface. Mode 3, another "fast" mode, is localized at the Al photoresist interface. (From Kirtley et. al.[26]).

plasmons" and modes close to the light line as "surface plasmon polaritons". Mode 1 is sometimes referred to as the "junction" mode. For the sake of simplicity we will refer to all of the tunnel junction electromagnetic resonances as surface plasmons.

Lambe and McCarthy suggested that tunnel junctions emitted light when electrons tunneling inelastically excited the "slow" mode, which radiated light. The coupling to the "slow" mode should be fairly efficient, since this mode has high field strengths and field gradients in the barrier region. DAVIS[18] has calculated coupling efficiencies as high as 1 percent. However, since this mode is so slow, very short wavelength roughness is required to cause it to radiate[21,22]. While such roughness may in fact be present in randomly roughened junctions, it is very hard to control or measure, making quantitative comparison with theory difficult.

These difficulties are avoided if the roughness required is supplied by fabricating the junctions on gratings (Fig. 3)[25,26,27,28,29]. We produced substrates with grating roughness by holographically exposing a thin layer of photoresist on a silicon wafer and

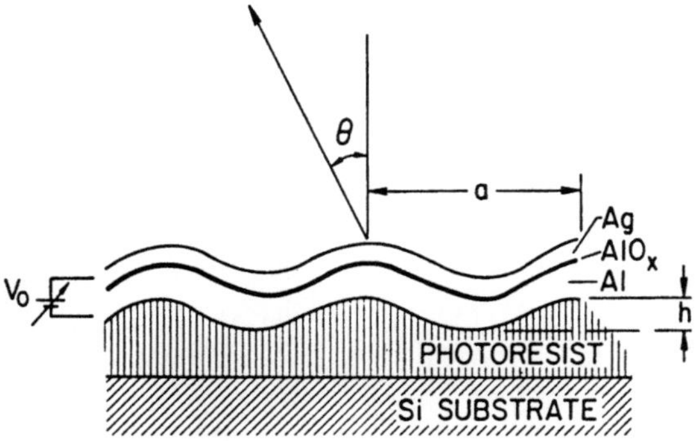

Fig. 3. Geometry of our tunnel junctions on photoresist gratings. The junction films are evaporated directly onto the photoresist, which has a nearly sinusoidal groove depth h and periodicity a. The emission is observed at an angle θ with respect to the substrate normal in the plane defined by the surface normal and the grating wavevector.

then developing it[30]. The resulting photoresist had nearly sinusoidal profiles with highly uniform grating periodicities (set by the exposure geometry) and grating amplitudes (set by the exposure and development conditions). Tunnel junctions were fabricated on the substrates as usual[25].

The junctions were mounted in a closed cycle refrigerator and cooled to about 25 K for measurement. The light emission process is relatively insensitive to temperature, but we found that junctions with silver counterelectrodes were more stable at low biases if operated at low temperatures. The junctions were biased at a given voltage, the emitted light was collected through an aperture (about 1.4 degrees wide) parallel to the grating grooves, focussed onto the entrance slit of a grating monochromater, and detected using photon counting.

The emission from an Al-Al_2O_3-Ag junction on a 8200 Å periodicity grating (grating amplitude approximately 50 Å), is shown in Fig. 4. As the observation angle (measured with respect to the substrate normal, and in the plane defined by the substrate normal and the grating wavevector) was increased, the sharp surface plasmon emission peak moved to higher energies. A second peak, due to second order scattering, moved in the opposite direction. The peak positions as a function of emission angle for a junction on a 8200 Å periodicity, 815 Å amplitude grating (Fig. 5) were well represented by the resonance condition:

$$(\omega/c) \sin(\theta) = k_{||} \mp nQ, \qquad (1)$$

where $Q = 2\pi/a$, a is the groove spacing, the integer n gives the scattering order, and the dispersion curve is that for the Ag-vacuum interface mode. The observed dispersion curves were slightly lower than calculated for several reasons: 1) The curves were calculated for real dielectric constants. Damping due to losses in the metal films tends to slow the plasmons down. 2) Plasmons on gratings with larger grating amplitudes have to travel farther along the surface to phase match with emitted photons at a given emission angle. 3) Radiative damping at high grating amplitudes also slows down the plasmons[31,32]. Junctions on gratings with higher grating amplitudes radiated more efficiently (proportional to the grating amplitude squared), had broader emission peaks, and slightly lower dispersion curves[26].

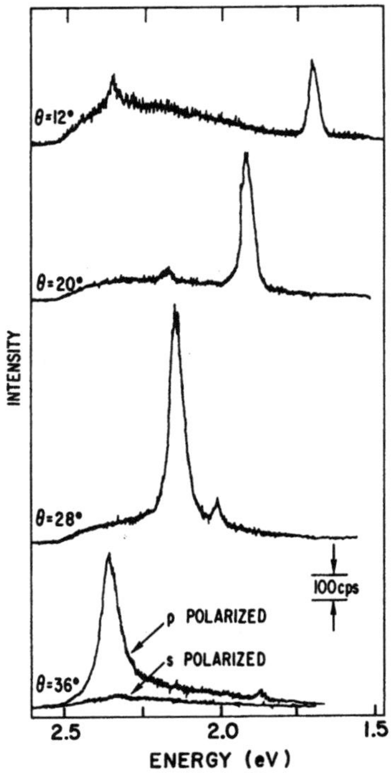

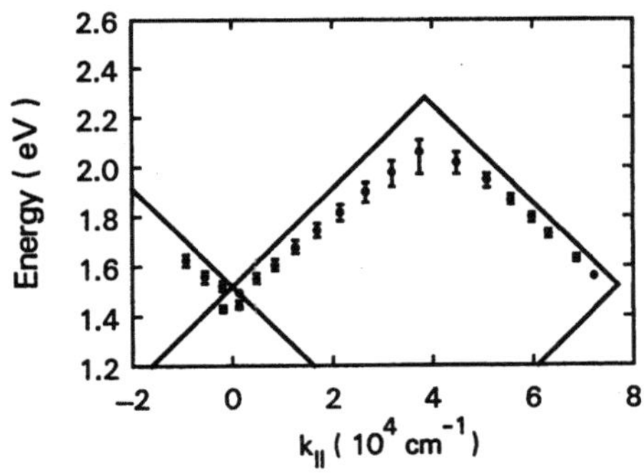

Fig. 4. Light emission observed at various angles θ from an Al-Al_2O_3-Ag junction on a 50 Å amplitude, 8200 Å periodicity photoresist grating. The sharp, highly polarized emission peaks move in energy with θ. The large peak is due to scattering to first order in the grating wavevector of the Ag-vacuum "fast" surface plasmon. The small peak that moves in the opposite direction is due to second order scattering in the grating periodicity. (From Kirtley et. al.[25].)

Fig. 5. Experimental vs. theoretical dispersion curves for the emission from an Al-Al_2O_3-Ag tunnel junction on an 815 Å amplitude, 8200 Å periodicity photoresist grating. The error bars in the experimental points represent half intensity points. As the grating amplitudes were made larger, the observed peaks became broader, more intense, and shifted to slightly lower energies. The solid curves are the calculated dispersion curves shifted by ± 1 and ± 2 grating wavevectors. (From Kirtley et. al.[26]).

It is obvious from Fig. 4 that the radiation from tunnel junctions on gratings is dominated by photons phase matched to surface plasmons by the grating periodicity. Fig. 5 demonstrates that the important surface plasmon mode is that localized at the Ag-vacuum interface. Radiation from the "slow" mode is relatively weak in these samples, presumably because there is insufficient short-scale roughness in our junctions on gratings to cause it to efficiently radiate. It should be emphasized that the "slow" mode should be driven efficiently by the tunneling electrons but should radiate inefficiently. The "fast" mode is driven inefficiently by the tunneling electrons but radiates quite well if surface roughness of the right scale is present.

The next question one might ask is: Where does the coupling between the tunneling electrons and the surface plasmons occur? If the coupling occurs primarily in the insulator, the emission process may be thought of as inelastic. However, if the coupling occurs primarily in the metal electrodes, it may be more useful to think of it as a hot-electron process.

We have several reasons for thinking the latter: 1) The primary radiation mode, at least on our gratings, is localized at the Ag-vacuum interface[25]. 2) Measurements as a function of Ag film thickness show that the emission intensities fall off exponentially with a length that agrees with the hot electron mean free path in metals measured using other techniques. In fact, this data can only be understood if the electrons couple most strongly to the surface plasmons at or near the Ag-vacuum interface[26,33] 3) Measurements of the dependence of emission efficiency on temperature show structure that tracks with the resistivity of the Ag film, and not with the junction resistance[11]. 4) We have observed similar emission efficiencies and linear cutoffs when hot electrons are excited in the junction structures by laser irradiation[11]. No tunneling is present in this case. 5) Similar emission efficiencies and spectral features are observed when hot electrons are injected into a rough metal film with electron injection devices[34,35], as will be described below. Direct tunneling across the insulator is not present in these experiments either.

Just as the positions in energy of the emission peaks tell us about the identity and dispersive properties of the radiative mode, an analysis of emission intensities using the hot electron model tell us about dynamic processes in the junction structure. Figure 6 shows the integrated intensities for a junction on an 8150 Å periodicity grating as a function of peak emission energy, for a number of junction bias voltages V. The integrated peak intensities cut off linearly at high energies, and decrease nearly exponential-

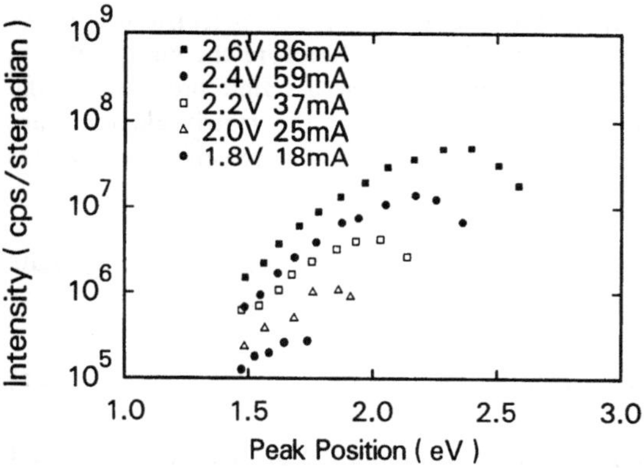

Fig. 6. Integrated peak intensities vs. peak energies for light emission from an Al-Al_2O_3-Ag tunnel junction with a 200 Å Ag film on a 8150 Å periodicity photoresist grating, for a number of different bias conditions. A series of emission spectra were measured by varying the observation angle, and the intensity under each peak was integrated after subtracting out the smooth background which was presumably due to residual roughness. As the bias voltage was increased the junctions became brighter and the linear cutoffs at high frequencies moved out to higher energies. The intensities fell off nearly exponentially as the peaks moved into the red. (From Kirtley et. al. [26]).

ly at low energies. The linear cutoff at high energies occurs because tunneling electrons cannot have more energy relative to the Fermi energy of the second metal electrode than the bias voltage energy. The exponential decrease in intensities at low photon energies occurs because of decreased radiation efficiencies for the fast plasmon at these energies[29].

We have developed a semi-classical theory of surface plasmon emission from tunnel junctions, directly analogous to the semi-classical theory of radiation from atoms[36]. It is not appropriate to present the details of this theory here. Fortunately, the important physical properties of the theory can be presented quite simply. The emission intensity from a rough film with a given non-thermal distribution function of hot electron occupied states f(E) can be written as:

$$I(\omega) = C\int_0^\infty A(\omega)\,|M(E,\omega)|^2 f(E)(1-f(E-\hbar\omega))dE \qquad (2)$$

where $A(\omega)$ is the " antenna " factor which describes the radiative properties of the film, $|M(E,\omega)|^2$ is the matrix element which describes coupling between the hot electrons and the surface plasmon fields, and C is a constant. The antenna factor has been worked out by Laks and Mills[29] for the model of light emission by inelastic tunneling; the same factors occur in the semi- classical theory of hot electron surface-plasmon mediated emission. The matrix elements are sensitive to the details assumed for the Ag-vacuum interface: we will treat them as unknown. Our modeling indicates, independent of the details of the Ag-vacuum interface, that at these energies the matrix elements are relatively insensitive to the absolute hot electron energy, but depend only on the energy loss. If one assumes that the matrix elements depend only on ω equation 2 can be rewritten as:

$$A(\omega) = I(\omega)/\{C\,|M(\omega)|^2\int_0^\infty f(E)(1-f(E-\hbar\omega))dE\}. \qquad (3)$$

The " antenna " factor $A(\omega)$ only depends on the geometry and dielectric properties of the junction; it does not depend on the bias conditions. Therefore the curves of Fig. 6 should collapse to a universal curve if we correctly model the dynamic processes in the junction. For the sake of this analysis we assume further that the matrix elements are independent of the surface plasmon energy. There is no particular reason to make this assumption, except that it seems to work, as will be seen below.

We model the dynamical processes of the junction by expanding the steady state occupation function f(E) in a prudently chosen basis set,

$$\int_0^\infty f(E)(1-f(E-\hbar\omega))dE = \sum_{n=1}^{N} A_n(1-\hbar\omega/eV)^n. \qquad (4)$$

The best fit collapse of the data of Fig. 6 to the antenna factor of Laks and Mills (using the appropriate dielectric properties for Ag, Al_2O_3 and Al, grating periodicities, and film thicknesses) using N=4, is shown in Fig. 7. The compression of the data using this basis set is quite good; significantly better than that for the original model of Laks and Mills. This is not surprising, since the simple model has no adjustable parameters. More important than the quality of the fit, however, is that it allows us some insight into the shape of the steady-state hot electron distributions.

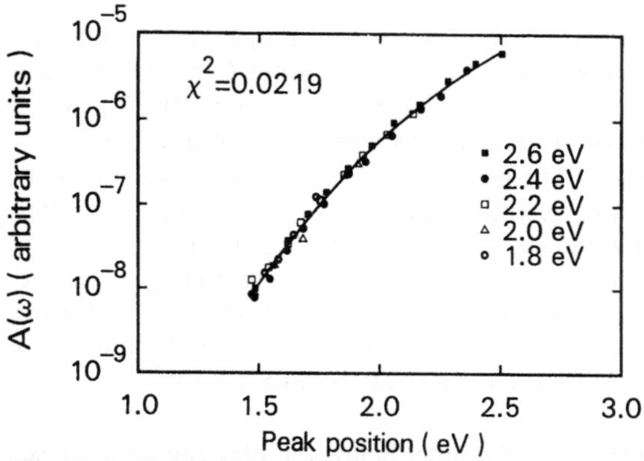

Fig. 7. Collapse of the data of Fig. 6 to $A(\omega)$, the "antenna factor" of Laks and Mills[29] by fitting the steady state occupation function f(E) as described in the text. The data all collapse to a common line, as they should, since the "antenna" factor, which is set by the geometry and dielectric constants of the junctions, should not depend on the bias conditions. By fitting such curves, we can infer the hot electron dynamics in the tunnel junctions.

To a very good approximation, f(E) is proportional to the derivative of the integral in (4). The resulting best fit non-equilibrium functions are displayed as the solid curves in Fig. 8. Also displayed are dashed curves giving the results for a Monte-Carlo simulation of the hot-electron scattering process. In this simulation, the electrons were taken to be injected into the Ag film at z=0 with an energy distribution given by Simmons[37], assuming a trapezoidal tunneling barrier of height 2.4 eV at the Al-Al_2O_3 interface and 3.5 eV at the Al_2O_3-Ag interface[38]. The electrons were assumed to be injected into

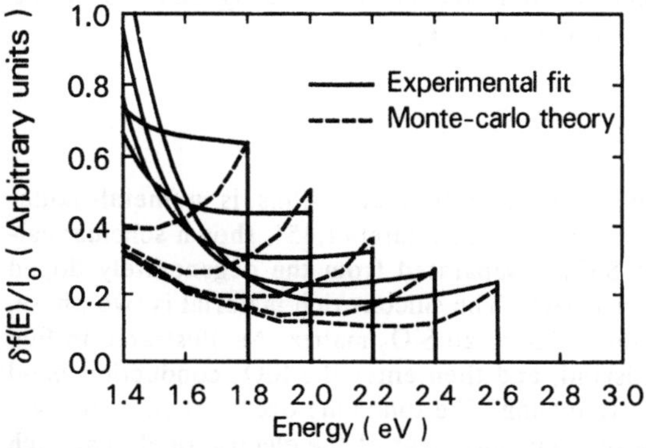

Fig. 8. The solid lines are the deviation from thermal equilibrium $\delta f(E)$ of the occupation function of states an energy E above the Fermi level of the silver, derived from a fit of the data of Fig. 6 to the "antenna factor" of Laks and Mills[29], for bias voltages of 1.8, 2.0, 2.2, 2.4, and 2.6 volts. The dashed curves are the predictions of a Monte-Carlo calculation as described in the text. The theory is fit to the data for the 1.8 volt curve at an energy of 1.8 eV. The theoretical peak amplitudes agree well with the experimental fits, but the theoretical peak shapes fall off more rapidly with energy than the fit, indicating either that the tunnel injection is less strongly energy dependent than predicted by simple tunneling theory, or that the scattering to intermediate energies of the hot electrons is stronger than predicted by the Monte-Carlo calculations.

the metal normal to the interface. The electrons were allowed to propagate through the film, scattering from phonons and other electrons using the Monte-Carlo techniques described by Fitting and Freiman[39]. During electron-electron scattering, the kinematics were retained, but afterwards one of the final particles was discarded and the effect of the other multiplied by a factor of two. The electron's scattering angle was randomized after a phonon scattering event. The electrons were allowed to elastically scatter from both the Ag-Al_2O_3 and Ag-vacuum interfaces. The occupation function plotted in Fig. 8 was calculated by recording the electron's energy at each scattering from the Ag-vacuum interface. The electron-electron and electron- phonon mean free paths were taken from Crowell and Sze[40]. The experimental fits are matched to the Monte-Carlo simulations at one point, where the bias and photon energies are both 1.8 eV.

The Monte-Carlo occupation functions fall off rapidly below the bias energy because less energetic electrons are less likely to tunnel. They turn up again at low energies because of the longer lifetimes of hot electrons at low energies[40]. The experimental fits do not dip as deeply as the simulations. This could occur either because the tunneling is less sensitive to electron energy than predicted by the simple tunneling theory, or because there is more effective scattering to intermediate energies in the Ag than predicted by the Monte-Carlo simulation.

Double junction experiments have measured hot electron distributions in tunnel junctions at energies of a few milli-electron volts above the superconducting gap edge [41,42]. Tunnel emission measurements have measured hot electron distribution functions in junctions in the Fowler-Nordheim regime at energies more than 4 eV above the Fermi level[43]. Photoemission measurements (either into a semiconductor or into vacuum) have also been used to measure hot electron mean free paths in metal films[40,43]. However, surface plasmon emission gives unique spectroscopic information on the energy dependence of tunnel injection and hot electron dynamics over the important energy range of 1-3 eV above the Fermi energy.

III. Electron Injector Structures

Yet another method for injecting hot electrons into metal films is a metal-oxide-semiconductor capacitor called an electron injector structure[44,45], shown schematically in Fig. 9. In this device, a layer of SiO_2 is separated from the degenerately doped n-type Si substrate by a layer of silicon-rich SiO_2. The silicon- rich material is two phase, containing many small (~ 50 Å) inclusions of Si in an SiO_2 matrix. As illustrated in Fig. 9, electrons tunnel from Si island to Si island, and then enter the SiO_2 conduction band at thermal energies by Fowler-Nordheim tunneling. The tunneling occurs at comparatively low bias voltages because of the local field enhancement of the electric field near each island. Electrons in the SiO_2 conduction band are quickly swept to the opposite interface and injected into the Al electrode. There they relax in energy, partially through the emission of surface plasmons, which in turn radiate. As current is run through the SiO_2, charge trapping occurs. This tends to bend the conduction band up. The current through the device is set by the field at the Si-rich SiO_2 - SiO_2 interface. In order to keep this current constant, the gate bias V_g is increased. This increases the average field across the oxide and the field at the SiO_2-Al interface (the anode field). Electrons injected at the SiO_2 conduction band edge E_c can generate a surface plasmon of maximum energy E_c-E_f, where E_f is the Fermi energy. Surface plasmon radiation at higher energies is evidence for electron heating in the SiO_2. In addition, as the anode field increases, the electrons

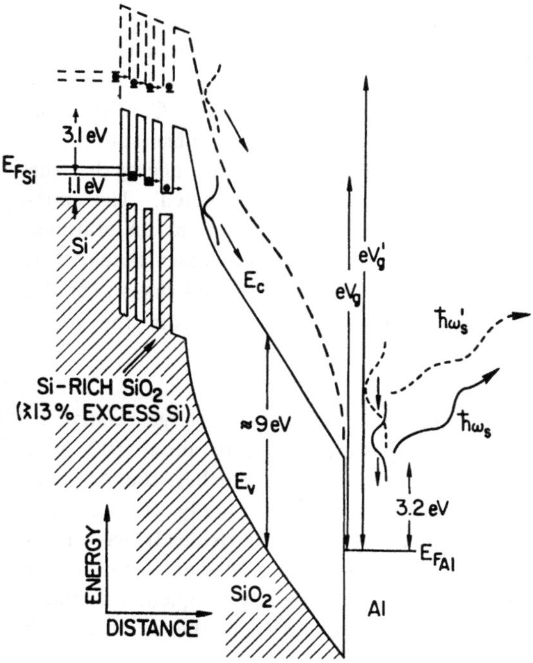

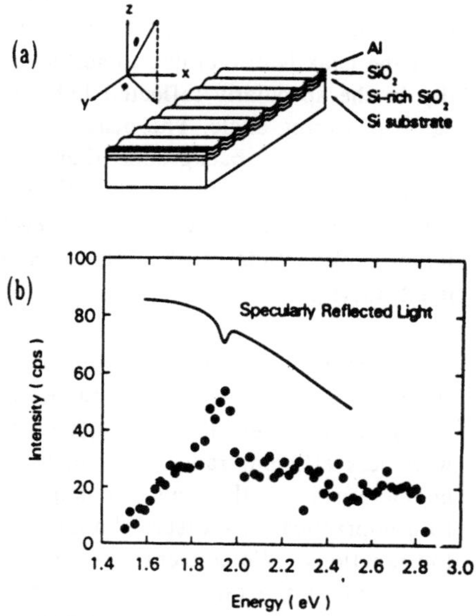

Fig. 9. Schematic of the conduction and light emission properties of surface-plasmon mediated emission from electron injector structures. When a bias voltage V_g is applied across the device, electrons from the conduction band of the Si tunnel from silicon island to silicon island in the silicon rich SiO_2. The high fields generated at the interface between the silicon islands and the stoichiometric SiO_2 cause field emission into the conduction band of the SiO_2. The electrons are quickly swept through the SiO_2 and injected into the Al film, where they relax in energy, partially through surface plasmon mediated light emission. As current is passed by the oxide, trapping occurs. This bends the silicon conduction band up as indicated. The bias voltage V_g is increased to keep the current through the device constant. This results in higher average fields across the oxide and at the Al-SiO_2 interface, causing hotter electron distributions. The surface plasmon mediated emission is a measure of the amount of heating occuring.

Fig. 10. Emission spectrum from an Al gate electron injection structure on a silicon substrate. The silicon substrate was holographically patterned and etched to form an 8150 Å periodicity, 360 Å amplitude grating etched on the surface (a). A narrow line (b) at the appropriate energy for surface plasmon mediated emission is superimposed on a broad background. A simultaneous measure of specularly reflected light from a tungsten lamp (b) shows a surface plasmon mediated dip at the same energy. The background is relatively larger than in Al-Al_2O_3-Ag tunnel junctions both because of the larger dielectric losses in Al than in Ag, and because of luminescence from the oxide itself. (From Theis et. al.[34]).

injected into the Al gate become more energetic, and photons with correspondingly higher energies are emitted.

We have unequivocally demonstrated that surface plasmon mediated emission does occur in the electron injector structures by fabricating them on Si substrates with a holographically exposed grating etched into the wafer[34]. Fig. 10 shows the emission from an Al gate (250 Å thick) electron injector on a 8150 Å periodicity, 360 Å amplitude Si grating. At an observation angle of 13 ° there is an emission peak at just the right energy for the n=1 branch of the Al-air surface plasmon mode. Confirmation that the peak is in fact due to surface plasmon coupling comes from the observation of a dip in the specularly reflected light from a tungsten lamp at the same energy for the same scattering angle. The peak emission energy varied with θ as expected. The emission has a larger background than was observed for the tunnel junctions on gratings. There are two reasons for this. One is that Al has larger dielectric losses than Ag at these energies, raising the relative height of the background. The second reason for the large backgrounds is that there is a component to the emission from the oxide itself[35,46].

Although the currents passed by electron injector structures were large relative to standard metal-oxide-semiconductor capacitors, they were still only about 100μA, a thousand times less than through light emitting tunnel junctions. Further, the Al gates were less efficient radiators than the Ag used in our junctions. Finally, the emission characteristics of the devices changed as charge was trapped in the oxide. It was therefore impractical to measure the emission spectra for a series of angles as was done for the junctions. We concentrated instead on emission from randomly roughened samples. This lead to the difficulty of distinguishing between emission from the oxide and Si islands and surface plasmon emission.

The luminescence from the Si islands and the SiO_2 was very complicated[46]: we do not have the space to treat it here. Fortunately, we were able to take advantage of the dependence of surface plasmon mediated luminescence on surface roughness to eliminate the oxide luminescence. We prepared two sets of charge injector samples. One set was fabricated on substrates intentionally roughened by oxidizing and then chemically removing the oxide from a polysilicon layer on the silicon substrate. The other set was identically prepared, but on smooth substrates. We obtained the surface plasmon mediated luminescence from these samples by subtracting the spectra of the smooth samples from the rough ones. The integrated intensity from the roughened samples was about 10 times larger than that for the smooth samples, so little error was involved in the subtraction procedure.

The results from such a subtraction are showed in Fig. 11. The spectra were taken by focusing the emitted light on the entrance slit of a monochromator with a spherical mirror. The aperture of the mirror was adjusted to just fill the gratings of the monochromator, which had an "f" number of 5.8. The light was detected using photon counting techniques at an emission angle normal to the substrate. As successive emission spectra were taken, charge trapped in the SiO_2 , the applied voltage was raised to keep the current through the device constant, and the junctions became more intense, with linear cutoffs at high photon energies that increased as the anode field increased. By varying the operating conditions so that different average fields corresponded to the same anode field, we were able to show that it was the anode field, and not the average field, that controlled the emission characteristics.

Since the samples were fabricated on randomly roughened samples, the "antenna" factor was not known and no compression of the data analogous to the analysis of the

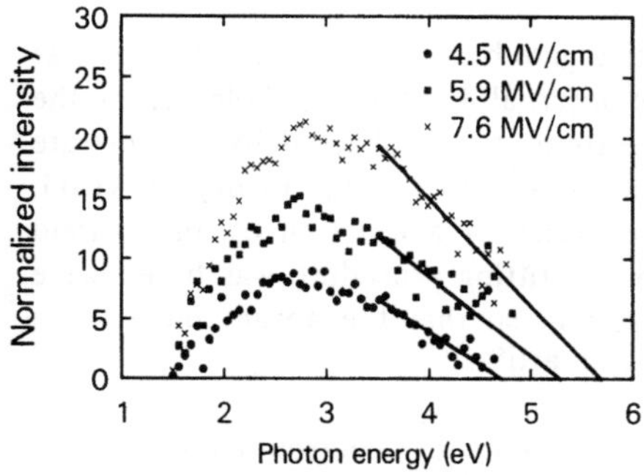

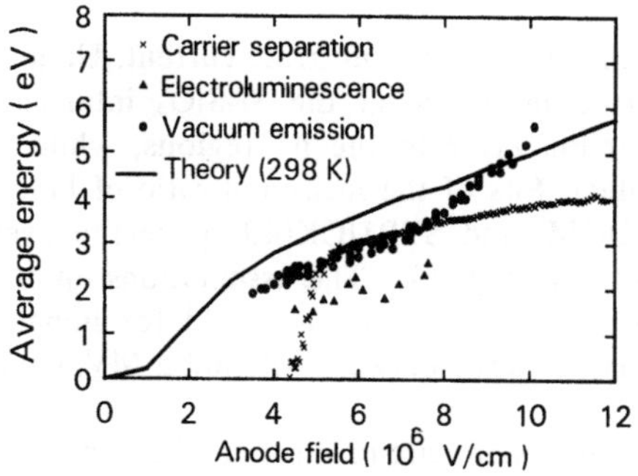

Fig. 11. Emission from an Al gate electron injector structure fabricated on an intentionally roughened polysilicon film on a silicon substrate. A background from a smooth control sample has been subtracted out. The excess emission from the rough sample, due to surface plasmon mediated emission, becomes more intense and has a linear cutoff at higher photon energies as the field at the anode increases. The solid lines are fits to the data above 3.6 eV. The increase in the cutoff energies implies more electron heating at higher anode fields. (From Theis et. al.[48]).

Fig. 12. Comparison of the average energy of hot electrons in SiO_2, measured with respect to the bottom of the SiO_2 conduction band or the Al Fermi level, for surface plasmon mediated emission (electroluminescence) measurements, carrier separation, vacuum emission, and a theoretical prediction using a Monte-Carlo simulation. There is some scatter in the absolute average energies given by the different methods, but they all indicate that the average energy increases with anode field, but does not " run away ". The Monte-Carlo calculation invokes a combination of umklapp scattering from acoustic phonons with energy loss to optical phonons to stabilize the hot electrons in the silicon dioxide.

tunnel junction data was possible. However, since it is probable that the emission processes are the same, just as in the tunnel junction case the energy at which electrons are injected into the Al films should be roughly equal to the cutoff energy of the emission spectra. The solid curves in Fig. 11 are linear fits to the data above 3.6 eV. The intercepts of these lines increased with anode field, implying that the injected electrons, and therefore the electrons in the SiO_2, became more energetic as the anode field was increased.

We plot in Fig. 12 the inferred electron energies, measured relative to the bottom of the conduction band of the SiO_2, for a series of electroluminescence measurements. Also included for comparison in this figure are the results of two other experimental techniques and a Monte-Carlo simulation of electron heating in SiO_2.

The x's in Fig. 12 represent the results from carrier separation measurements[47,48]. These experiments were performed on field-effect transistors based on the electron injector structure. The deposition sequence of the SiO_2 and Si-rich SiO_2 layers was the reverse of that for the luminescence devices: the gate electrode was n^+ polycrystalline Si, and the n-type Si substrate had p^+ source and drain diffusions. The Si- rich layer allowed enhanced current flow from the gate, through the SiO_2 and into the Si substrate. Energy loss in the Si was by the competing mechanisms of phonon generation and impact ionization. The ionization rate was monitored by separating the electron and hole components of the substrate current. Under negative gate bias a hole inversion layer was formed in the Si at the Si-SiO_2 interface. The internal field swept holes from the inversion layer to the p^+ regions, while the electrons were collected by a substrate contact. Fits of the measured ratio of hole to electron currents to the theory of ALIG, BLOOM, and STRUCK[49] gave the average energies as a function of anode field plotted in Fig. 12. The assumptions upon which this fitting procedure was based were not reliable near the threshold for impact ionization, so that the abrupt increase in average energy between 4.5 and 5 MV/cm may be an artifact.

The open circles in Fig. 12 represent measurements of electron emission into vacuum[48]. The electron injector structures used for these experiments were identical to those used for electroluminescence, except that no roughened n^+ polysilicon layer was used. The devices were biased and electrons emitted through a 250 Å Al gate electrode were energy analyzed using an electron focusing and energy analyzing system similar to that described by Solomon[50]. The data of Fig. 12 represent average energies of the electron distributions measured in this way.

Each of the three experimental techniques described above had uncertainties associated with them. We assumed that the linear cutoffs in the electroluminescence data acted in a way analogous to the tunnel junction data. It is not obvious that this should be true. For example, hot electron mean free paths at these energies become quite short[40], and the energy dependence of the transmission through the film should influence the shape of the spectral cutoff. The same difficulty exists in the vacuum emission data. How are the measured distributions skewed by passage through the Al films? What is the role of pinholes? The carrier separation data must be analyzed in terms of a theoretical model. Given these uncertainties, it is remarkable how well the three techniques agree with each other. Although there are shifts of a volt or so in the curves, the agreement in the slopes of the curves is excellent.

Theoretical studies of transport in polar insulators in general[51] and SiO_2 in particular[52,53,54,39] have emphasized longitudinal-optical (LO) phonon scattering as the dominant energy-loss mechanism for conduction band electrons. This has been demonstrated experimentally for SiO_2 for fields up to 1 MV/cm[55], but the results of Fig. 12 show that it is not the case for fields from 4-12 MV/cm. Because the rate of energy loss to the lattice is peaked near the dominant LO- phonon energy $\hbar\omega_{LO}$ (about 0.15 eV), there exists a threshold electric field F_{th} above which the electronic energy would be unstable, were optical phonon scattering the only loss mechanism[39]. It has often been assumed that above this threshold field electrons are accelerated to impact ionization energies[51,52,53,54,55]. However, acoustic-phonon scattering must become dominant at higher energies, and RIDLEY[56] has suggested that this loss mechanism may stabilize the electronic energy distributions at fields above F_{th} but below impact ionization energies. The data of Fig. 12 show that the energy distribution is indeed stabilized.

The solid curve in Fig. 12 represents the predictions of a Monte-Carlo simulation[57] of hot-electron transport in SiO_2, including energy loss to LO phonons, but also including umklapp scattering from acoustic phonons. The addition of the acoustic phonon scattering stabilizes the predicted electron distributions in a way that does not happen for optical phonon scattering alone. Aside from a constant shift, the agreement between the Monte-Carlo simulation and experiment is remarkable, especially when one considers that there are no adjustable parameters in this calculation.

Carrier separation experiments as a function of SiO_2 thickness[47] show that the electron distribution is stabilized after travelling about 30 Å through the SiO_2. These measurements, combined with the successful Monte-Carlo calculation, show that electron-heating in SiO_2 can be quantitatively understood in a straightforward way if one includes the effects of acoustical phonon scattering.

The results from the electron injector devices combined with those from the light emitting tunnel junctions show that surface plasmon mediated luminescence can be a useful spectroscopic tool for understanding hot electron dynamics in metals and electron transport in insulators. Since the emission characteristics are independent of the pumping mechanism, the ideas we have outlined should be applicable to a number of other systems.

Acknowledgement

We would like to acknowledge the technical assistance of A.M Torreson, J.A. Tornello, and F.L. Pesavento in this work. A number of the electron injector samples were made by D.W. Dong (deceased). We would like to thank D.L. Mills for his help in working out the "antenna" factors and pointing out to us the literature on acoustic phonon umklapp scattering in insulators. P.J. Price pointed out an important reference on impact ionization in Si to us. We would also like to acknowledge useful discussions with P.K. Hansma, D.J. Scalapino, J.J. Chang, and P.M. Solomon on hot electron and light emission effects.

REFERENCES

1. R.A. Ferrell: Phys. Rev. 111, 1214 (1958).
2. R.H. Ritchie and H.B. Eldridge: Phys. Rev. 126, 1935 (1962).
3. M.S. Chung, T.A. Callcott, E. Kretschmann, and E.T. Arakawa: Surf. Sci. 91, 245 (1980).
4. J. Bodesheim and A. Otto: Surf. Sci. 45, 441 (1974).
5. D-L Hornauer: Opt. Commun. 16, 76 (1976).
6. E. Kretschmann and H. Raether: Z. Naturf. 23a, 2135 (1968).
7. Ye-Yung Teng and E.A. Stern: Phys. Rev. Lett. 19, 511 (1967).
8. D. Heitman: J. Phys. C 10, 397 (1977).
9. J.C. Tsang, J.R. Kirtley, and T.N. Theis: Solid State Commun. 35, 667 (1980).
10. A. Girlando, W. Knoll, and M.R. Philpott: Solid State Commun. 38, 895 (1981).
11. J.R. Kirtley, T.N. Theis, J.C. Tsang, and D.J. DiMaria: Phys. Rev. B 27, 4601 (1983).
12. J. Lambe and S.L. McCarthy: Phys. Rev. Lett. 37,923 (1976).
13. S.L. McCarthy and J. Lambe: Appl. Phys. Lett. 30, 427 (1977).
14. D.L. Mills, M. Weber, and Bernardo Laks: in Tunneling spectroscopy, edited by P.K. Hansma (Plenum, New York 1982), p. 121.
15. P.K. Hansma and H.P. Broida: Appl. Phys. Lett. 32, 545 (1978).
16. A. Adams and P.K. Hansma: Phys. Rev. B 23, 3597 (1981).
17. A. Adams, J.C. Wyss, and P.K. Hansma: Phys. Rev. Lett. 42, 912 (1979).
18. L.C.Davis: Phys. Rev. B 16, 2482 (1977).
19. D. Hone, B. Muhlschlegel, and D.J. Scalapino: Appl. Phys. Lett. 33, 203 (1978).
20. R.W. Rendell, D.J. Scalapino, and B. Muhlschlegel: Phys. Rev. Lett. 41, 1746 (1978).

21. B. Laks and D.L. Mills: Phys. Rev. B 20, 4962 (1979).
22. B. Laks and D.L. Mills: Phys. Rev. B 21, 5175 (1980).
23. J. Moreland, A. Adams, and P.K. Hansma: Phys. Rev. B 25, 2297 (1982).
24. E.N. Economou: Phys. Rev. 182, 539 (1969).
25. J.R. Kirtley, T.N. Theis, and J.C. Tsang: Appl. Phys. Lett. 37, 435 (1980).
26. J.R. Kirtley, T.N. Theis, and J.C. Tsang: Phys. Rev. B 24, 5650 (1981).
27. N. Kroo, Zs. Szentirmay, and J. Felszerfalvi: Phys. Lett. 81A, 399 (1981).
28. N. Kroo, Zs. Szentirmay, and J. Felszerfalvi: Phys. Lett. 86A, 445 (1981).
29. B. Laks and D.L. Mills: Phys. Rev.B 22, 5723 (1980).
30. I. Pockrand: Phys. Lett. 49A, 259 (1974).
31. D. Heitman and H. Raether: Surf. Sci. 59, 17 (1976).
32. R.H. Ritchie, A.T. Arakawa, J.J. Cowan, and R.N. Hamm: Phys. Rev. Lett. 21, 1530 (1968).
33. N. Kroo, Zs. Szentirmay, and J. Felserfalvi: Phys. Status Solidi 102, 227 (1980).
34. T.N. Theis, J.R. Kirtley, D.J. DiMaria, and D.W. Dong: Phys. Rev. Lett. 50, 750 (1983).
35. T.N. Theis, J.R. Kirtley, D.J. DiMaria, and D.W. Dong: in Insulating films on semiconductors, edited by J.F. Verweij and D.R. Wolters, (North-Holland, Amsterdam 1983), p. 134.
36. L.I. Schiff: Quantum Mechanics, 3rd edition, (McGraw-Hill, New York 1968), p. 397.
37. J.G. Simmons: J. Appl. Phys. 34, 1793 (1983).
38. J.C. Lau and R.V. Coleman: Phys. Rev. B 24, 2985 (1981).
39. H-J. Fitting and J-U. Friemann: phys. stat. sol. (a) 69, 349 (1982).
40. C.R. Crowell and S.M. Sze: in Physics of Thin Films, vol.4, edited by George Hass and Rudolf E. Thun (Academic, New York 1967) p. 325.
41. H.W. Willemsen and K.E. Gray: Phys. Rev. Lett. 41, 812 (1978).
42. J.R. Kirtley, D.S. Kent, D.N. Langenberg, S.B. Kaplan, J.J. Chang, and C.C. Yang: Phys. Rev. B. 22, 1218 (1978).
43. H. Kanter: Phys. Rev. B. 1, 522 (1970).
44. D.J. DiMaria, R. Ghez, and D.W. Dong: J. Appl. Phys. 51, 4830 (1980).
45. D.J. DiMaria, K.M. DiMeyer, C.M. Serrano and D.W. Dong: J. Appl. Phys. 52, 4825 (1981).
46. D.J. DiMaria, J.R. Kirtley, E.J. Pakulis, D.W. Dong, T.S. Kuan, F.L. Pesavento, T.N. Theis, J.A. Cutro, and S.D. Brorson: J. Appl. Phys. (June 15, 1984).
47. D.J. DiMaria, T.N. Theis, J.R. Kirtley, F.L. Pesavento, D.W. Dong, and S.D. Brorson: accepted for publication in J. Appl. Phys..
48. T.N. Theis, D.J. DiMaria, J.R. Kirtley, and D.W. Dong: Phys. Rev. Lett. 52, 1445 (1984).
49. R.C. Alig, S. Bloom, and C. W. Struck: Phys. Rev. B. 22, 5565 (1980).
50. P.M. Solomon: in The Physics of SiO_2 and Its Interfaces, edited by S.T. Pantelides, (Pergamon Press, New York 1978), p.35.
51. K.K. Thornber and Richard P. Feynmann: Phys. Rev. B 1, 4099 (1970).
52. W. T. Lynch: J. Appl. Phys. 43, 3274 (1972).
53. D.K. Ferry: Appl. Phys. Lett. 27, 689 (1975).
54. D.K. Ferry: J. Appl. Phys. 50, 1422 (1979).
55. R.C. Hughes: Phys. Rev. Lett. 35, 449 (1975).
56. B. K. Ridley: J. Appl. Phys. 46, 998 (1975).
57. M. Fischetti: submitted to Phys. Rev. Lett.

Spectroscopy of Two Dimensional Plasmas in Semiconductor Heterostructures

A. Pinczuk

AT&T Bell Laboratories, Murray Hill, NJ 07974, USA

This contribution reviews optical studies of two dimensional plasmas confined in semiconductor heterojunctions and superlattices. Experiments in modulation doped GaAs-(AlGa)As heterostructures are considered.

1. INTRODUCTION

The electronic structure of semiconductor heterojunctions and superlattices can show a reduction in dimensionality that leads to properties in common with those of an idealized two dimensional electron gas. The combined effects of the discontinuities in the energy band structure and space charge electric fields act to *quantize* the electron motion normal to the plane of the *heterojunctions* into *discrete* energy levels. In these states the in-plane motion is free. Therefore, each of the discrete energy levels gives origin to a *two dimensional subband*. The proposal of ESAKI and TSU [1] of synthetic thin layer semiconductor superlattices was followed by extensive research of *quasi* two dimensional behavior in semiconductor heterojunctions and superlattices. The recent discovery of fractional quantization of the Hall effect in high mobility 2D electron gases at GaAs-(AlGa)As heterostructures [2,3] is one of the most exciting new developments in solid state and low temperature physics.

This presentation covers a few relatively recent developments in the *optical spectroscopy* of two dimensional electron systems[1] in semiconductor heterostructures and superlattices. Optical spectroscopy plays a foremost role in the field. In the early experimental efforts, in the GaAs-(AlGa)As system, photocurrent measurements [4], optical absorption [5] and resonant Raman scattering [6] were used to identify the discrete quantization of energy levels in quantum wells and superlattices. In 1978 BURSTEIN et al. [7] proposed inelastic light scattering as a spectroscopic method for the investigation of the elementary excitations of 2D electron systems in semiconductors. It was emphasized that with *resonant* enhancements of the light scattering cross-section there is sensitivity to observe elementary excitations of 2D electron or hole gases with areal densities lower than $5\times10^{11}cm^{-2}$. The proposal led to the

1. The expression two dimensional electron systems is used here in a general sense and refers to both electrons or holes.

first observations of light scattering by *intersubband* excitations of electrons confined in GaAs- (AlGa)As heterostructures [8,9]. During the last five years extensive light scattering investigations of 2D electron systems in semiconductors gave insight into energy levels and collective electron-electron interactions.

This short review succinctly considers light scattering research in GaAs- (AlGa)As heterostructures. Recent work [10,11,12] on optical absorption and emission processes by 2D electron plasmas confined in these heterostructures is also discussed. The reader interested in light scattering scattering by free carrier excitations in semiconductors, including those of 2D systems, will find extensive discussions in a recently published review chapter [13]. The optical spectroscopy of semiconductor heterostructures is of great current interest. Some of the most exciting recent work has been presented at two international conferences held this summer [14,15].

The results described here have been obtained in *modulation doped* [16,17] GaAs- (AlGa)As *quantum-well* heterostructures fabricated by *molecular beam epitaxy*. For the case of n-type systems, the multilayers have the structure shown in Fig. 1. The free carriers occupy *conduction* subband states in the GaAs quantum wells, where they constitute a two dimensional electron gas. The undoped (AlGa)As spacer, with thickness d_3, separates further the mobile electrons from the ionized donors. This results in a substantial enhancement of the electron mobility [18]. A similar multilayer structure can be modulation doped with Be acceptors leading to a p-type system. In this case the free carriers constitute 2D hole gases in valence subband states of the GaAs quantum wells [19].

2. LIGHT SCATTERING SPECTROSCOPY

Within the framework of the effective-mass approximation the mechanisms and selection rules for resonant light scattering by two

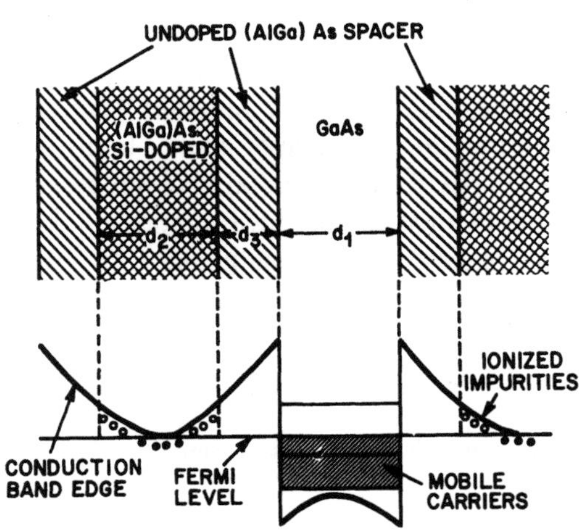

Fig. 1. Sequence of layers and relative position of the conduction band edge in, *n*-type, modulation-doped multiple quantum-well GaAs-(AlGa)As heterostructures

dimensional plasmas in semiconductors are similar to those of 3D systems [7,20]. Resonant enhancements, that are essential in these experiments, are expected for photon energies near the optical gaps that involve states occupied by the free carriers. Of interest to us are the *direct* gap zincblende semiconductors. In these cases the relevant resonances occur at optical gaps near the Brillouin Zone center.[2] In GaAs we are concerned with the fundamental and spin-orbit split-off optical gaps of $E_0 \simeq 1.5\,eV$ and $E_0 + \Delta_0 \simeq 1.9\,eV$. For 2D electron gases the majority of the experiments have been carried out with photon energies near the $E_0 + \Delta_0$ gap. The difficulties in obtaining light scattering spectra in resonance with the E_0 gap, due to intense luminescence background, have been recently resolved for the case of 2D hole gases in modulation doped GaAs- (AlGa)As quantum well heterostructures [21].

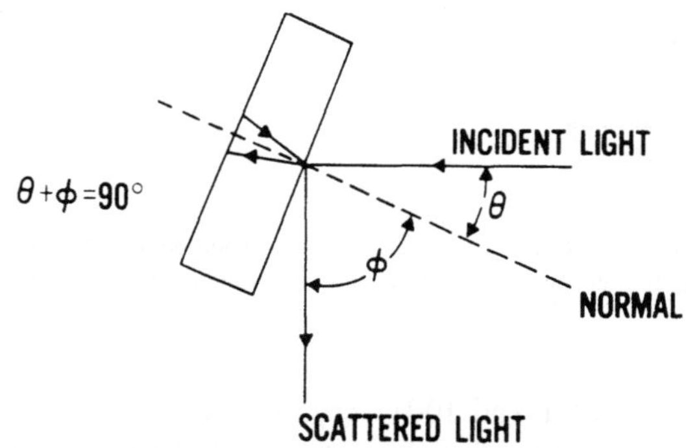

Fig. 2. Nearly backscattering geometry frequently used in inelastic light scattering

The "backscattering" geometry shown in Fig. 2 has been used in experiments described here. The in-plane and normal components of the scattering wavevector are given by

$$k_{\parallel} = \frac{2\pi}{\lambda}\,(\sin\theta - \cos\theta) \tag{1}$$

and

$$k_{\perp} = \frac{4\pi}{\lambda}\,\eta\,(1 - \frac{1}{4\eta^2}) \tag{2}$$

where λ is the laser wavelength and η is the refractive index. With this pseudo-backscattering geometry, $k_{\parallel}$ can be varied from a small value of $\sim 10^4 cm^{-1}$, at $\theta \approx 45°$, up to a maximum value of $\sim 10^5 cm^{-1}$. In GaAs typical values of $k_{\perp}$ are $\sim 7\times10^5 cm^{-1}$ near the $E_0 + \Delta_0$ gap and $5\times10^5 cm^{-1}$ at the E_0 gap.

Figure 3 shows spectra of intersubband excitations of electrons confined in GaAs quantum wells. z and $\bar{z}$ are the propagation directions of the

2. Figure 5 illustrates a case of optical transitions in resonant inelastic light scattering.

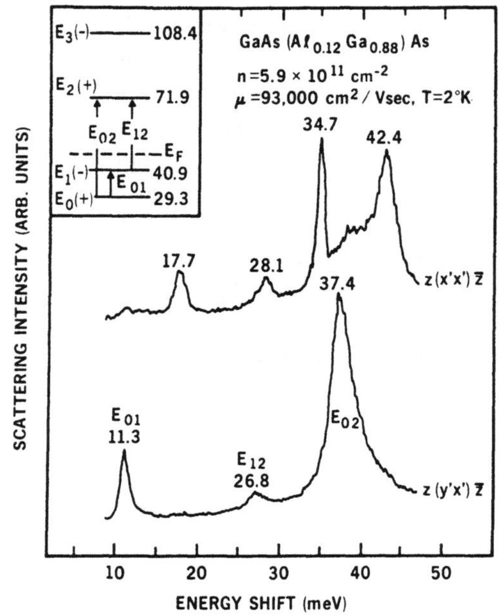

Fig. 3. Polarized and depolarized spectra of 2D electrons confined in a modulation-doped multiple quantum-well sample. The inset shows calculated energy levels of the conduction states and the assignment of intersubband transitions

incident and scattered light, close to (001) and $(00\bar{1})$ crystal axes. x' and y' are (110) and $(1\bar{1}0)$ in-plane directions of polarizations of light. The *depolarized* $z(y'x')\bar{z}$ spectrum shows a set of three bands that are assigned to, *single particle, spin-density* intersubband excitations associated with transitions shown in the inset. These spectra were obtained with $\theta \simeq 45°$. Thus, $k_{\parallel}$ is small and the intersubband transitions can be assumed to be *vertical* in the 2D wavevector space of the conduction subbands. Therefore, since final state interactions for these excitations are small [22,23], the energies measured in $z(y'x')\bar{z}$ spectra are very close to subband spacings. The peaks in the *polarized* $z(x'x')\bar{z}$ spectrum are assigned to, *collective, charge-density* intersubband excitations. In zincblende semiconductors these excitations couple to longitudinal optical phonons [20].

The differences between the energies of single particle and collective excitations are direct evidence of the macroscopic electric fields (or depolarization effect) due to *resonant screening* phenomena. These effects were first predicted by CHEN et al. [24], and were invoked in the interpretation of subband optical absorption [25,26]. The separate measurement of light scattering by single particle and collective excitations have for the first time made possible simple determinations of depolarization field effects [27]. Collective intersubband excitations were considered by DAHL and SHAM [28]. BURSTEIN et al. [20] have included the coupling to optical phonons that occurs in polar semiconductors. These treatments have been extended to incorporate the specific conditions that exist in quantum-well heterostructures [27,29]. In the analysis of light scattering spectra of collective intersubband excitations the only adjustable parameters are the matrix elements of the electron-electron interactions. The values of the matrix elements determined from the experiments are in good agreement with the ones evaluated with simple model calculations [27,30].

One of the important questions in the spectroscopy of intersubband excitations is the relation to transport properties. A correlation has been identified between lineshapes in resonant light scattering spectra and the in-plane mobilities [31]. These studies have been carried out in four similar modulation doped quantum well heterostructures in which the electron mobilities are directly related to d_3, the thickness of the undoped (AlGa)As spacer (see Fig. 1) [18]. The spectral features obtained in light scattering measurements from these samples are similar to those seen in Fig. 3, which correspond to the sample with the widest d_3 and greatest electron mobility. The widths of the bands in the spectra from the four samples where found to increase dramatically with decreasing Hall mobility and d_3. The observed correlation is also associated with an unexpected dependence of the linewidths on incident photon energy in the range of the $E_0 + \Delta_0$ optical gap [31]. This behavior is most marked in samples having Hall mobilities lower than $\sim 25{,}000\ cm^2/Vsec$.

To interpret these results it was proposed that resonant inelastic light scattering is subject to wavevector relaxation processes related to those that limit the electron mobility. In modulation-doped samples these processes, which involve scattering of electrons by the Coulomb potential of the ionized donors, depend crucially on d_3. Consideration of these relaxation processes leads to breakdown of wavevector conservation for resonant light scattering [13,31]. Thus, *non vertical* intersubband transitions with an in-plane wavevector transfer $q \neq k_{\parallel}$ become active. The spectral bands of non-vertical transitions have widths of $\sim \hbar q v_F$ (where v_F is the Fermi velocity). The stronger resonant behavior of light scattering with relaxation of wavevector conservation [13,32] is invoked to explain the dependence of linewidth on photon energy.

The resonant enhancement of light scattering by intersubband excitations has been reported in single [8,33,34] and multiple [9,35] GaAs-(AlGa)As heterostructures. The results at photon energies near the $E_0 + \Delta_0$ optical gaps of the GaAs layers display the general behavior anticipated by BURSTEIN et al. [7,20]. Figure 4 shows the results that correspond to E_{01} and E_{02} transitions in the $z(y'x')\bar{z}$ spectra like that of Fig. 3. The integrated intensities of the *even* parity transition E_{02} are much larger than those of the *odd* parity transition E_{01}. Scattering by the odd parity excitation, that breaks the parity selection rule for dipole transitions in quantum wells [5], has been assigned to $k_{\perp}$-dependent contributions to the optical matrix element [13].

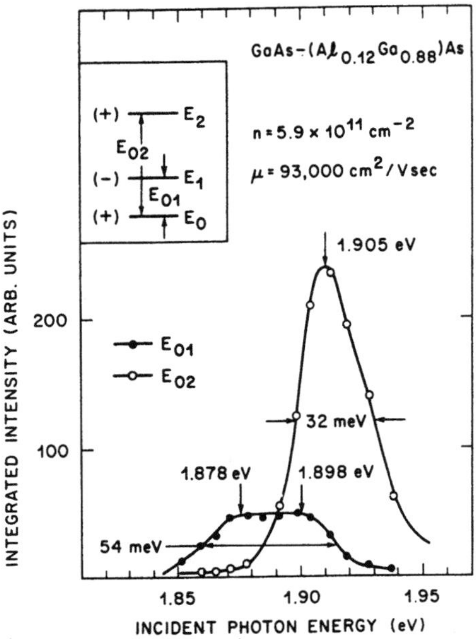

Fig. 4. Resonant enhancement of light scattering by single-particle, spin-density, intersubband excitations of 2D electrons confined in modulation-doped multiple quantum wells. The $z(y'x')\bar{z}$ spectra are similar to those in Fig. 3

Resonant inelastic light scattering by very high mobility 2D hole gases has been recently reported in p-type, modulation-doped, GaAs- (AlGa)As quantum wells [21]. These *multilayer* 2D hole systems display striking phenomena like the quantized Hall effect (integral and fractional) [19,36]. They also offer unique opportunities for the investigation of the intricate structure of the valence subband states in the quantum wells. The complexities are related to the degenerate structure of the valence band states in zincblende-type semiconductors that in quantum wells leads to separate, but coupled, light and heavy valence subbands. The light scattering spectra display the complex character of the subband structure through spectral lineshapes that are very different from those from 2D electron gases in the n-type quantum-well heterostructures.

For 2D hole gases in GaAs the spin-orbit split-off gap $E_0 + \Delta_0$, that is widely used in the case of electron gases, does not give resonant enhancements of the light scattering cross section [7,20]. The spectra were excited with tunable infrared dye lasers with photon energies (1.6-1.7 eV) chosen to be in resonance with one of the optical gaps that involve higher lying conduction subband states of the GaAs quantum wells. The intersubband transitions of interest are shown in Fig. 5(a). We consider transitions from the ground heavy subband h_0 to the lowest light subband l_0 and to the first excited heavy subband h_1. Figure 5(b) shows the two optical transitions that

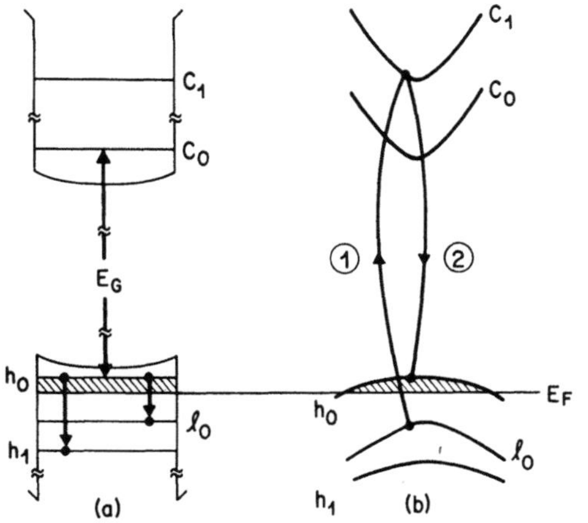

Fig. 5. Energy levels and subband structure in the GaAs quantum wells in p-type GaAs-(AlGa)As multiple quantum wells. (a) Shows the energy gap E_G and indicates intersubband transitions of holes. (b) Optical transitions in resonant inelastic light scattering by $h_0 \rightarrow l_0$ excitations

participate in resonant light scattering by $h_0 \rightarrow l_0$ excitations. This procedure for obtaining spectra by 2D holes gases has the advantage of large resonant enhancements and at the same time minimizes the background due to optical emission at the fundamental gap E_G.

Figure 6 shows spectra measured in one of the lower concentration samples. All the features that appear in the spectra have the resonant enhancements that are associated with the processes represented in Fig. 5. The structures labeled $h_0 \rightarrow l_0$ and $h_0 \rightarrow h_1$, that occur in depolarized as well as polarized spectra, are assigned to the intersubband transitions shown in Fig. 5(a). The polarized $z(y'y')\bar{z}$ spectrum has two additional structures. One is the band at the energy of the LO phonon of GaAs (at 36.6 meV). The other is the intense and broad structure at energies below those of conventional intersubband excitations. Higher resolution spectra show that this low energy scattering actually has a maximum at an energy shift of ~1 *meV*. The absence of large differences between peak positions measured in $z(y'y')\bar{z}$ and $z(y'x')\bar{z}$ spectra suggest that depolarization feld effects, if any, are much weaker than in the 2D electron gases discussed above. This interpretation is consistent with the observation of optical lattice vibrations at the LO phonon energy, uncoupled to the free holes.

The striking difference in spectral lineshapes of $h_0 \rightarrow l_0$ and $h_0 \rightarrow h_1$ excitations can be interpreted in terms of the different dispersion of the corresponding subbands in the two dimensional space of the in-plane wavevector $k_{\parallel}$. The $h_0 \rightarrow h_1$ excitations appear in the spectra of Fig. 6 as relatively sharp peaks, indicating that close to the Brillouin zone center the subbands are nearly parallel. The $h_0 \rightarrow l_0$excitations appear as broad structures because the two subbands have very different dispersions. The spectral lineshapes of both transitions are found to depend on incident photon

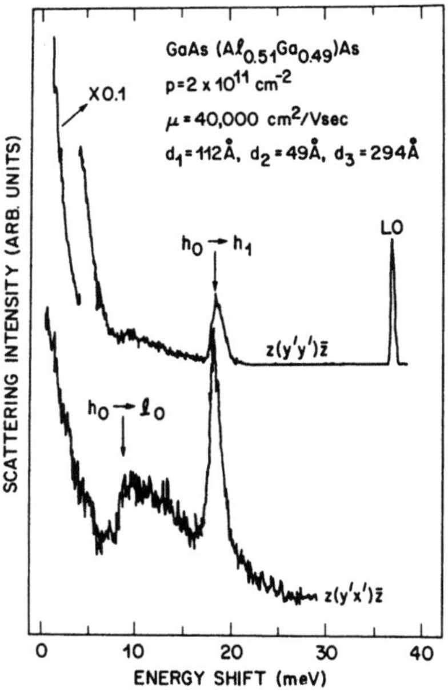

Fig. 6. Polarized and depolarized spectra of 2D holes confined in p-type modulation doped GaAs-(AlGa)As quantum wells. The structures labeled $h_0 \rightarrow l_0$ and $h_0 \rightarrow h_1$ (see Fig. 5) are assigned to the corresponding transitions

energy. These behaviors reflect the *non-parabolicity* of the valence subbands. The measured energies of $h_0 \rightarrow h_1$ excitations are in excellent agreement with the subband spacings calculated with a square well model using the energy band discontinuities and heavy hole mass proposed by MILLER et al. [37].

The low energy scattering observed in polarized $z(y'y')\bar{z}$ spectra is interesting because it occurs below the conventional intersubband excitations. It was also determined that its position and lineshape do not have a significant dependence on $k_{\parallel}$ for the values attainable in a nearly backscattering geometry (see Eqs. (1) and (2)). This indicates that the low energy scattering is not due to superlattice plasmons of the kind reported by OLEGO et al. [38] in the n-type multiple quantum wells. An interesting interpretation of these unexpected observations is that the low energy scattering is related to transitions of holes within a split ground valence subband h_0. This interpretation is supported by the temperature dependence of the low energy scattering that shows broadening and energy shifts like those of intersubband excitations even at temperatures $T \lesssim 20K$. These results are intriguing. Although a lifting of the Kramers degeneracy has been reported in single heterojunctions [19], it should not exist in quantum wells in which the potentials have mirror symmetry.

3. OPTICAL PROCESSES OF 2D PLASMA

Optical processes, emission and absorption, have been extensively investigated in semiconductor heterostructures and superlattices. Undoped quantum-wells show the sharp structure due to quasi-2D excitons[5,37,39-42]. Band to band recombination was observed in InAs-GaSb superlattices[43]. Photoexcited systems were studied in GaAs-(AlGa)As heterostructures[44,45] and in doping superlattices of GaAs (n-i-p-i)[46]. In contrast to previous work, the optical spectra from modulation-doped heterostructures show the distinctive behavior expected from a degenerate 2D electron plasma in a semiconductor[10,11]. The spectral lineshapes of optical recombination at the energy gap have simple interpretations in terms of densities of states and equilibrium occupation factors. The optical gap for emission is subject to substantial renormalizations, comparable to the Hartree energies of the 2D plasmas. Luminescence excitation spectra allowed the study of optical absorption processes[40]. In the higher mobility samples they show well defined absorption onsets, with relatively weak additional structure that suggests the persistence of final-state electron-hole interactions even at large plasma density ($n > 5 \times 10^{11} cm^{-2}$).

Figure 7 compares the emission (luminescence) spectrum with the excitation spectrum (equivalent to absorption) near the energy gap of a modulation doped GaAs-(AlGa)As multiple quantum well heterostructure.

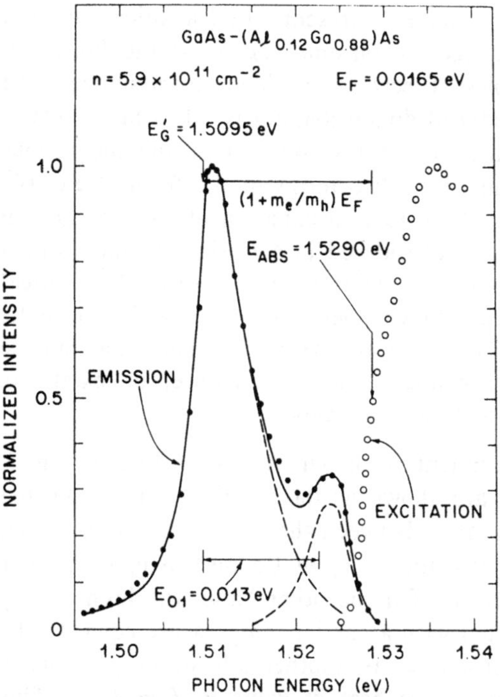

Fig. 7. Spectra of optical emission (full dots) and luminescence excitation (open dots) of the 2D electron plasma confined in GaAs-(AlGa)As quantum wells. The lines are a fit to the emission spectra as explained in the text. Dashed lines are the contributions due to each of the two occupied conduction subbands (see Fig. 3). The full line is the total fit. Inelastic light scattering spectra from the sample are shown in Fig. 3

This is the high mobility sample that has the inelastic light scattering spectrum shown in Fig. 3. We can see a characteristic separation between the emission and excitation spectra. The energy difference between the maximum in the emission spectrum, E_G', and the onset of absorption, E_{ABS}, has been represented as

$$E_F(1 + \frac{m_e}{m_h}) \tag{3}$$

where E_F is the Fermi energy of the 2D plasma. In the case of parabolic conduction and valence subbands with effective masses m_e and m_h Eq. (3) represents the separation between the optical gap and the onset of absorption in a degenerate semiconductor electron plasma[47]. Quantitative agreement with experiment has been achieved by using $m_e = 0.068m_0$[48] and considering m_h as an adjustable parameter. We obtain $m_h = (0.39 \pm 0.03)m_0$.

The emission spectra were fitted with the lineshape function

$$L(\hbar\omega) = A\int_0^\infty \rho(E)f_e(E)f_h(E)\delta(E-\hbar\omega)dE \tag{4}$$

where ω is the emitted photon energy and E represents the transition energies. $f_e(E)$ is the Fermi function of the electron plasma. $f_h(E)$ is the Boltzman factor of the weakly photoexcited holes. Electrons and holes are assumed to be at the same temperature T. The reduced dimensionality is taken into account by introducing step-like 2D densities of states $\rho(E)$ with inhomogeneous broadening Γ_i. In the case shown in Fig. 7 two terms of the form of Eq. (4) are considered since, as seen in Fig. 3, there is occupation of two conduction subbands. Homogeneous broadening is considered in $\delta(E-\hbar\omega)$ by means of a damping parameter Γ_h (in high mobility samples $\Gamma_h = 1$ meV). The optical matrix element is taken as energy independent together with parabolic conduction and valence subbands. These are approximations since the valence subbands are known to be highly nonparabolic. Final-state, or excitonic, electron-hole interactions are excluded from this analysis.

The best fit to the experiment is shown as a solid line in Fig. 7. The two terms required for the fit are shown as dashed lines. The adjusted value of the energy gap and the separation between the two terms are indicated in the figure. An inhomogeneous broadening of $\Gamma_i = 1.5$ meV is used in order to fit the low energy tail of the spectra. On the other hand, the high energy tail of the spectrum could not be fitted with a carrier temperature equal to the lattice temperature. Thus, T is considered as another adjustable parameter. For $m_e = 0.068m_0$ and $m_h = 0.39m_0$, the best fit value is $T = 7$K . This behavior may be a consequence of valence subband mixing[12] as well as indication of final-state interaction and nonparabolicity in the ground valence subband. Other effects apparent in this analysis are large renormalizations of the energy gap and breakdown of the parity selection rule for the optical matrix element in quantum wells[11]. The determination of E_{01} made from the emission spectra can be compared with the one from inelastic light scattering spectra shown in Fig. 3. The difference of ~2 meV is an indication of final-state interactions associated with intersubband excitations[49].

4. CONCLUDING REMARKS

The intensive work in modulation doped GaAs-(AlGa)As heterostructures has established optical spectroscopy as a valuable tool for studies of semiconductor plasmas under the conditions of reduced dimensionality that exist in semiconductor heterostructures. Inelastic light scattering can be used in the elucidation of the energy level structure and collective interactions. One of the most interesting applications of the optical emission of 2D plasmas is in the capability to study valence to conduction subband transitions in 2D wavevector space. As shown recently[50], this allows an unexpected test of effective-mass theories of the electronic states in semiconductor heterostructures.

REFERENCES

[1] L. Esaki and R. Tsu: IBM J. Res. Develop. *14,* 61 (1970).

[2] D. C. Tsui, H. L. Störmer and A. C. Gossard: Phys. Rev. Lett. *48,* 159 (1982).

[3] H. L. Störmer, A. Chang, D. C. Tsui, J. C. Hwang, A. C. Gossard and W. Wiegmann: Phys. Rev. Lett. *50,* 1953 (1983).

[4] R. Tsu, L. L. Chang, G. A. Sai-Halasz and L. Esaki: Phys. Rev. Lett. *34*, 1509 (1975).

[5] R. Dingle, W. Wiegmann and C. H. Henry: Phys. Rev. Lett *33*, 827 (1974); also R. Dingle, A. C. Gossard and W. Wiegmann: Phys. Rev. Lett. *34,* 1327 (1975).

[6] P. Manuel, G. Sai-Halasz, L. L. Chang, C. A. Chang and L. Esaki: Phys. Rev. Lett. *35*, 1701 (1976).

[7] E. Burstein, A. Pinczuk and S. Buckner: *Physics of Semiconductors 1978* (Institute of Physics, London 1979), p. 1231.

[8] G. Abstreiter and K. Ploog: Phys. Rev. Lett. *42*, 1308 (1979).

[9] A. Pinczuk, H. L. Störmer, R. Dingle, J. M. Worlock, W. Wiegmann and A. C. Gossard: Solid State Commun. *32*, 1001 (1979).

[10] A. Pinczuk, J. Shah, H. L. Störmer, R. C. Miller, A. C. Gossard and W. Wiegmann: Surface Sci. *142*, 492 (1984).

[11] A. Pinczuk, J. Shah, R. C. Miller, A. C. Gossard and W. Wiegmann: Solid State Commun. *50,* 735 (1984).

[12] R. Sooryakumar, D. S. Chemla, A. Pinczuk, A. C. Gossard and W. Wiegmann: J. Vac. Sci. Technol. *B2*, 349 (1984).

[13] G. Abstreiter, M. Cardona and A. Pinczuk: *Light Scattering in Solids IV* (Springer Verlag, Berlin-Heidelberg 1984), pp. 5-150.

[14] Proceedings of the 17th International Conference on the Physics of Semiconductors, San Francisco 1984, to be published.
[15] Proceedings of the 1st International Conference on Superlattices, Microstructures and Microdevices, Urbana 1984, to be published.
[16] R. Dingle, H. L. Störmer, A. C. Gossard and W. Wiegmann: Appl. Phys. Lett. *33*, 665 (1978).
[17] H. L. Störmer: J. Phys. Soc. Japan *49* Suppl. A., 1013 (1980).
[18] H. L. Störmer, A. Pinczuk, A. C. Gossard and W. Wiegmann: Appl. Phys. Lett. *38*, 691 (1981).
[19] H. L. Störmer, A. M. Chang, Z. Schlesinger, D. C. Tsui, A. C. Gossard and W. Wiegmann: Phys. Rev. Lett. *51*, 126 (1983).
[20] E. Burstein, A. Pinczuk and D. L. Mills: Surface Sci. *98*, 451 (1980).
[21] A. Pinczuk, H. L. Störmer, A. C. Gossard and W. Wiegmann: to be published in Ref. [14].
[22] T. Ando and S. Mori: J. Phys. Soc. Japan, *47* 1518 (1979).
[23] S. Katayama and T. Ando: to be published in Ref. [14].
[24] W. P. Chen, Y. J. Chen and E. Burstein: Surface Sci. *58*, 263 (1976).
[25] S. J. Allen, Jr., D. C. Tsui and B. Vinter: Solid State Commun. *20*, 425 (1976).
[26] T. Ando: Solid State Commun. *21*, 133 (1977).
[27] A. Pinczuk, J. M. Worlock, H. L. Störmer, R. Dingle, W. Wiegmann and A. C. Gossard: Solid State Commun. *36*, 43 (1980).
[28] D. Dahl and L. J. Sham: Phys. Rev. *B16*, 651 (1977).
[29] A. Pinczuk and J. M. Worlock: Surface Sci. *113*, 69 (1982).
[30] G. Fishman: Phys. Rev. *B27*, 7611 (1983).
[31] A. Pinczuk, J. M. Worlock, H. L. Störmer, A. C. Gossard and W. Wiegmann: J. Vac. Sci. Technol. *19*, 561 (1981).
[32] A. Pinczuk, G. Abstreiter, R. Trommer and M. Cardona: Solid State Commun. *30*, 429 (1979).
[33] G. Abstreiter: Surface Sci. *98*, 117 (1980).
[34] A. Pinczuk, J. M. Worlock, H. L. Störmer, A. C. Gossard, W. Wiegmann and R. Dingle: Bull. Am. Phys. Soc. *26*, 254 (1981).
[35] A. Pinczuk, H. L. Störmer, R. Dingle, J. M. Worlock, W. Wiegmann and A. C. Gossard: Surface Sci. *98*, 126 (1980).
[36] H. L. Störmer, D. C. Tsui, A. M. Chang, A. C. Gossard and W. Wiegmann: work in progress.
[37] R. C. Miller, D. A. Kleinman and A. C. Gossard: Phys Rev. *B29*, 3740 (1984).
[38] D. Olego, A. Pinczuk, A. C. Gossard and W. Wiegmann: Phys. Rev. *B25*, 7867 (1982).
[39] R. C. Miller, D. A. Kleinman, W. A. Nordland and A. C. Gossard: Phys. Rev. *B22*, 863 (1980).
[40] C. Weisbuch, R. C. Miller, R. Dingle and A. C. Gossard: Solid State Commun. *37*, 219 (1981).

[41] E. E. Mendez, G. Bastard, L. L. Chang, L. Esaki, H. Morkoc and R. Fischer: Phys. Rev. *B26*, 7101 (1982).
[42] D. A. B. Miller, D. S. Chemla, D. J. Eilenberger, P. W. Smith, A. C. Gossard and W. T. Tsang: Appl. Phys. Lett. *41*, 679 (1982).
[43] P. Voisin, G. Bastard, C. E. T. Gonsalves da Silva, M. Voos, L. L. Chang and L. Esaki: Solid State Commun. *39*, 79 (1981).
[44] R. C. Miller, D. A. Kleinman, O. Munteanu and W. T. Tsang: Appl. Phys. Lett. *39*, 1 (1981).
[45] N. Holonyak, Jr., W. D. Laidig, B. A. Vojak, K. Hess, J. Coleman, P. D. Dapkus and J. Bardeen: Phys. Rev. Lett. *45*, 1703 (1980).
[46] G. H. Döhler, H. Kunzel, D. Olego, K. Ploog, P. Ruden, H. J. Stolz and G. Abstreiter: Phys. Rev. Lett. *47*, 864 (1981).
[47] E. Burstein: Phys. Rev. *93*, 632 (1954).
[48] J. M. Worlock, A. Pinczuk, Z. J. Tien, C. H. Perry, H. L. Störmer, R. Dingle, A. C. Gossard, W. Wiegmann and R. L. Aggarwal: Solid State Commun. *40*, 876 (1981).
[49] R. Sooryakumar, A. Pinczuk, A. C. Gossard and W. Wiegmann: submitted for publication.
[50] R. Sooryakumar, D. S. Chemla, A. Pinczuk, A. C. Gossard, W. Wiegmann: submitted for publication.

Surface Coherent Anti-Stokes Raman Spectroscopy

Rance M. Fortenberry

Optical Sciences Center, University of Arizona, Tucson, AZ 85721, USA

Application of integrated optics to Coherent Anti-Stokes Raman Spectrosocpy (CARS) results in Surface CARS, a highly sensitive analytical tool for the study of adsorbed molecules on surfaces. In CARS two input laser beams at frequencies ω_1 and ω_2 interact non-linearly producing an output signal at the frequency $2\omega_1 - \omega_2$. This process is greatly enhanced when a vibrational resonance of a molecule occurs at the frequency $\omega_1 - \omega_2$. Thus, by tuning one of the two input frequencies, one can obtain a spectrum corresponding to an ordinary Raman spectrum of the interacting molecules.

While Raman spectroscopy is a relatively weak incoherent process, the coherent aspects of CARS lead to Raman signal levels proportional to the cube of the input optical power and the square of the interaction length, provided phase-matching constraints are observed. Trapping this power in a thin film waveguide results in increased power density and longer surface interaction lengths. Film signal levels of 10^9 photons per Q-switched laser pulse and conversion efficiencies of 0.2% have been obtained from polystyrene films. In addition, the coherence of CARS can be exploited to totally eliminate the bulk CARS signal from the waveguide itself, while leaving the CARS signal from adsorbed surface molecules unaffected. Detailed analysis indicates that surface coverages of as low as 0.01 monolayer should be measurable in this way.

Other advantages include high resolution (limited only by laser linewidth) and obtaining of orientation information by varying polarization of input laser beams. Applications to fast transient phenomena, using picosecond pulses, and excited state spectroscopy appear promising.

Part VI

Magnetic Properties

Surface Magnesium by Spin Polarized Electrons

H.C. Siegmann
Laboratorium für Festkörperphysik ETH Zürich, CH-8093 Zürich, Switzerland

Spin polarized electrons are an ideal tool for studying bulk and surface magnetism. Effects of quantum mechanical exchange in elastic and inelastic electron scattering are readily observable with spin modulated electron beams from semiconductor photocathodes, and provide an extremely rich variety of interesting and new phenomena pertinent to open questions in metallic magnetism. Applications include viewing of magnetic domains with unprecedented spatial and time resolution, and non-destructive depth profiling of remanence and coercive force.

1. Introduction

The purpose of this paper is to show that one obtains good and new information on magnetism and particularly surface magnetism by directing an electron beam onto a magnetic material and by observing how the various elastic and inelastic scattering processes depend on the relative direction of the electron spin and the magnetization. An alternative, probably even more powerful method is to measure the spin polarization of electrons photoemitted from magnetic materials. The spin polarized electronic structure in the bulk and at the surface as observed in photoemission often discloses directly the causes of magnetic phenomena.

Historically, neutron beams have been the most important probe of magnetism revealing the magnetic structure. On the knowledge gained from neutron scattering,whole new industries could be built employing ferrites and garnets. However, to the solid state physicist, neutrons behave like classical particles, interacting with magnetism simply by the dipole force. This is completely different with electrons.

Spin polarized electrons display quantum character in a pure form. They allow direct probing of Heisenbergs and Diracs famous exchange interaction. Quantum mechanical exchange leads to many surprising and new phenomena that are not encountered in the classical world of neutron scattering. Another important difference between neutrons and electrons is that electrons can have a very short probing depth of up to one layer only opening up the field of surface magnetism.

The tools employing electron spin polarization come just at an exciting new stage of magnetism. Today, it is possible to engineer very thin new magnetic materials by epitaxial growth. If one deposits say Fe on a GaAs substrate, one can manipulate the lattice parameters of Fe, and one expects to obtain something very different from ordinary Fe. We will see that electron beams or photoelectrons are ideal tools for studying these and other phenomena at surfaces. These latter include surface magnetochemistry, hysteresis loops in very thin films, and dynamic magnetic domain observation with high spatial resolution.

2. Elastic exchange scattering of electrons

Elastic scattering of electrons from atoms or ions with a magnetic moment depends on the orientation of the electron spin s with respect to the atomic spin S. The spin-dependence is relatively weak since the exchange energy $J \vec{S} \cdot \vec{s}$ is generally small compared to the Coulomb energy. The energy-dependent exchange constant J can be calculated rigorously only in the simplest possible case, which is the scattering of 2 electrons on each other. Already the scattering of one electron on an H-atom is very difficult, since it requires a solution to a three body problem [1]. If one now thinks of an Fe-atom embedded in the sea of valence electrons, one can see there is little hope at present to treat exchange scattering from magnetic materials in a rigorous theoretical way.

However, CELOTTA and coworkers have shown that the elastic exchange scattering can readily be measured with a semiconductor spin polarized electron source [2].

The principle of this technique is shown in Fig. 1. It relies on producing a spin polarized electron beam by photoemission of electrons from a semiconductor such as GaAs, GaAsP, Ge or others [3]. The electrons are spin polarized, if circularly polarized light is used for photoemission. This arises as a consequence of the selection rules in optical transitions. The emission intensity is large, the energy spread of the electrons very small, pulsed operation is easy, and most importantly, the direction of spin polarization can be reversed simply by switching the sense of the circular polarization of the incident light. This switching from right to left circularly polarized light does not affect the position and focus of the electron beam. Therefore, one obtains a spin-modulated electron beam.

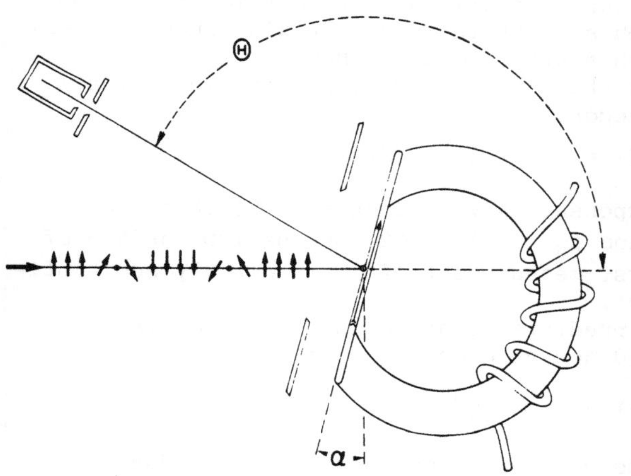

Fig. 1
The spin-modulated electron beam from a GaAs-photocathode is incident from the left onto a sample magnetized in the direction of the arrow by a horse-shoe electromagnet. The angle of incidence α as well as the scattering angle Θ can be varied. The intensity of the electrons scattered elastically into the Faraday cup is measured for electron spin parallel (i↑) and antiparallel (i↓) to the magnetization with lock-in-techniques. In the spin asymmetry $A = (i\uparrow - i\downarrow)/(i\uparrow + i\downarrow)$, the Coulomb-term is eliminated, and the electrons effectively are "surface neutrons".

The spin-modulated electron beam hits the target which is magnetized by a horseshoe-electromagnet. The magnetic stray fields have to be small in order to avoid excessive Lorentz-deflection of the electron beam. A Faraday cup is positioned at the scattering angle Θ and measures the current $i\uparrow(\downarrow)$ of electrons scattered elastically from the target with spin parallel (antiparallel) to the magnetization. The spin-dependence $A = (i\uparrow - i\downarrow)/(i\uparrow + i\downarrow)$ of this scattering is caused by the exchange interaction alone in the geometry shown in Fig. 1. Spin-orbit coupling cannot contribute, because the electron beam polarization lies in the scattering plane. With many ferromagnetic materials, $A \sim 10^{-2}$ at electron energies of $\sim$ 100 eV. This can easily be measured with the spin-modulated electron beam. Even spin dependences as small as 10^{-5} have accurately been determined with this technique. It was common belief that the effects of exchange in elastic scattering of electrons are a small, mostly negligible correction to the Coulomb interaction and, therefore, almost impossible to observe. This is not true any more, since spin modulated electron beams are available.

The main virtue of the elastic scattering technique lies in the very short probing depth given by $1/2\ \lambda$, where λ is the mean free path of the electrons. λ is shortest at electron energies ranging from 50 - 100 eV. The factor 1/2 arises since the electron has to travel, on the average, one half λ into and out of the sample. This amounts to little more than one atomic layer in most materials. Hence one can test magnetism of the very first layer.

The main drawback of elastic electron scattering has the same physical reason, namely, the strong interaction of electrons with solid matter. Due to multiple scattering, low-energy electron diffraction (LEED) data are difficult to interpret even with nonmagnetic materials. Loss of symmetry at the surface, reconstruction and/or contraction of the first layer can cause new types of magnetic order, such as antiferromagnetic or ferrimagnetic rearrangement of spins at the surface of a ferromagnet. If one superimposes the complexities of surface magnetism to LEED, it becomes evident that spin polarized LEED with magnetic surfaces will need some time to develop. In the following, we mention some of the most striking phenomena observed with elastic electron scattering up to now.

Scaling laws [4] predict that the surface magnetization μ must change with temperature T according to

$$\mu(T) = \text{const}\ (1 - T/T_c)^{\beta_s}$$

where β_s is the critical exponent. ALVARADO et al. [5] obtained $\beta_s = .79 \pm .02$ for (110)- and $\beta_s = .81 \pm .02$ for the 100-surface of Ni. This represents the first measurement of β_s on a truly semi-infinite crystalline ferromagnet.

MILLS and MARADUDIN [6] predicted that in the low T-regime, the surface magnetization should decrease as

$$\mu(T) = 1 - c_s T^{3/2},$$

where $c_s = 2\ c_B$, with c_B the corresponding bulk constant, due to the excitation of additional of magnons at the surface. D.T. PIERCE et al. [7] confirmed this theory again by using a low-energy spin modulated electron beam, however it turned out that $c_s = 3c_B$.

D. WELLER, S.F. ALVARADO, and M. CAMPAGNA [8] observed a surface-enhanced magnetic order transition at the surface of (0001)-Gd epitaxially grown on W(110). The critical temperature of the surface as detected by elastic scattering of low-energy spin polarized electrons lies 22 K higher compared to the bulk which orders at 293 K. J. KIRSCHNER [9] investigated the effects of adsorbates on the elastic exchange scattering of electrons from a (110)-Fe surface. A strong dependence from O and S coverage was found, but no magnetic dead layer as long as the coverage is small. KIRSCHNER [10] also studied the effects of temperature on the exchange asymmetry A. These data show that A can be a complicated function of T due to top layer expansion and other thermal effects. However, these findings do not affect the interpretation of the experiments [5], [7] and [8].

Another interesting experiment to observe (quasi-)elastic exchange scattering of electrons on paramagnetic atoms was done by F. MEIER, G.L. BONA, and S. HUEFNER [11]. Since paramagnetic atoms do not define a quantization axis, one needs a source and a detector of spin polarized electrons. The source was a Ge-crystal, irradiated with circularly polarized light of an energy of 3.05 eV. The vacuum level of Ge is adjusted to lie within 2.4 and 2.8 eV by deposition of potassium. Under this condition, the emerging photoelectrons have a polarization of 23% parallel to the light beam. If now paramagnetic atoms such as Ce, Gd or Ni are deposited on top of this structure, one observes depolarization of the photoemitted electrons.

The depolarization is due to spin exchange scattering of the electrons with the randomly-oriented atoms with spin S [12]. One obtains from this experiment the effective exchange constant J(E), as it varies with the energy of the electron with spin s. Hence Heisenbergs and Diracs famous exchange interaction $J\,\vec{S}\cdot\vec{s}$ becomes directly accessible to experimental investigation. If an unoccupied $4f^{n+1}$ level, for instance, is available, into which the electron can scatter, the interaction with the spin of the $4f^n$-moment becomes very large as expected. This happens e.g. in Ce-metal ~ 2.5 eV above the Fermi energy.

A very elegant way to probe surface magnetism by elastic exchange scattering of electrons was found by S.F. ALVARADO, D. WELLER, B. ACKERMANN and R. FEDER [13]. It requires neither a source nor a detector of polarized electrons. In fact, the joint action of magnetic exchange and spin orbit interaction gives rise to an asymmetry for unpolarized electron scattering upon reversal of the sample magnetization. Naturally, the Lorentz deflection in the stray magnetic fields from the sample can cause such an asymmetry too. However, such stray magnetic fields can be made sufficiently small. One can think of this as an experiment using a source of polarized electrons, this source being the spin orbit scattering, and the magnetic target being integrated with the source. The observed spin asymmetries A are surprisingly large, up to 6% for a Gd film on W(110) or a (001)-Ni surface, depending on the electron energy, the angle of incidence, and the diffraction spot. Hence, one only needs an ordinary LEED apparatus to obtain information on surface magnetization.

3. Inelastic exchange scattering of electrons

The scattering of two electrons takes place in the singlet state $|s = 0, s_z = 0\rangle$ with complex amplitude f_1 or in either of the triplet states $|s = 1, s_z = 0, \pm 1\rangle$ with amplitude f_3. Let us assume that the spins of the 2 electrons are antiparallel before the collision. The initial state is then a superposition of singlet and triplet states $1/2(|1,0\rangle + |0,0\rangle)$ and leads to a direct scattering amplitude $f_d = 1/2(f_3 + f_1)$ and an exchange scattering amplitude $f_e = 1/2(f_3 - f_1)$ on account of the indistinguishability of the electrons. If the spin direction of one electron before the collision and of both electrons after the collision is known, the spin of the other electron before the collision is given by conservation of spin momentum in the collision. Hence, the quantum mechanical exchange, which arises from the indistinguishability of electrons, becomes directly observable through spin polarization measurements. According to Ref. [14], this requires only a detector of spin polarization if one uses a ferromagnet as target. $Fe_{83}B_{17}$ is particularly suited, because it is a strong ferromagnet with pure minority spin states in the unoccupied part of the 3d-valence bands [15].

The incident unpolarized electrons suffer a characteristic loss of ~ 57 eV due to the resonant excitation of a 3p electron into the 3d - states. Since there are only down spin states available in the 3d - bands, the 3p electron that was excited is down spin as well. The polarization P of the primary electron after excitation is measured. It depends, as expected, on the energy E of the incident beam. $P \cong 30\%$ for $E/57 = 2$ and ~ 8% for $E/57 = 10$ [14]. This decrease of P is expected within the simplest theoretical model, in which the interaction of the electrons is given by an unscreened Coulomb potential. It also shows that in the case of large energy-transfer, the cross-section for exciting a minority spin is larger for a majority spin primary electron compared to a minority spin electron. It immediately follows that one can create spin polarized core holes even in a nonmagnetic material, by using a spin polarized incident beam. Spin polarized core holes lead to spin polarized Auger-electrons [16]. The exchange in inelastic electron - electron collision makes it then possible to perform spin polarized Auger - studies, even in nonmagnetic materials.

It is also possible to study the spin dependence of core hole excitations using a spin polarized incident electron beam and a ferromagnetic target. KIRSCHNER performed spin dependent appearance potential spectroscopy on the 2p - level of ferromagnetic iron [17]. One finds that minority spin electrons are somewhat more effective in producing a 2p - hole at threshold. This arises, because the final state of the incident electron is close to the Fermi-energy at threshold where the minority spin states are more abundant. However, it is not possible to decide from this experiment whether the core hole is up or down. KIRSCHNER was able to fit his data to a band model neglecting the effects of quantum mechanical exchange. This is in contradiction to the findings of Ref. 14. The agreement with a single particle bandstructure may be fortuitous. The application of single particle densities of states is hardly justified in the presence of 2 extra electrons in the 3d-band and a hole in the 2p - shell.

A very important application of inelastic exchange scattering of electrons is concerned with Stoner excitations. The net result of this process is that an electron from the occupied part of the majority spin band flips its spin and occupies a minority spin state above the Fermi-level. The minimum energy required for this process is called the Stoner gap, and the maximum energy is the Stoner gap plus the width of the occupied part of the majority spin bands and the width of the empty minority bands. Hence, Stoner excitations typically extend over an energy range of a few eV. They cannot be observed with neutrons, yet their temperature-dependence and threshold enery are quite important issues in the theory of itinerant magnetism. KIRSCHNER, REBENSTORFF and IBACH [18] have observed the spectrum of Stoner excitations in Ni for vertical electron transitions (momentum transfer q = 0) measuring characteristic energy losses of a spin polarized electron beam. HOPSTER, RAUE and CLAUBERG [19] used an unpolarized primary electron beam and measured the polarization of inelastic electrons, having created a Stoner excitation in the ferromagnetic glass $Fe_{80}B_{20}$. The first group showed that a Stoner excitation happens with a primary minority spin electron, and the second group showed that electrons having lost a few eV emerge predominantly, depending on the energy of the primary beam as majority spin electrons. This is exactly what one expects. GLAZER and TOSATTI [20] have predicted and analyzed the Stoner excitations theoretically. HOPSTER et al. [21] also observed pronounced structures in the low-energy range of the secondary cascade which they attributed to Stoner excitations as well. However, according to D. MAURI and M. LANDOLT [22], the high spin polarization of the low-energy cascade may also be explained simply by the band structure involving no spin flip transitions. This explanation has the additional advantage that no large difference in the mean free path of up spins compared to down spins is required, in agreement with electron scattering on magnetic glasses [7, 23].

4. Spin Polarization of Auger-electrons

Auger electron emission is frequently used for elemental and chemical analysis of surfaces. M. LANDOLT and D. MAURI [24] have shown on the example of $Fe_{83}B_{17}$ that the spin polarization of Auger electrons can also be measured. The polarization of Auger electrons from the decay of various core holes in Fe is large, of the order of 20%, and can have either direction parallel or antiparallel to the magnetization. The study of spin polarized Auger-electrons offers several new possibilities. First one can test some basic issues in metallic magnetism. The key idea here is that one creates, upon excitation of a core hole, a magnetic impurity with nuclear charge (Z + 1), e.g. Co in the case of Fe. This magnetic impurity is instable, and will give information on magnetism in alloys when it decays via an Auger process. For the information to be realistic with respect to real impurities, we have to postulate that the life time of the deep core hole is long enough such that the spectator electron orbitals can adjust themselves to the new situation created by removing one electron, and also such that even the external conduction electrons have time to flow to the impurity and reach an equilibrium in screening the additional charge. The existing experience with core level spectroscopies, particular X-ray photoemission, shows that this may be the case depending on which core hole was excited. Hence, one can even think of studying the actual process of screening a magnetic impurity by exciting core holes with different lifetimes.

The observation of the kinetic energy and spin of the emitted Auger electron will yield various types of information. When the valence states are involved in the Auger decay, we will learn about the valence bands around the impurity. When deeper lying levels are involved, we will learn about the sign and magnitude of the exchange coupling of these states to the 3d-moment [25]. Interestingly, there are actually 2 possibilities to create a hole in, say, the 3p - shell of a 3d - transition element: The first one occurs if a large energy is transferred so that the 3p - electron is removed to infinite distance from the atom. The binding energy E_B may be determined by X-ray photoemission. The core hole left behind is unpolarized, and the impurity is initially charged electrically. It will then be stabilized by screening of this charge. Most effective in the screening will be conduction electrons of 3d - parentage, because they can draw closer to the nucleus. The second excitation occurs if exactly the energy of a 3p to 3d transition is transferred. This resonant energy E_R may be determined from characteristic energy-loss structures [26] or from photoemission with synchrotron radiation [27]. This type of impurity is not charged electrically. It may be called a self-screened impurity. The core hole is spin polarized because the empty 3d - states to which the core electron was excited is predominantly of minority spin.

Depending on whether $E_B \gtrsim E_R$ or $E_B < E_R$, a magnetic impurity will acquire an additional 3d - electron, that is reduce its magnetic moment, or it will screen its charge by conduction electrons and keep a large magnetic moment. In intensity studies of the Auger-electrons, the difference between E_B and E_R is hardly detectable, but the spin polarization of Auger - electrons from the decay of the 2 types of core holes is very different, because in one case the core hole is polarized and in the other it is not. It turned out that $E_R - E_B \cong 4$ eV in the case of metallic Fe and for a 3p core hole. These remarks demonstrate the great potential of spin polarized Auger - electrons for the study of magnetism in alloys.

5. Low-energy cascade and threshold photoelectrons

In many applications of spin polarized electrons, it is important to have a relatively large probing depth. Technical materials are mostly complicated alloys, and their structure and composition is disturbed at the very surface by ion bombardment, oxidation, or deposition of hydrocarbons. In principle, one has large probing depths at high (10^4 eV) or at very low electron energies. Since most spin-dependent phenomena are weak at electron energies above 10^3 eV, the best choice for a reduced surface sensitivity is to go to very low electron energies. Technically, it is of course difficult to produce and scatter very low energy electrons. This difficulty can be avoided by measuring the spin polarization of very low energy electrons emerging from the sample upon excitation with a primary (unpolarized) high energy electron beam, or in the case of photoemission, upon irradiating the sample with light of an energy close to photoelectric threshold. If the work function is lowered to about 2 eV by deposition of alkali such as Cs, the average depth from which threshold electrons emerge exceeds 50 Å even in most metals. Systematic variation of the photoelectric threshold by deposition of chemicals leads to non-destructive depth profiling of the magnetization. The results can be compared to the magneto-optical Kerr rotation, which has a probing depth of ~ 200 Å, and therefore yields the magnetic properties of the bulk [28]. Such non-destructive depth profiling is essential. The commonly used ion milling cannot be applied to depth profiling of magnetic properties, since it severely alters remanence and coercive force [29]. Hysteresis loops depend largely on the defect structure, and therefore surface hysteresis loops represent fingerprint marks of the defects at the surface that are difficult to obtain with other techniques.

The shape, size and arrangement of magnetic domains and domain walls and their dynamic response to external influences, like stress, magnetic fields, annealing, corrosion, ion bombardment etc. characterize a magnetic material in a rather complete way under a variety of conditions, in particular with respect to applications. KOIKE and HAYAKAWA [30] have now built a conceptually simple scanning electron microscope that displays magnetic domains with extraordinary contrast,using spin selective detection of the electron beam-induced low-energy cascade. The ultimate resolution for observation of magnetic domains in the spin polarized reflection mode is far ahead of conventional reflection methods using Lorentz-deflection of the secondaries. Whereas the measurement of the spin polarization of cascade electrons provides highly detailed domain structural information, it is limited to changes in the domain pattern slower than $\sim 10^{-3}$ sec. If domain switching behavior at a much faster rate (equivalent and superior to rates used in present magnetic recording devices) is to be studied, pulsed laser-induced threshold photoemission has to be considered [31]. Here, however, the spatial resolution is limited by the size of the laser focus ($\sim 10^{-6}$ m).

Threshold photoemission has additional unique features. Whereas the spin polarization of the low-energy cascade represents some still poorly understood average magnetization of the valence bands, compare section 3, the spin polarization of threshold photoelectrons is well understood,and contains clear information on the electronic structure [29]. This arises, because threshold photoelectrons are excited from well-defined states close to the Fermi energy in metals or from the highest lying occupied electron states in magnetic semiconductors or insulators. For example, in the ferrite family, γ-Fe_2O_3 yields positive and Fe_3O_4 negative polarization at threshold [32]. This additional information can be very handy in applications to technical materials. The time needed to measure the direction of a magnetic domain is given by the duration of the laser pulse. Hence it can be as short as 10^{-12} sec.

The best information on spin polarized electronic structure, its changes at the magnetic phase-transitions and other issues of fundamental interest are of course obtained using angle, energy and spin-resolved photoemission,and a synchrotron radiation light source as demonstrated by M. CAMPAGNA and collaborators [33]. With undulators, the intensity may become high enough to even exploit the time structure of the synchrotron radiation for studying dynamics of surface magnetic phenomena in the greatest detail.

6. Conclusion

Spin polarization in elastic and inelastic scattering of electrons and in photoemission has a remarkable potential for advancing our understanding of surface and bulk magnetism. Of course, other techniques,like the recently developed scattering of polarized positrons [34] or the spin polarized metastable He - deexcitation [35] will also contribute to this field, but spin polarized electron beam techniques are certainly much simpler. Even in this review, limited to the impact of spin polarized electrons on magnetism, we have not been able to treat such important fields as surface magnetochemistry, or such simple techniques as the spin-dependence of the current absorbed by the surface of a ferromagnet [23]. Future progress in the field of applied magnetism may well be enhanced by these and similar techniques employing spin polarization of electrons.

Acknowledgement

The author is much indebted to his colleagues for discussions, and for communicating results prior to publication. He also wishes to thank IBM San José for support and hospitality while this paper was prepared for publication.

References

1 G.D. Fletcher, M.J. Alguard, T.J. Gay, V.W. Hughes, C.W. Tu, P.F. Wainwright, M.S. Lubell, W. Raith and F.C. Tang, Phys. Rev. Lett. 48, 1671 (1982)

2 R.J. Celotta, D.T. Pierce, G.C. Wang, S.D. Bader, and G.P. Felcher, Phys. Rev. Lett. 43, 728 (1979)

3 C.S. Feigerle, D.T. Pierce, A. Seiler, and R.J. Celotta, Appl. Phys. Lett. 44, 866 (1984); J. Kirschner, H.P. Oepen, and H. Ibach, Appl. Phys. A30, 177 (1983) and ref. cited

4 H.W. Diel and S. Dietrich, Z. Physik B42, 65 (1981), and ref.

5 S.F. Alvarado, M. Campagna, F. Cicacci, H. Hopster, J. Appl. Phys. 53, 7920 (1982), and Phys. Rev. Lett. 48, 51 (1982), and ibidem 48, 1768 (1982)

6 D.L. Mills, Comm. on Sol. State Phys. 4, 28 (1971) and 4, 95 (1972) and ref. cited

7 D.T. Pierce, R.J. Celotta, J. Unguris, and H.C. Siegmann, Phys. Rev. B26, 2566 (1982)

8 D. Weller, S.F. Alvarado, and M. Campagna, preprint

9 J. Kirschner, Surf. Sci. 138, 191 (1984)

10 J. Kirschner, Phys. Rev. B30, 415 (1984)

11 F. Meier, G.L. Bona, and S. Hüfner, Phys. Rev. Lett. 52, 1152 (1984)

12 J.S. Helman and H.C. Siegmann, Solid State Commun. 13, 891 (1973)

13 S.F. Alvarado, D. Weller, B. Ackermann, and R. Feder, preprint

14 D. Mauri, R. Allenspach, and M. Landolt, Phys. Rev. Lett. 52, 152 (1984)

15 E. Babić, Ž. Marohnić, and E.P. Wohlfarth, Phys. Lett. 95A, 335 (1983)

16 L.A. Feldkamp and L.C. Davis, Phys. Rev. Lett. 43, 151 (1979)

17 J. Kirschner, Solid State Comm. 49, 39 (1984)

18 J. Kirschner, D. Rebenstorff, and H. Ibach, Phys. Rev. Lett. 53, 698 (1984)

19 H. Hopster, R. Raue, and R. Clauberg, Phys. Rev. Lett. 53, 695 (1984)

20 J. Glazer and E. Tosatti, preprint

21 H. Hopster, R. Raue, E. Kisker, G. Güntherodt, and M. Campagna, Phys. Rev. Lett. 50, 70 (1983)

22 D. Mauri and M. Landolt, to be publ, and D. Mauri, Diss. ETH 1984

23 H.C. Siegmann, D.T. Pierce, and R.J. Celotta, Phys. Rev. Lett. 46, 452 (1981)

24 M. Landolt and D. Mauri, Phys. Rev. Lett. 49, 1783 (1982)

25 K.H. Bennemann, Phys. Rev. B28, 5304 (1983); A. Kotani and H. Mizuta, Solid State Commun. 51, 727 (1984)

26 S.D. Bader, G. Zajac, and J. Zak, Phys. Rev. Lett. 50, 1211 (1983), and ibidem 50, 1713 (1983)

27 R. Clauberg, W. Gudat, E. Kisker, E. Kuhlmann, and G.M. Rothberg, Phys. Rev. Lett. 47, 1314 (1981) and ref. cited

28 H.C. Siegmann and E. Kay, Appl. Phys. Lett. 44, 927 (1984)

29 H.C. Siegmann, F. Meier, M. Erbudak, and M. Landolt, Adv. El. and El. Phys. 62, 1 (1984)
30 Kazuyuki Koike and Kazunobu Hayakawa, Jap. J. Appl. Phys. 23, L187 (1984)
31 E. Kay and H.C. Siegmann, IBM Techn. Discl. Bull. 27, 317 (1984)
32 S.F. Alvarado, W. Eib, H.C. Siegmann, and J.P. Remeika, J. Magn. Magn. Mater. 3, 121 (1976)
33 M. Campagna, Physica Proc. (1984), in press
34 D.W. Gidley, A.R. Köymen, and T. Weston Capehart, Phys. Rev. Lett. 49, 1779 (1982)
35 M. Onellion, M.W. Hart, F.B. Dunning, and G.K. Walters, Phys. Rev. Lett. 52, 380 (1984)

Magnetism at Interfaces and Surfaces, as Probed by Neutron Scattering

G.P. Felcher

Materials Science and Technology Division, Argonne National Laboratory
Argonne, IL 60439, USA

The diffraction of polarized neutrons from metal-metal superlattices, with one of the two metal magnetic, has yielded significant information on the microscopic distribution of magnetism, although the ultimate possibilities of the technique are yet sample-limited. Fresnel reflection of polarized neutrons from superconductors and ferromagnets is providing the depth profile of the magnetization close to the surface of these materials.

1. Introduction. Why Neutrons?

The study of interfaces and surfaces by neutron scattering is relatively novel, and for this reason it is worthwhile to examine briefly the main properties of this probe, by which the configurations and the dynamics of magnetism in the bulk have been successfully explored. Neutrons produced in modern sources are available in a thermal spectrum, with wavelengths ranging from 0.5 to 10 Ångstroms, or roughly of the same length as the interatomic distances in the solids, from which they can be scattered or diffracted. In view of their mass, neutrons are rather slow, their corresponding velocities ranging from 1.2 to 25 μsec/cm: hence their energy can be sorted out by time of flight. Neutrons interact with matter in two ways.[1,2] First, they interact with the nuclei, with a scattering amplitude b, which varies from nucleus to nucleus, but it is always of the order of 10^{-12} cm, or of the size of the nuclear radius. Secondly the neutrons, having spin 1/2, interact with the magnetic fields in the material with an energy $\vec{\mu}_n \cdot \vec{B}$, where $\vec{\mu}_n$ is the neutron moment and $\vec{B}$ is the local induction field. From this expression of the energy, a magnetic scattering amplitude per atom p, can be derived; in absolute value, $|p|=0.27\times10^{-12}$cm for an atomic moment of one Bohr magneton. Magnetic and nuclear scattering amplitudes have comparable sizes; this property has made the neutron quite a unique tool for examining the magnetic configurations in materials.

In reference to the study of interfaces and surfaces, the size of the neutron-matter interaction has been at the same time the curse and the blessing of the neutron research. The interaction is so weak that neutrons penetrate through meters of materials, and therefore they are relatively surface-insensitive. However, the weakness of the interaction assures that the scattered intensities can be very well understood, since single scattering processes are by far prevalent. In other words, if the neutrons scattered by interfaces and surfaces were sufficient to obtain a reasonable signal, the conversion of the measured intensities into the magnetic configuration in real space could be carried out in the framework of the first Born approximation, without having to recur to simplifying models in order to solve the matrix elements of the general n-beams scattering theory.

There are two ways in which the intensity scattered from surfaces and interfaces can be enhanced: the first is to increase the amount of viewed surface, the second to select a particular geometry, in which surface scattering is enhanced. Magnetic superlattices are an ideal way of obtaining a large amount of interface area, simply by stacking a sizeable quantity of individual sandwiches. For neutron diffraction studies it is equally important that the samples are truly superlattices, in the sense that the stacking should be exactly repeated over a large number of layers.

A second way to enhance the surface signal is to send the neutron beam at a grazing incidence to the surface, and to measure the intensity the Fresnel reflected beam. While in this geometry the first Born approximation is no longer valid (the reflected beam can be as strong as the incoming beam), the two beams scattering theory is still simple enough to provide a rigorous expression of the reflected intensity as a function of the magnetic profile from the surface in the material. I will review the work done and what is still ahead in the experimentation with the two methods.

2. Magnetic Superlattices

The preparation and the crystalline form of metal-metal superlattices have been presented in another lecture of this course.[3] In magnetic superlattices, at least one of the two metals is in an ordered magnetic state, which is in most of the cases studied of ferromagnetic nature. The ultimate purpose of the preparation of magnetic superlattices is to create materials with novel magnetic properties. Film thicknesses of a few tens of Ångstroms are of the same order of the range of the magnetic interactions, hence, whatever magnetic order exists in the bulk material is deeply modifed by the presence of the interfaces. The perturbations stem not only from the electronic mismatch of the two metals but also as a result of their different crystalline structure.[4] The thickness of the non-magnetic layer controls the amount and the type of magnetic interaction between different layers, or the confinement of the magnetically ordered state. Such is the variety of materials that can be artificially prepared - compared with the limited number of thermodynamically stable-ordered alloys - that their study should provide a much deeper understanding of magnetism in solids.

The first characterization of the samples is usually done by magnetization measurements. These give the average ordered moment per atom $\bar{\mu}$: the extent of the fluctuations of magnetization within each layer may only be inferred from the behavior of $\bar{\mu}$ as a function of the applied field H and the temperature T. By polarized neutron diffraction methods, it is instead possible to identify the state of a magnetic atom as a function of its position in the crystalline structure. This wealth of potential information contrast with the limited information that can be gathered from disordered alloys, in which the magnetic atoms are set in a variety of magnetic and chemical environments, so that their properties have to be statistically averaged. For disordered alloys, only the average magnetization, and the mean-square fluctuation of the magnetic moments can be obtained by diffuse neutron scattering.[1] A detailed behavior of the role of a non-magnetic in a ferromagnetic metal can be obtained only when the amount of impurity is very small, so that different impurities do not interact with each other. In particular it is known that non-magnetic impurities in nickel[5] affect generally the magnetic moments of nickel up to the nearest-neighbours, while impurities in iron do not affect appreciably the magnetization of the matrix.

A polarized neutron diffraction experiment is schematically illustrated in Fig. 1. The neutrons from a continuous source are monochromatized at a wavelength λ. The monochromatized beam is also polarized, either parallel (+) or antiparallel (-) to the magnetization of the sample, which is taken to be in the plane of the superlattice. By rotating the angle θ between the beam and the sample surface, diffraction lines are revealed by a detector at an angle 2Θ. The intensities of these lines are proportional to a square of an amplitude, which is simply one of the Fourier components of the superlattice modulation. Conversely, the modulation of the superlattice can be obtained by a Fourier transformation of the experimental amplitudes of the Bragg reflections up to an index n. The de-

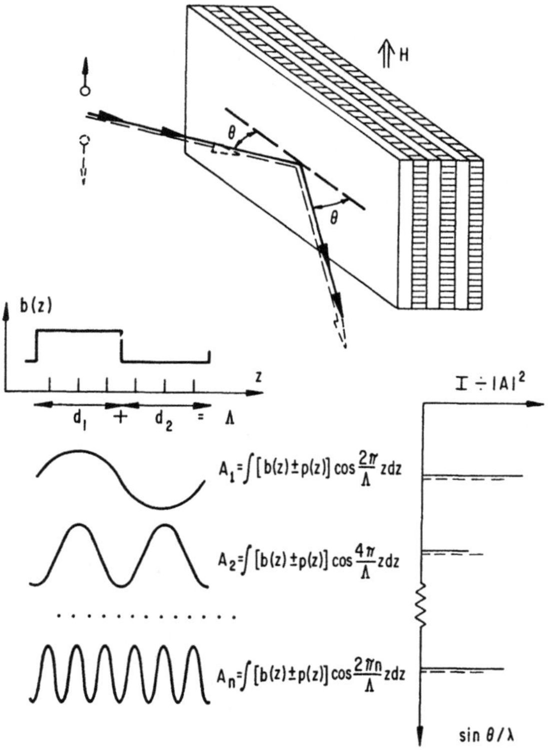

Fig. 1. Scheme of a polarized diffraction experiment. Top: diffraction geometry, with the orientations of neutrons polarized parallel (continuous line) and antiparallel (dotted lines) to the magnetization H. The intensities of the diffracted neutrons of the two spin states are measured sequentially. Bottom: analysis of the diffracted intensities. The modulation of the scattering amplitude of the bilayer is schematically represented as a square wave, but in fact the scattering centers are in atomic planes. Bragg reflections of ascending order have a structure factor A_n which is the product of the amplitude for the phase factor.

tail of the calculated modulation depends on the truncation of the Fourier serie. For instance,to learn about the composition of each atomic plane (with interplane spacing a), the intensities must be measured at least up to the n-th reflection, where $n = \Lambda/a$. The role of polarized neutrons is evident in separating the modulation of the nuclear amplitude b(z), that provides (as for x-rays) the structural and chemical composition in the material, and p(z), its magnetic counterpart, due to the magnetic moments aligned along the magnetization axis, or an external magnetic field H. Such a complete analysis has not been yet performed on any actual sample; the presence of some kind of disorder (of which a few examples have been given elsewhere[3]) has prevented it.

Figure 2 presents the intensities obtained with polarized neutrons of a superlattice Fe-Ge, for low scattering angles.[6] Seven satellites are well visible, with decreasing intensity. the fact that the ratio of the intensities I+/I- (usually called the flipping ratio) is not always larger than one implies that the magnetization is not constant across the thickness of the Fe layer. In the center of it, the moment of iron is close to the value for b.c.c. iron (2.2 μ_B), and decreases to about half that value at the iron/germanium interface. However, this finding must be qualified by a close look at the real crystalline

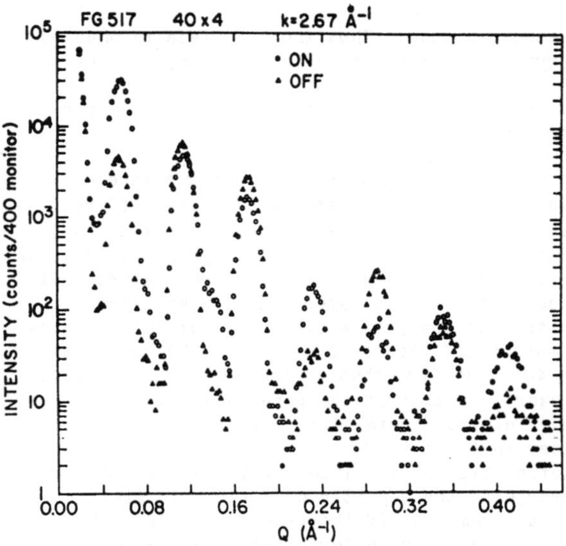

Fig. 2. Θ:2Θ polarized neutron diffraction scan ($Q = 4\pi\sin\Theta/\lambda$) of a multilayer of Fe-Ge. Λ = 108 Å; thickness of the iron layer D_1 = 36 Å. The "ON" ("OFF") data points correspond to incident neutrons in the + (-) eigenstate (from reference 6).

structure as determined by x-rays and neutron diffraction[7] for this sample of Fe-Ge. As shown in Fig. 3, germanium is deposited in the amorphous state, whilst Fe grows in platelets along the <110> direction of its body-centered-cubic structure. The amorphous state of germanium causes the disappearance of the diffraction lines of higher order, while those at small scattering angle are so beautifully visible. Even worse, germanium interdiffuses in the iron lattice for

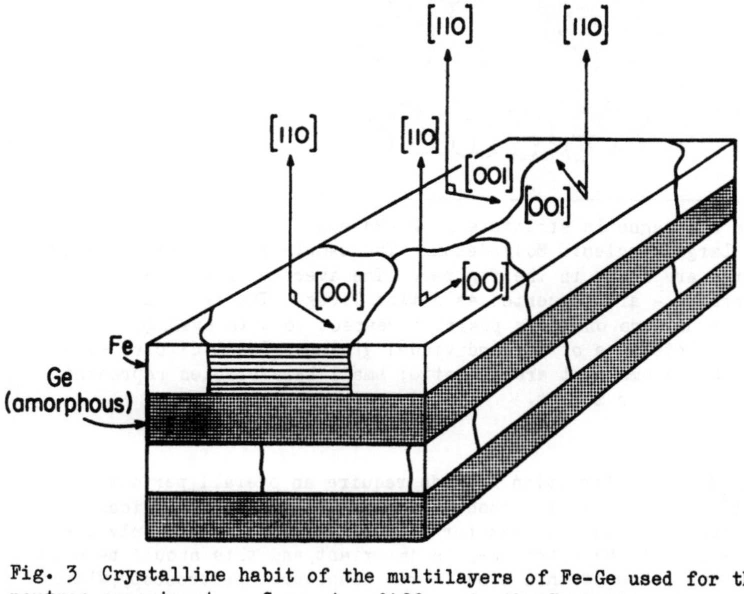

Fig. 3 Crystalline habit of the multilayers of Fe-Ge used for the polarized neutron experiments. Germanium differs in the Fe lattice for a thickness of a few atomic planes, "blurring" the interface (from reference 7).

a distance of approximately three atomic planes. Thus, the interface is smeared into a region of Fe-Ge alloy of varying composition; in this region, the lattice spacing is contracted with respect to that of the pure iron. The reduced moment of iron is found in such an alloy, in which the magnetic atoms do not have a perfectly defined chemical and magnetic environment.

A different kind of disorder is encountered in nickel-molybodemun superlattices. In these layered materials, composed of elements of quite different atomic radii (~ 10%), rather perfect crystallites grow [8] but with axes misaligned compared to each other and to the normal to the lamina by several degrees (Fig. 4). These superlattices are ferromagnetic, with a Curie temperature that rapidly increases with the thickness of the nickel layer, and with an average nickel moment which is approaching that of the pure nickel only for layers thicker than 80 Ångstroms. Polarized neutron diffraction experiments, [9] conducted on thin layered samples (less than 40 Å) confirm that at the center of the layer nickel has a moment considerably lower than that of the pure metal (0.6 μ_B/atom). The orientational disorder of the crystalline grains prevents a detailed description of magnetism close to the interface. However, from a careful comparison of the spin-dependent intensities of the lines at low and large scattering angles, an unusual picture emerges: the magnetic modulation is flatter than the chemical modulation across the lamina. If the lamina is divided (as with the dotted lines in Fig. 4) in layers, which alternatively can be labeled "nickel sites" and "molybdenum sites", the nickel atoms are magnetically active only if they are on the nickel sites, irrespective of the details of the local configuration. This picture is highly simplified; the data do not exclude a smoother variation of the moments across the nickel-molybdenum sites boundary.

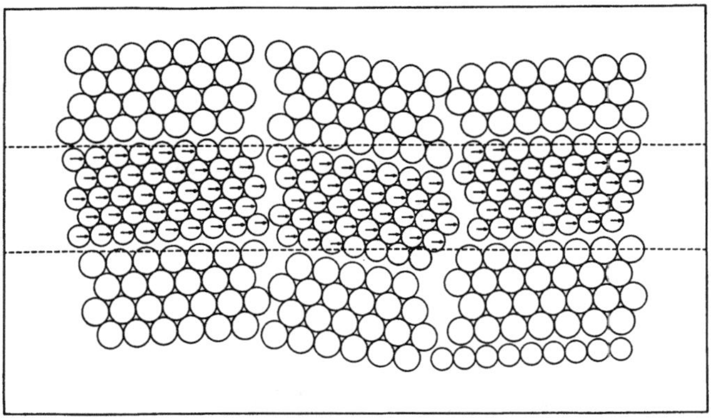

Fig. 4 Crystalline and magnetic structure in Ni-Mo superlattice. Small circles: nickel. Large circles: Molybdenum. The sample is made of different perfect grains (three are shown in the figure). The average interfaces - averaged over the grains - are presented as dotted lines. The ferromagnetic moments of nickel depend more on their position respect to this average interface than to the interface of the individual grains. Small circles with arrows represent nickel atoms that are magnetic; small bare circles represent nickel atoms without magnetic moment.

It might be argued that diffraction methods require an overall perfection of the crystalline edifice, which in the case of the artificial superlattices is quite difficult to achieve. In contrast, for a local magnetic probe, only the local order (for example at the interface) is important, and this should be more easily obtainable. One probe of this kind is the Mössbauer spectroscopy, which has been used to measure the magnetic moments of iron layered with different elements: the most recent work being done on Fe-V and Fe-Sb superlattices

[10,11,12] The measurements on these systems, which comprise limited polarized neutron observations as well as Mössbauer spectroscopy, have been interpreted in terms of an iron moment which has a reduced value only in the atomic plane adjacent to the interface. But how well defined is the interface in these systems? An independent study[13,14] of Fe-V superlattices suggests that some interdiffusion occurs even in this system, and that the moment reduction at the interface is at least partially due to this alloying process.

The record of the results obtained up to now on magnetic superlattices might appear not entirely positive. Still, the basic object of the research, to find novel magnetic materials, is too important to be lightly dismissed: and several groups are laboring to refine the technique of growth to produce better materials. The microscopic analysis of the magnetization within the layers by polarized neutron diffraction proceeds in parallel with such an effort. Even with the present quality samples, it should be at least possible to measure accurately the lower harmonics of the modulation of magnetization as a function of the temperature. This might help in understanding the findings of conventional magnetization measurements. $\overline{M}$ (T) has a form similar to that of the bulk material for thick layered superlattices; when the magnetic material is confined to a few atomic planes, $\overline{M}$ (T) has an almost linear behavior, reminiscent of that expected of a 2-D ferromagnet.[15]

3. Magnetism at the Surface

The ground magnetic state at the surface of ferromagnets is generally altered from the bulk for a thickness of a few atomic planes; the altered state is extensively studied by a variety of surface-sensitive probes, such as polarized electron diffraction and spin analyzed photoemission. However, when the temperature is raised, and brought close to the temperature of magnetic order T_c, the surface becomes thicker, in the sense that the region in which the magnetization is perturbed from the bulk value has the length characteristic of the critical fluctuations. "Thick" magnetic surfaces are also encountered in superconducting materials; an external magnetic field penetrates the material for a characteristic depth of the order of a few hundred Ångstroms.

The magnetic depth profile can be studied, especially in the case of "thick" surfaces, by polarized neutron reflection. The neutron beam is considered here as a conventional optical probe, which is Fresnel reflected from a surface of a material material, characterized by a refractive index n. For most materials, the neutron refractive index is less than one: the materials are optically less dense than the vacuum. An important consequence is that, if the angle of incidence is smaller than a critical value Θ_c, the beam is totally reflected; even at angles slightly larger than Θ_c the reflectivity is sizeable. Suppose that the refractive index n varies as a function of the depth from the surface z. The measurement of the reflectivity as a function of incidence angle, or equivalently (as shown below) the neutron wavelength, allows a determination of n(z); for magnetic materials and polarized neutrons this quantity is directly proportional to the depth profile of the magnetic induction B(z).

Suppose that the material is magnetized in a direction parallel to the surface. Neutrons polarized parallel (+) or antiparallel (-) to the direction of the applied magnetic field have refractive indices:

$$n^{\pm}(z) = 1 - \frac{\lambda^2}{2\pi} \left[\frac{b}{v} \pm c\ (B(z) - H)\right] \qquad (1)$$

where λ is the neutron wavelength, b the average nuclear scattering amplitude in the atomic volume v, c is a constant, and B(z) is the magnetic induction in the material. The refractive index for neutrons is, for wavelengths of the order of a few Ångstroms, only slightly different from unity, since for all materials

$|b/v|<10^{-5}Å^{-2}$. However, the constant $c=2\pi\mu_n m/h^2=2.3\times10^{-10}Å^{-2}\ Oe^{-1}$: even modest magnetic fields are sufficient to yield a sizeable spin-dependence of n-1.

The reflectivity as a function of the refractive index n(z) can be given in a form exact but cumbersome. However, a more visual expression can be given in an approximate form, when the neutron beam is only weakly reflected from the surface, and the magnetic component of the refractive index is considerably smaller than the nuclear component. In this case:

$$R^{\pm} \sim R_n[(1 \pm c \int_0^{\infty} B'(z) \exp(i K_n z)\, dz] \qquad (2)$$

R_n is the reflectivity due to the nuclear part only, and K_n is the neutron momentum transfer in the material. From eq. (2) it appears that $\Delta R = R^+ - R^-$ is a Fourier transform of the derivative of the magnetic profile. If the magnetic induction reaches a constant value B_o in the bulk, ΔR at the critical angle (K_n=o) is the same irrespective of the shape B(z). Only for larger angles of incidence does the shape of the magnetic profile become important; for instance, the effect of a smoother variation of B(z) is to decrease ΔR. Since the the refractive index n is only slightly different from unity, the critical angles are only a fraction of a degree; high collimation and very flat surfaces are required in order to perform reasonable measurements.

The first experimental equipment set up at the Intense Pulsed Neutron Source at Argonne (Fig. 5) has been primarily used to perform measurement on

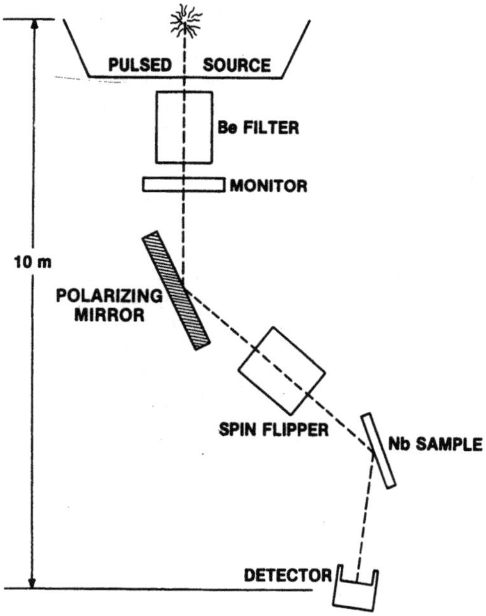

Fig. 5 Layout of the polarized neutron reflection experiment. A pulsed beam of cold neutrons (λ>3.8Å) is polarized by reflection of a magnetized mirror. The polarized beam is partially reflected by the sample surface, which is kept in a magnetic field; neutrons of different wavelength arrive at the detector at different times, and are sorted out by time-of-flight. The magnetic fields (and the quantization axis of the neutron spins) are normal to the plane of the page. Reflectivity measurements for the two states of the neutron spins are taken in rapid alternation, by switching the spin flipper.

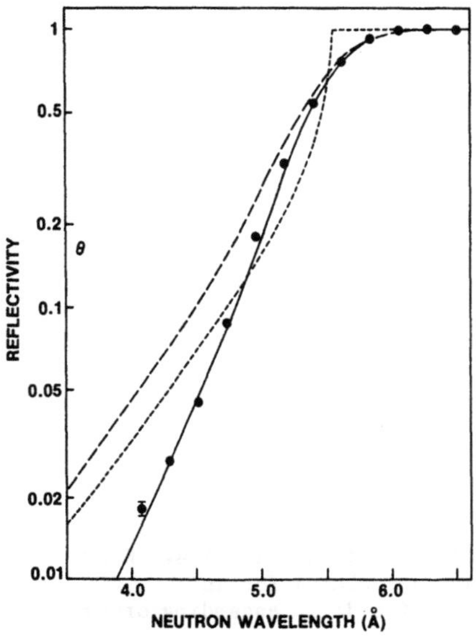

Fig.6. Reflectivity of niobium (10 K). Dotted line, calculated reflectivity for an angle of incident $\theta_i = 0.36°$. Dashed line, the same, but with a Gaussian spread of angles $\Delta\theta_i = 0.02°$. Continuous line, the effect of a surface roughness $(\langle z_D^2\rangle^{1/2} = 70$ Å) has been added.

superconductors. It was well known that a superconductor completely expels from its bulk an applied magnetic field H, provided H is below a critical value H_{c1}. However, the magnetic induction becomes zero only deeper than the penetration depth Λ; the variation of B with the distance from the surface is usually approximated as an exponential function.

Measurements were done on niobium, pure lead and an alloy $Pb_{0.992}Bi_{0.008}$. Fig. 6 presents a typical reflectivity curve as obtained for niobium[16] at a temperature of 10K and in a field of 500 Oe. At this temperature, above the superconducting temperature $T_c = 9.2$K, the reflectivity is spin-independent; the reflectivity in this region is solely used to calibrate experimental and sample parameters. The reflectivity calculated for niobium has to be averaged over the effective angular spread of the neutron beam, which includes the divergence of the beam incident on the sample and the undulations of the surface, which is considered only locally flat. If locally the surface is not entirely smooth, its roughness modifies the reflectivity as a Debye-Waller factor. Using these two adjustable parameters (angular spread, and surface roughness) a good fit is obtained with the experimental reflectivity.

The measurement of the magnetic field penetration depth of superconducting niobium was carried out at different field and temperatures. A typical result[16] is presented at Fig. 7 in the form of the ratio R^+/R^-, or "flipping ratio". The results are compared with the flipping ratios calculated for different penetration depths, assuming that the field decays in the solid as in an exponential function. The best fit at T = 4.6 K, H = 500 Oe gives Λ = 430 ± 40 Å. From these and other data (at different fields and temperatures) the value of Λ at zero-temperature is estimated to be Λ=410±40Å. The value of $\Lambda(o)$ found by this new, and direct technique is in comfortable agreement with that inferred from magnetic induction measurements, as well as that obtained from band calculations.

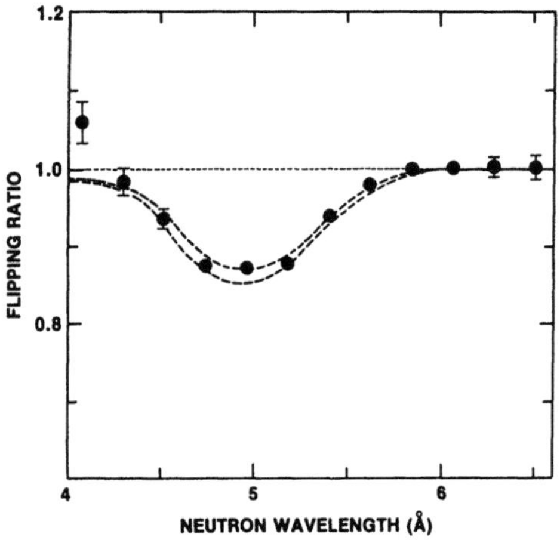

Fig. 7. The flipping ratio for a superconducting niobium film at 4.6 K and 500 Oe. θ_i = 0.34±0.02°. The dotted lines are calculated for an exponential field decay of the external magnetic field; in ascending order, the penetration depth is Λ = 380 and 480Å.

A second, and more intriguing possibility, is to directly observe the diamagnetism (field expulsion) associated with the surface sheath of superconductivity predicted[17] to occur between H_{c2} (the critical field, at which the bulk becomes normal) and $H_{c3} \simeq 1.7H_{c2}$. Possibly the best system to measure the effect is represented by the lead-bismuth alloys. Alloying minute amounts (0.8%) of bismuth to lead produces a material, with an incipient type II behavior, and with a low value of the Ginzburg-Landau parameter ($\kappa \sim 1/\sqrt{2}$). In these circumstances is expected[18] a deep magnetic disturbance at the surface. The preliminary results[19] confirm the presence of the surface superconductivity; however, more experiments are needed in order to obtain a detailed profile of B(z) and thus furnish an experimental test for the validity of the wavefunctions used to calculate the super-conducting surface sheath.

The polarized neutron reflection technique is still in its infancy, and it is difficult to assess its ultimate limitations. Roughly speaking, it should be possible to observe magnetic perturbations for thicknesses between 10 and 5000Å, the lower boundary being due to the neutron counting statistics, and the upper boundary to the angular resolution. Magnetic depth profiling in this way is useful to test technologically important materials in a non-destructive way. For instance, measurements on thin films of the ferromagnetic iron oxides used for magnetic tapes have alredy been started, and have revealed large differences in the magnetic profiles for materials prepared in different ways.[20]

The next challenge will be the determination of the critical exponents of ferromagnets. Both the bulk and surface magnetizations has a temperature dependence close to T_c of the form $M \propto (1-{}^{T}/T_c)^{\beta}$; however, the critical exponent at the surface (normally distinguished as β_1) is different from the β for the bulk and generally higher: the surface becomes less magnetic than the bulk. While the critical magnetic behavior at the surfaces is vigorously researched theoretically, [21,22] on the experimental side only few systems have been investigated: for example the value of β_1 at the surface of nickel has been determined by polarized electron diffraction (PLEED).[23] Polarized neutron reflection may offer a complementary method of analysis: even if the results have less resolution in z than those obtained by PLEED, in a single experiment the surface and bulk magnetization, as well as the correlation length, are simultaneously obtained.

4. Conclusions

I have reviewed the present status of two polarized neutron techniques that are being used in order to study magnetism at interfaces and at surfaces. The first of these techniques (diffraction) is quite ready; unfortunately, for a thoroughly accurate investigation it requires samples more perfect than those presently available. The second of these techniques (Fresnel reflection) imposes a less taxing requirement on the sample, (a single surface is sufficient) but does not have yet the resolution of a single atomic plane. In principle it is possible to bridge the gap between Fresnel reflection and diffraction from a single atomic plane, as indeed it has been done in X-ray scattering.[24] The matter is that modern synchrotron light sources are much brighter in X-rays than nuclear reactors or spallation sources are for neutrons. The advent of the next generation of neutron sources, which promise neutron fluxes ten to one hundred times higher than these presently available, may make these experiments more feasible.

Acknowledgements

The present lecture was given during a period which I spent as a guest of the Second University of Rome at Tor Vergata. To Prof. F. Sacchetti, my host, and Prof. I. Modena, Chairman of the Physics Department, my thanks for their support. I would like also to thank Dr. Majkrzak for informing me of his findings and for providing me with visual material prior to publication.

References

1. W. Marshall and S. W. Lovesey, Theory of thermal neutron scattering, Oxford 1971

2. G. E. Bacon, Neutron Diffraction, Third Edition, Oxford 1976

3. C. M. Falco, W. R. Bennett and A. Boufelfel, Metal-Metal Superlattices, in this volume

4. A. J. Freeman, Electronic Structure of Surfaces, Interfaces and Superlattices, in this volume

5. J. M. Comly, T. M. Holden and G. G. Low, J. Phys. C., 1, 458 (1968)

6. C. F. Majkrzak, J. D. Axe and P. Böni, Proceedings of the 30th Conference on Magnetism and Magnetic Materials, to be published in Journal of Applied Physics

7. C. F. Majkrzak and L. Passell, Acta. Cryst., to be published

8. M. R. Khan, C. S. L. Chun, G. P. Felcher, M. Grimsditch, A. Kueny, C. M. Falco and I. K. Schuller, Phys. Rev. B 27, 7186, (1983)

9. J. W. Cable, M. R. Khan, G. P. Felcher, and I. K. Schuller, to be published

10. T. Shinjo, N. Hosoito, K. Kawaguchi, T. Takada, Y. Endoh, Y. Ajiro and J. M. Friedt, J. Phys. Soc. Japan, 52, 3154 (1983)

11. N. Hosoito, K. Kawaguchi, T. Shinjo, T. Takada and Y. Endoh, J. Physical Soc. Japan 53, 2659 (1984)

12. Y. Endoh, N. Hosoito, and T. Shinjo, J. Magnetism and Magnetic Materials 35, 93 (1983)

13. H. K. Wong, H. Q. Yang, B. Y. Jin, Y. H. Shen, W. Z. Cao, and J. B. Ketterson, J. Appl. Phys. 55, 2494, (1984)

14. M. K. Jaggi, L. H. Schwartz, H. K. Wong, and J. B. Ketterson, Journal of Magnetism and Magnetic Materials, 49, 1 (1985)

15. N. K. Flevaris, J. B. Ketterson, and J. E. Hilliard, J. Appl. Phys. 53, 2439, (1982)

16. G. P. Felcher, R. T. Kampwirth, K. E. Gray and Roberto Felici, Phys. Rev. Letters 52, 1539, (1984)

17. D. Saint-James and P. D. DeGennes, Phys. Letters 7, 306, (1964)

18. Roberto Felici and K. E. Gray, Phys. Rev. B 29, 6129 (1984)

19. G. P. Felcher, Roberto Felici, R. T. Kampwirth, and K. E. Gray, 30th Conference on Magnetism and Magnetic Materials, San Diego, November 1984. To be published in J. Appl. Phys.

20. S. S. Parkin, R. Sigsbee, Roberto Felici and G. P. Felcher, to be published

21. H. W. Diehl, J. Appl Phys. 53, 7914 (1982)

22. K. Binder and D. P. Landau, Phys. Rev. Letters 52, 318 (1984)

23. S. F. Alvarado, M. Campagna, and H. Hofster, Phys. Rev. Letters 48, 51 (1982)

24. P. Eisenberger and W. C. Marra, Phys. Rev. Letters 46, 1081, (1981)

Contribution

Long Wavelength Bulk and Surface Spin Waves in Metallic Superlattices

Talat S. Rahman

Cardwell Hall, Department of Physics, Kansas State University, Manhattan, KS 66506, USA

At long wavelength where exchange interactions play a small role, and the dominant contribution to the excitation energy of the spin waves in a ferromagnet comes from the external Zeeman field and the demagnetizing field set up by the motion of the spins, the surface spin waves, also called the Damon-Eschbach mode, [1] have unique characteristics. Unlike the short wavelength, exchange-dominated surface spin waves, these dipolar surface magnons lie in frequency above the bulk bands and are unidirectional in nature. This effect shows up very well in Brillouin scattering measurements on ferromagnetic materials for both the case of the film [2] and the semi-infinite solid [3], where the surface magnon appears on either the Stokes or the anti-Stokes part of the spectrum, depending on whether the Zeeman field is parallel or antiparallel with respect to the saturation magnetization of the ferromagnet. The calculated light scattering spectra [4], in agreement with the experimental results, show the surface magnon peak to be much more intense than those due to the bulk magnons.

Related theoretical studies [5] on a semi-infinite stack of ferromagnetic films of thickness d_1, separated from each other by a layer of non-magnetic metal of thickness d_2, show that the dispersion of the dipolar surface and bulk spin waves and their light scattering spectra are quite distinct, albeit related, from the features in each individual ferromagnetic film. Here, the demagnetizing field generated by the spin fluctuations is treated in the magnetostatic limit, and the usual boundary conditions are applied at each interface. In order to keep the calculations tractable, exchange interaction is ignored, which is a reasonable assumption when the films are either very thick or very thin.

The calculated dispersion of the collective spin wave excitations of the stack as a function of the ratio d_1/d_2 is presented in Fig. 1 (taken from reference [5]). There appear two types of surface modes and a broad band of bulk states sandwiched in between. The surface magnon at the higher frequency arises from surface mode in each ferromagnetic film, and its frequency is exactly that of the Damon-Eschbach mode for a semi-infinite ferromagnet. The set of surface modes at the lower frequencies are composed of standing bulk modes in each ferromagnetic film. Note that unlike the bulk modes, the frequency of the surface modes is independent of the ratio d_1/d_2 except that the higher frequency surface mode does not exist if d_1 is smaller than d_2.

The thermal fluctuations in the spin system lead to modulations of the dielectric tensor of the medium, thus facilitating the inelastic scattering of light from the spin waves. The spin-spin correlation functions are calculated using linear response theory and Bloch's equations. The electric field of the scattered light is then obtained, using standard Green's function technique for Maxwell's equations. The calculated [5] Brillouin scattering also displays the characteristic surface, and bulk modes of the stack, discussed above. In particular it is seen that the high-frequency surface mode appears only in those cases where $d_1 > d_2$ and that it is unidirectional, just like the Damon-Eschbach mode in each ferromagnetic film. In addition, the light scattering spectrum features the bulk band. These results are in qualitative agreement with data on Brillouin scattering from Molybdenum-Nickel layered structure [6].

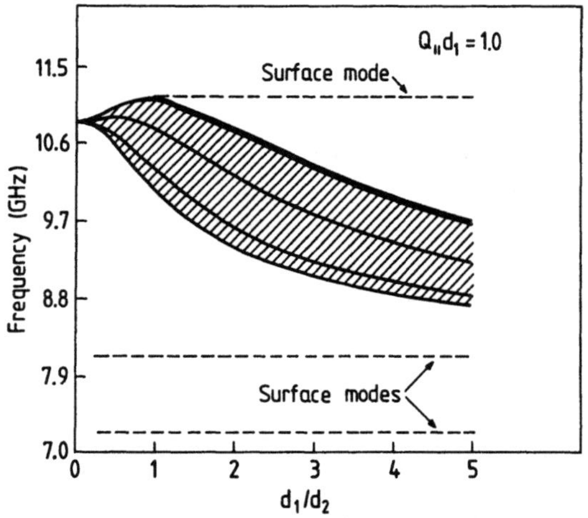

Figure 1. Frequency of collective spin wave excitation of the metallic superlattice vs the ratio d_1/d_2. The shaded region represents the bulk band and the surface modes are the dashed lines

References

1. R.W. Damon and J.R. Eschbach: J. Phys. Chem. Solids 19, 308 (1960)
2. M. Grimsditch, A. Malozemoff and A. Brunsch: Phys. Rev. Lett 43, 711 (1979)
3. J. Sandercock: J. Appl. Phys. 50, 7784 (1979)
4. R.E. Camley, T.S. Rahman and D.L. Mills: Phys. Rev. B23, 1226 (1981)
5. R.E. Camley, T.S. Rahman and D.L. Mills: Phys. Rev. B27, 261 (1983)
6. M. Grimsditch, M. Khan, A. Kueny and I. Schuller, Phys. Rev. Lett. 51, 498 (1983).

Index of Contributors

Theory of Chemisorption

Editor: **J. R. Smith**

1980. 116 figures, 8 tables. XI, 240 pages. (Topics in Current Physics, Volume 19). ISBN 3-540-09891-7

Contents: *J. R. Smith:* Introduction. - *S. C. Ying:* Density Functional Theory of Chemisorption of Simple Metals. - *J. A. Appelbaum, D. R. Hamann:* Chemisorption of Semiconductor Surfaces. - *F. J. Arlinghaus, J. G. Gay, J. R. Smith:* Chemisorption on d-Band Metals. - *B. Kunz:* Cluster Chemisorption. - *T. Wolfram, S. Ellialtioğly:* Concepts of Surface States and Chemisorption on d-Band Perovskites. - *T. L. Einstein, J. A. Hertz, J. R. Schrieffer:* Theoretical Issues in Chemisorption.

M. A. Van Hove, S. Y. Tong

Surface Crystallography by LEED

Theory, Computation and Structural Results

1979. 19 figures, 2 tables. IX, 286 pages. (Springer Series in Chemical Physics, Volume 2). ISBN 3-540-09194-7

"... This is an excellent book for anyone seriously interested in LEED who would like to be able to perform his own sophisticated calculations. It is clearly and carefully written, and all the nuances of the techniques are thoroughly covered. If any book will make the practical calculation of LEED intensity profiles widely available, this is the one."
American Scientist

Vibrational Spectroscopy of Adsorbates

Editor: **R. F. Willis**

With contributions by numerous experts
1980. 97 figures, 8 tables. XII, 184 pages. (Springer Series in Chemical Physics, Volume 15). ISBN 3-540-10429-1

Contents: Introduction. - Theory of Dipole Electron Scattering from Adsorbates. - Angle and Energy Dependent Electron Impact Vibrational Excitation of Adsorbates. - Adsorbate Induced Optical Phonons. - Inelastic Electron Tunnelling Spectroscopy. - Inelastic Molecular Beam Scattering from Surfaces. - Neutron Scattering Studies. - Reflection Absorption Infrared Spectroscopy: Application to Carbon Monoxide on Copper. - Raman Spectroscopy of Adsorbates at Metal Surfaces. - Vibrations of Monatomic and Diatomic Ligands in Metal Clusters and Complexes - Analogies with Vibrations of Adsorbed Species on Metals. - Coupling Induced Vibrational Frequency Shifts and Island Size Determination: CO on Pt {001} and Pt {111}.

Springer-Verlag
Berlin
Heidelberg
New York
Tokyo

MIX
Papier aus verantwortungsvollen Quellen
Paper from responsible sources
FSC® C105338

If you have any concerns about our products,
you can contact us on
ProductSafety@springernature.com

In case Publisher is established outside the EU,
the EU authorized representative is:
Springer Nature Customer Service Center GmbH
Europaplatz 3, 69115 Heidelberg, Germany

Printed by Libri Plureos GmbH
in Hamburg, Germany